Phage Nanobiotechnology

RSC Nanoscience & Nanotechnology

Series Editors:
Professor Paul O'Brien, *University of Manchester, UK*
Professor Sir Harry Kroto FRS, *University of Sussex, UK*
Professor Harold Craighead, *Cornell University, USA*

Titles in the Series:

How to obtain future titles on publication:
A standing order plan is available for this series. A standing order will bring delivery of each new volume immediately on publication.

For further information please contact:
Book Sales Department, Royal Society of Chemistry, Thomas Graham House, Science Park, Milton Road, Cambridge CB4 0WF, UK
Telephone: +44 (0)1223 420066, Fax: +44 (0)1223 420247,
Email: books@rsc.org
Visit our website at http://www.rsc.org/Shop/Books/

Phage Nanobiotechnology

Edited by

Valery A. Petrenko
Auburn University, Auburn, AL, USA

George P. Smith
University of Missouri, Columbia, MO, USA

RSCPublishing

RSC Nanoscience & Nanotechnology No. 17

ISBN: 978-0-85404-184-8
ISSN: 1757-7136

A catalogue record for this book is available from the British Library

Published by The Royal Society of Chemistry,
Thomas Graham House, Science Park, Milton Road,
Cambridge CB4 0WF, UK

Registered Charity Number 207890

For further information see our web site at www.rsc.org

Preface

Bacteriophage, viruses that infect bacteria, are in many ways ideal nanomaterials. Their structure and infection cycle are encoded in small genomes that can be readily manipulated using simple recombinant DNA techniques. Naturally occurring virions (virus particles) exhibit a great diversity of shapes in the nanometer range, with well-defined geometry and uniformity favorable for nano-fabrication. Moreover, they have turned out to be remarkably tolerant of structural alteration. The atomic structures of many viruses have been determined, allowing precise engineering of their modified forms with predetermined shapes and functions and a precise spatial distribution of virus-fused functional peptides at a nanoscale level. In this book, we will outline progress in the development of the phage-derived nanomaterials and their use as bioselectable interfaces in medical and technical devices.

Phage nanobiotechnology has evolved from phage display, a suite of techniques for surveying vast populations of peptides for rare structures with some desired target behavior. The surveys are accomplished by selective strategies that exploit the ability to display phage-encoded peptides on the outer virion surface by genetic fusion to phage coat proteins. The phage libraries that are the initial input to selection can comprise billions of distinct phage clones, displaying billions of distinct peptides. The most common survey strategy is affinity selection, which enriches for phage whose displayed peptides bind an immobilized selector. The selector can be a conventional biomolecule such as an antibody or receptor; it can be a complex biological structure such as whole cells in culture or whole tissues in living animals; and it can be a non-biological surface such as a semiconductor. It is increasingly possible to enrich for properties other than affinity for a target selector. In most phage display applications, the desired end-product is the displayed peptide; the virion serves only a vehicle for discovery. In phage nanotechnology, in contrast, it is the entire ensemble of peptide and virion that is the goal of discovery. Successful

RSC Nanoscience & Nanotechnology No. 17
Phage Nanobiotechnology
Edited by Valery A. Petrenko and George P. Smith
© Royal Society of Chemistry 2011
Published by the Royal Society of Chemistry, www.rsc.org

selections deliver virions with highly desirable emergent properties that depend on the displayed peptides, but may well not be exhibited by those peptides in isolation. In summary, it is in the ability to select nanomaterials with favorable behavior from vast initial nanomaterial libraries that phage nanotechnology differs most dramatically from other modes of nanotechnology.

The commercial success of the phage display libraries over the last 25 years has brought the technology to the bench of many innovative researchers working in very diverse disciplines, greatly expanding its technological reach. At the same time, new modes of peptide display designed to modify the properties of the entire virion have come to the fore. In 'landscape' libraries, for instance, almost one-quarter of the virion's surface area differs from one phage clone to another. These creative developments have already paid off in contributions to multiple areas of medicine and technology, including medical diagnostics and monitoring, molecular imaging, targeted drug and gene delivery, vaccine development and bone and tissue repair. Further advances in these areas will require the collective efforts of specialists in diverse fields – medical doctors, microbiologists, structural biologists, chemists, pharmacologists and many others. We hope that the assembling in this book of a collection of up-to-date reviews on phage nanotechnology will help the members of the growing community of phage bioengineers to stay abreast of recent trends and inspire them to create new ones.

Valery A. Petrenko
George P. Smith

Contributors

Hélène Blois, *Pherecydes Pharma, Biocitech, Bâtiment Lavoisier, 102 avenue Gaston Roussel, 93230 Romainville, France*

Binrui Cao, *Department of Chemistry and Biochemistry, University of Oklahoma, Norman, OK 73019, USA*

Bryan A. Chin, *Materials Research and Education Center, Auburn University, Auburn, AL 36849, USA*

Susan L. Deutscher, *Biochemistry Department, 117 Schweitzer Hall, University of Missouri, Columbia, MO 65211 and Harry S Truman Medical Veterans Hospital, Columbia, MO 65212, USA*

Manuel Gea, *BIO-MODELING SYSTEMS SAS, 26 rue Saint Lambert, 75015 Paris, France*

François Iris, *BIO-MODELING SYSTEMS SAS, 26 rue Saint Lambert, 75015 Paris, France*

Prashanth K. Jayanna, *Department of Chemistry, Sam Houston State University, Huntsville, TX 77340, USA*

Kimberly A. Kelly, *Department of Biomedical Engineering and Robert M. Berne Cardiovascular Research Center, University of Virginia, Health System Box 800579, Charlottesville, VA 22904, USA*

Ramji S. Lakshmanan, *Department of Chemical and Biological Engineering, Drexel University, Philadelphia, PA 19104, USA*

Paul-Henri Lampe, *BIO-MODELING SYSTEMS SAS, 26 rue Saint Lambert, 75015 Paris, France*

Suiqiong Li, *Materials Research and Education Center, Auburn University, Auburn, AL 36849, USA*

Lee Makowski, *Department of Electrical and Computer Engineering and Chemistry and Chemical Biology, Dana Research Center, Northeastern University, 360 Huntington Avenue, Boston, MA 02115, USA*

RSC Nanoscience & Nanotechnology No. 17
Phage Nanobiotechnology
Edited by Valery A. Petrenko and George P. Smith
© Royal Society of Chemistry 2011
Published by the Royal Society of Chemistry, www.rsc.org

Karen Manoutcharian, *Departamento de Biología Molecular y Biotecnología, Instituto de Investigaciones Biomédicas, UNAM, México DF, México*

Chuanbin Mao, *Department of Chemistry and Biochemistry, University of Oklahoma, Norman, OK 73019, USA*

Christopher J. Noren, *New England Biolabs, 240 County Road, Ipswich, MA 01938, USA*

Stanley J. Opella, *Department of Chemistry and Biochemistry, University of California, San Diego, La Jolla, CA 92035, USA*

Valery A. Petrenko, *Department of Pathobiology, College of Veterinary Medicine, Auburn University, AL 36849, USA*

Flavie Pouillot, *Pherecydes Pharma, Biocitech, Bâtiment Lavoisier, 102 avenue Gaston Roussel, 93230 Romainville, France*

Steven Ripp, *University of Tennessee, 676 Dabney Hall, Knoxville, TN 37996, USA*

Lana Saleh, *New England Biolabs, 240 County Road, Ipswich, MA 01938, USA*

George P. Smith, *Division of Biological Sciences, University of Missouri, Columbia, MO 65211, USA*

Contents

RSC Nanoscience & Nanotechnology No. 17
Phage Nanobiotechnology
Edited by Valery A. Petrenko and George P. Smith
© Royal Society of Chemistry 2011
Published by the Royal Society of Chemistry, www.rsc.org

CHAPTER 1

The Phage Nanoparticle Toolkit

GEORGE P. SMITH

Division of Biological Sciences, University of Missouri, Columbia, MO 65211, USA

1.1 Introduction

The development of phage virions (particles) as novel nanoparticles has been closely tied to phage display technology. Since filamentous phages of the Ff class (wild-type strains f1, fd and M13) are the predominant phage display vectors, so too they have been the predominant type of phage nanoparticle – at least so far. Accordingly, this chapter will concentrate on Ff phages, although it will point out ways in which they differ fundamentally from other phages such as T4. For a summary of Ff phage biology in general, the reader is referred to an excellent review.[1] Here the focus will be more specifically on those aspects of the life cycle that are of direct practical concern to scientists and engineers seeking to develop the virions as nanoparticles.

1.2 Virion Structure and Purification

The wild-type Ff virion is 900 nm long and 6 nm in diameter. Almost all of the nanoparticle's surface and 87% of its weight is a tubular sheath composed of ~2760 copies of the 50-residue major coat protein pVIII in a geometrically regular array,[2,3] as depicted in the molecular model in Figure 1.1. Residues in the first third of pVIII are accessible from the outside, whereas the last third of the protein, including four lysines and no acidic side-chains, are exposed in the lumen. Inside the lumen is the single-stranded circular viral DNA and at the

RSC Nanoscience & Nanotechnology No. 17
Phage Nanobiotechnology
Edited by Valery A. Petrenko and George P. Smith
© Royal Society of Chemistry 2011
Published by the Royal Society of Chemistry, www.rsc.org

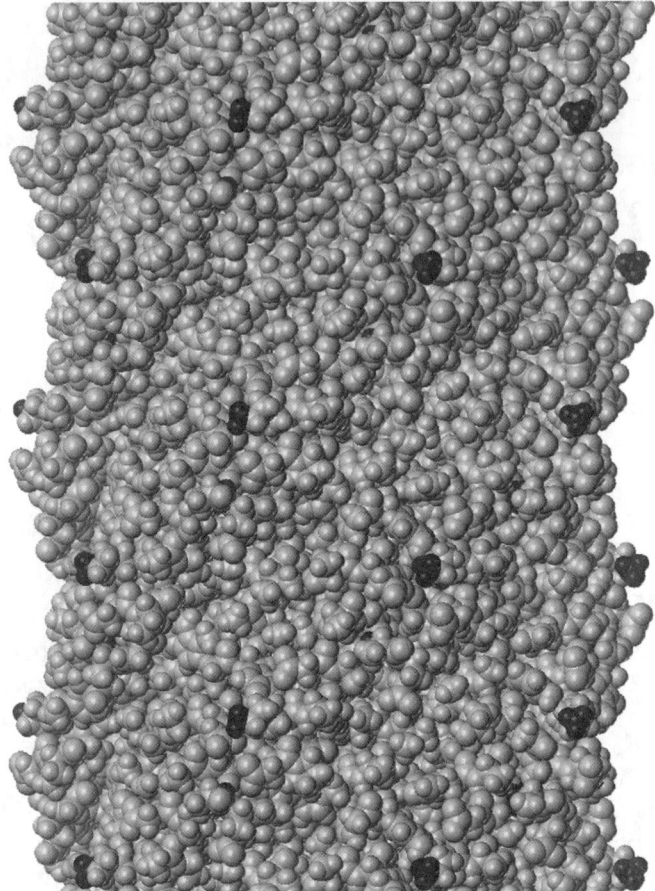

Figure 1.1   Space-filling model (including hydrogens; Protein Data Bank accession
number 2C0X) of a short section of the tubular sheath of filamentous
bacteriophage fd.[2,3] The section depicted includes all or parts of six rings
of five pVIII subunits, each ring rotated 36° relative to the ring above and
below it. The entire wild-type virion has ∼550 five-subunit rings. The
exposed α- and ε-amino groups are blackened, the former mostly obscured
by surrounding atoms.

tips are five copies each of the minor coat proteins: pVII (33 residues) and pIX
(32 residues) at the tip that is extruded first from the cell during assembly, pIII
(406 residues) and pVI (112 residues) at the other tip. The length of the virion is
dictated by the length of the viral DNA that it contains. If the viral DNA is
artificially lengthened or shortened by adding or subtracting segments, the
number of pVIII subunits and length of the virion increase or decrease in
proportion. The geometry of the pVIII array does not depend on a geometrically specific interaction between the pVIII subunits and the nucleotides of
the viral DNA. Instead, packing of the DNA in the lumen of the tubular sheath

seems to require only overall electrostatic balance between the positive charges on the four lumenal ε-amino groups of the pVIII polypeptide and the negatively charged phosphates.[4-7] Modifying that balance leads to a compensatory change in ratio of virion length to number of viral DNA nucleotides, without perceptibly changing the geometry of the pVIII array.[7]

Virions can be prepared to a high degree of purity in two complementary steps that are readily scaled to large volumes. First, since virions are released without lysing the host cells (see Section 1.5), simply removing cells by low-speed centrifugation eliminates all but a trace of intracellular components. A second low-speed centrifugation of the culture supernatant dramatically reduces contamination with residual intact cells. This simple, scalable way of removing the vast bulk of intracellular contaminants does not apply to phages such a λ and T4 that are released by lysis rather than extrusion. The second Ff virion purification step is precipitation from the culture supernatant with a low concentration (2% w/v) of poly(ethylene glycol) (PEG);[8] under these conditions, most residual contaminants, including DNA, RNA, proteins and ribosomes, remain in the supernatant after the virions have been sedimented by low-speed centrifugation. Purity can be substantially improved by re-centrifuging the pellet after pouring off the bulk of the supernatant and aspirating the residual supernatant. Spherical phages, including λ and T4, can also be precipitated with PEG,[8] but only at a concentration ($\sim 10\%$ w/v) that also precipitates many intracellular contaminants such as DNA, RNA and ribosomes.

Ff virions prepared by PEG precipitation from cleared supernatant are sufficiently pure for most purposes. Further purification, if necessary, is usually accomplished by CsCl density equilibrium ultracentrifugation, which accommodates multiple phage clones on a modest scale (up to $\sim 10^{15}$ virions), but achieves only modest reductions in contamination with proteins or non-particulate molecules. Size-exclusion chromatography effectively resolves virions from non-particulate proteins, but has very limited capacity ($\sim 10^{13}$ virions per 30 ml column).[9] Hydroxyapatite chromatography resolves virions from most protein contaminants and is easily scaled to any number of virions but, unlike the other two methods, is not highly reproducible and depends sensitively on the nature of displayed guest peptides.[10] Neither chromatography method conveniently accommodates multiple clones.

The virion is remarkably robust, retaining infectivity after ~ 10 min of exposure to pH 2.2, pH 12 or 6 M urea, indefinite exposure to 70 °C, and disulfide reduction; exposure to many proteases such as trypsin and chymotrypsin; and other harsh conditions. This robustness facilitates affinity selection, which depends on specific binding of phage to an immobilized target (the 'selector') and subsequent release by breaking the bonds between the phage and selector or between the selector and the immobilizing surface. Relatively harsh release conditions can be used without compromising infectivity – the ability of the released phages to be propagated by infecting fresh host bacteria. It should be noted, however, that some release conditions can subtly alter the physical characteristics of the virion even if they do not reduce infectivity.[11]

Virions remain physically intact almost indefinitely at refrigerator temperatures, even if they have not been purified. However, unpurified virions can lose
infectivity over a period of years, because two domains of pIII that are
necessary for infectivity (see the next section) but not for overall physical
integrity are relatively accessible to contaminating proteases. It is therefore best
to purify virions that must be stored for more than a few months. Virions can
be stored frozen, but they lose a substantial fraction of their infectivity with
each freeze–thaw cycle.

1.3   Intrusion

The filamentous phage infection cycle is shown schematically in Figure 1.2. Its
major stages will be discussed in this section and the two that follow.

Ff phages infect *Escherichia coli* cells carrying the F factor plasmid and
therefore displaying F pili on their surface. Infection, called 'intrusion' in
this chapter, is a two-stage process in which two domains, N1 and N2, at the
N-terminus of pIII play a critical role; it is summarized in the aforementioned
review.[1] In the first stage, domain N2 binds to the tip of the F pilus, which then
retracts, drawing the tip of the virion to the cell surface. In the second stage,
domain N1 interacts with the cell's Tol apparatus. This interaction somehow
triggers intrusion of the single-stranded viral DNA – the *plus* strand – into the
cytosol and concomitant transfer of the coat proteins into the cell's inner
membrane. When a large number of cells are mixed with a limiting number of
virions, infectivity – the number of successfully infected cells per virion – is very

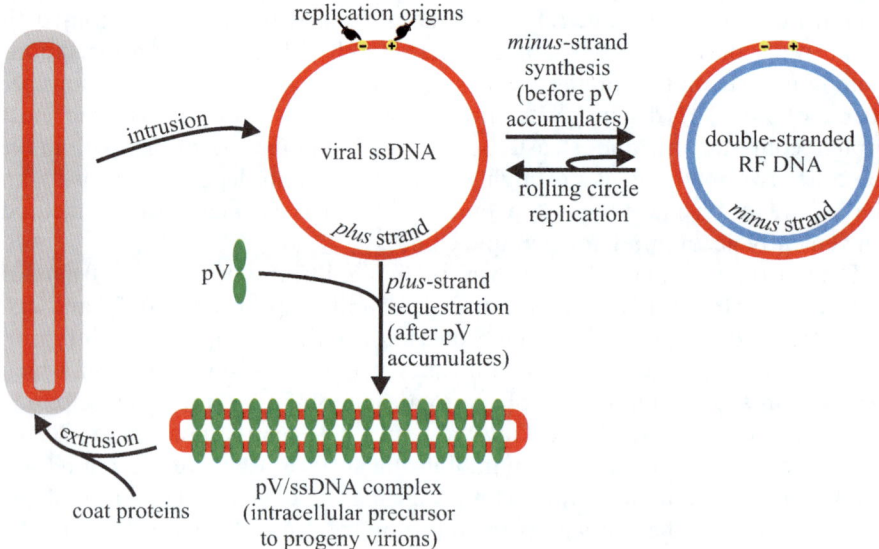

Figure 1.2   Schematic representation of the infection cycle.

high: typically around 50% for wild-type phage. This reflects both the virion's overall physical robustness (see the previous section) and the efficiency of *minus* strand synthesis, which converts the infecting *plus* strand to a double-stranded RF within a few minutes (see Figure 1.2 and the next section). Because the RF is considerably more stable than the viral single strand, once an infection reaches the RF stage it seldom aborts.

Neither the F pilus nor the Tol apparatus plays any role in phage production. Nor are pIII domains N1 or N2 required for phage assembly. Mutant phages in which those domains are mutationally inactivated or deleted altogether are produced in more or less normal yields, although of course they are non-infective.

1.4 DNA Replication Cycle and Gene Expression

Once the viral *plus*-strand DNA is inside the cell, cellular enzymes synthesize the complementary *minus* strand starting at an initiation site within a *minus*-strand origin, one of the functional elements in the intergenic region depicted in Figure 1.3.[12] The result is the circular double-stranded replicative form (RF), which is the template for mRNA transcription and eventually the starting point for further replication. That replication is initiated by nicking of the *plus* strand at the *plus*-strand origin by phage protein pII (Figure 1.3) and continues in the rolling circle mode until a complete progeny *plus* strand has been made. Early in infection, the progeny *plus* strand serves as template for *minus*-strand synthesis, just as did the original infecting viral DNA (Figure 1.2). Late in infection and in chronically infected cells, however, the phage single-stranded DNA binding protein pV sequesters most progeny *plus* strands to form the pV/ssDNA complex, which is the precursor to progeny virion assembly (see Figure 1.2 and the next section).

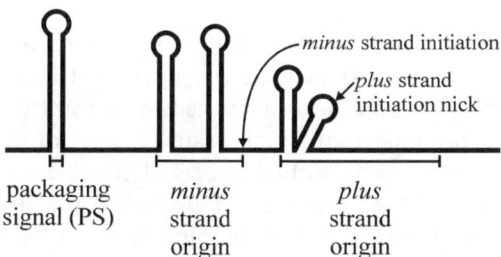

packaging   *minus*   *plus*
signal (PS)   strand   strand
      origin   origin

minus strand initiation
plus strand initiation nick

Figure 1.3 Schematic map of the functional elements of the intergenic region of the filamentous phage genome;[12] the 5′ to 3′ polarity of the *plus*-strand is from left to right. The hairpins are predicted, and in some cases confirmed, for the single-stranded viral *plus*-strand DNA. *Plus*-strand synthesis proceeds rightwards from the pII-generated nick; *minus*-strand synthesis proceeds leftwards from its initiation site. There is no gene transcription through the region depicted: gene IV mRNA terminates to the left and gene II mRNA initiates to the right.

The phage genes are expressed at very different levels. Some of the proteins in effect act catalytically in phage DNA replication and virion extrusion (see the next section), while five are consumed stoichiometrically by incorporation into the virion (see Section 1.2). The five virion proteins are trafficked to the cell's inner membrane, from which they are transferred into extruding virions. Two of them, pIII and pVIII, have signal peptides that are cleaved to form the mature polypeptides, whose N-terminal portions are exposed in the periplasm before transfer into virions. Chronically infected cells come to a steady state in which synthesis and consumption of viral DNA and proteins are balanced and do not kill the cell.

1.5   Extrusion of Progeny Virions

Most of the familiar phages, including λ and T4, are assembled in the cytosol of the infected cell and released by cell lysis. Ff phages, in contrast, are continuously extruded through the cell envelope without killing the cell, which continues to divide, albeit at a reduced rate. It is the reduction in growth rate, not lysis, that accounts for plaque formation.

The precursor for progeny virion assembly is the long, thin pV/ssDNA complex (Figure 1.2) produced late in infection and in chronically infected cells. At one tip of the complex the viral DNA forms a hairpin called the packaging signal (PS), one of the functional elements in the intergenic region (Figure 1.3). Assembly is initiated by interaction of the PS with a special pore through the cell envelope fashioned from three phage proteins. The complex's DNA is extruded through the pore, shedding its intracellular covering of pV molecules back into the cytosol and concomitantly acquiring its extracellular covering of the five coat proteins from the inner membrane. The process is extraordinarily productive; the wild-type yield reaches 10^{12} virions ml^{-1} under ordinary laboratory culture conditions and vigorous aeration can boost the yield to 10^{13} virions ml^{-1} – equivalent to 250 mg of pVIII molecules per liter.

The assembly process imposes constraints on the peptides or proteins that can be successfully displayed by genetic fusion to coat proteins. Since all the coat proteins are incorporated into the extruding virion from the inner membrane, peptides or proteins that block translocation of the fusion protein through the inner membrane cannot be displayed. Since the virion exits the cell through a narrow pore, large proteins are not efficiently displayed, although amazingly virions with up to 24 50-kDa immunoglobulin Fab domains fused to pVIII have been documented.[13] These constraints are lessened when the virions carry wild-type in addition to recombinant versions of the displaying coat protein. In particular, only very short peptides can be displayed on all copies of pVIII,[14] while proteins as large as the 50-kDa Fab domain can be displayed on a few copies of pVIII when the remaining pVIII subunits are wild-type.

Any substantial imbalance in the multiple convergent pathways of the infection cycle leads to cell killing and absence of phage production.[15] Even subtle impairments, such as those incurred by the presence of a small foreign peptide on a few copies of the pVIII coat protein, can compromise cell viability sufficiently to impose measurable selective pressure against the peptide.[16] The adverse effects of subtle imbalance, or even of gross defects in virion assembly, are largely eliminated by reducing the DNA copy number (the number of double-stranded RF molecules per cell), as will be explained in the final section.[17]

1.6 Display of Guest Peptides

Foreign ('guest') peptides and proteins can be fused genetically to the coat proteins and thereby displayed on the virion's outer surface. This is depicted schematically in Figure 1.4, where the recombinant peptide-bearing proteins

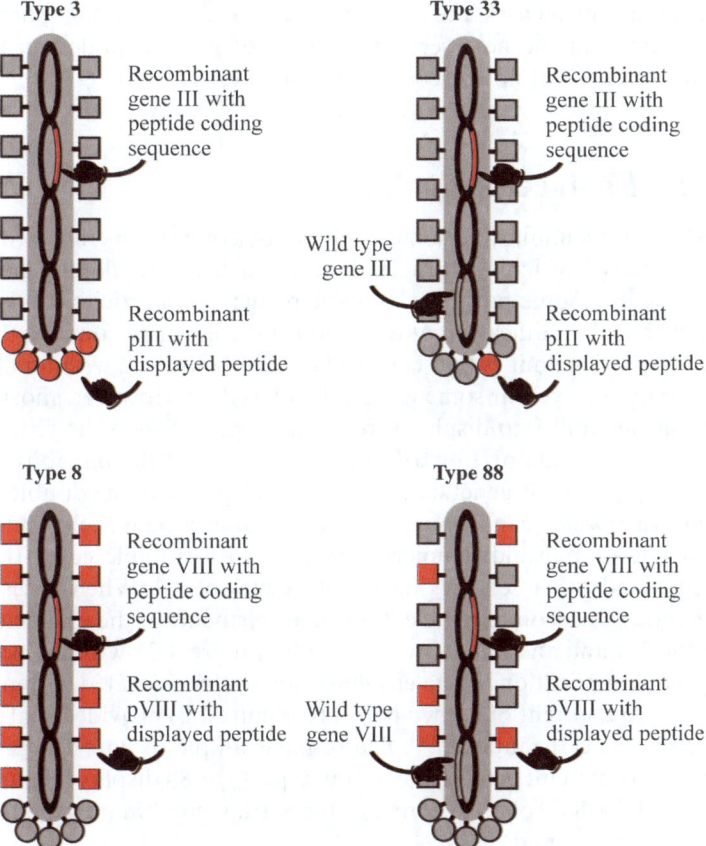

Figure 1.4 Modes of displaying guest peptides fused to pIII or pVIII as the host coat protein, as explained in the text. Recombinant coat-protein genes and their corresponding peptide-bearing recombinant proteins are colored red. Not shown are Type 3 + 3 and 8 + 8 display or display on pVI, pVII or pIX coat proteins.

and the recombinant genes that encode them are colored red. Usually (but not always) it is the pIII or pVIII coat protein that thus hosts the guest peptide or protein, as summarized in Figure 1.4. N-terminal display on these proteins requires the foreign coding sequence to lie between the coding sequences for the signal peptide and the mature coat protein, without disturbing the reading frame. Small peptides can be fused to all five copies of pIII (type 3 display) or to all copies (thousands of them) of pVIII (type 8 display). In these cases, the recombinant gene III or gene VIII encoding the fusion protein replaces the wild-type gene. Alternatively, peptides or proteins can be displayed on one copy of pIII or on a few dozen copies of pVIII, the remaining pIII or pVIII subunits being the wild-type (type 33 or 88 display). In these cases, the recombinant gene III or VIII is in addition to the corresponding wild-type gene (Figure 1.4). Not depicted in Figure 1.4 are type 3 + 3 and 8 + 8 displays, in which the wild-type and recombinant coat-protein genes are carried on separate genomes present in the same cell. In most cases, the recombinant gene is carried on a phagemid, as will be described in the next section. Display of guest peptides on the other three coat proteins will not be covered in this review.

1.7   The Engineer's Toolkit

This final section summarizes some of the most common methods for manipulating Ff phage to achieve practical goals such as developing new phage nanoparticles. For some purposes, suitable phage nanoparticles can be created by direct chemical modification of the virion surface, which contains thousands of surface-accessible amines that can be chemically modified without disturbing virion integrity, or even in some cases infectivity.[18,19] However, most applications exploit the ability to display foreign guest peptides on the virion surface by genetic fusion to the pIII or pVIII coat protein (see the previous section).

The starting point for genetic manipulation of phage is the double-stranded RF (Figure 1.2), which can be purified from stationary-phase cells and used like a typical bacterial plasmid. Numerous vectors with suitable gene-III or -VIII cloning sites and other features have been developed.[20] When necessary, the phage genome can accommodate extra genetic elements in the intergenic region between the PS and the *minus*-strand origin (Figure 1.3) without perceptibly affecting phage replication. Extra elements can also be tolerated within the *plus*-strand origin to the right of its two hairpins (Figure 1.3) provided that there is a compensating gene-II mutation.[21] The prime example of an extra genetic element is the recombinant gene III or VIII in type 33 or 88 display (Figure 1.4). In type 3 + 3 and 8 + 8 display, in contrast, the extra recombinant gene III or VIII is carried on a phagemid.

Phagemids are a special class of plasmids that carry the phage intergenic region, spanning the origins of replication and the nearby PS (Figure 1.3). When cells carrying a phagemid are superinfected by phage, the plasmid replicates in the phage mode and one of its strands is extruded in the form of a normal virion; the superinfecting phage in these circumstances is called a 'helper'. If the phagemid bears a gene encoding the N-terminal domains of pIII, that gene must be repressed before helper superinfection is possible, because those domains block deployment of the F pilus, the receptor for Ff phage infection.[22] In addition, because pIII and pVIII are toxic at high concentrations, expression of gene III or VIII on a high copy number phagemid must generally be greatly down-regulated in order to ensure cell viability even after superinfection has been successfully completed. When the recombinant gene III or VIII in type 3 + 3 or 8 + 8 display is carried on a phagemid, it is included in the phagemid virions released from the cell. This is the design of type 3 + 3 and 8 + 8 libraries, allowing phagemids bearing rare selector-binding peptides to be affinity selected along with the corresponding recombinant gene.

In type 33, 3 + 3, 88 and 8 + 8 display, virion production and infectivity do not depend on the recombinant pIII or pVIII. Although defects in the recombinant proteins might severely affect their incorporation into virions, overall virion production and infectivity usually remain normal. In particular, virions with a mixture of recombinant and wild-type pIII subunits retain infectivity even if the infection-mediating N-terminal domains are entirely missing from the recombinant subunits. For this reason, it will often be convenient use type 33 or 3 + 3 systems for the development of such recombinant pIIIs in order to take advantage of infection during the research process, even if the ultimate goal is a nanoparticle with no wild-type pIII subunits.

The *plus*-strand origin has two distinguishable functions: it is the site where *plus*-strand synthesis is initiated and the site where *plus*-strand synthesis terminates. If an infected cell contains an artificial plasmid with two *plus*-strand origins, both initiation and termination occur at both of them, creating a separate circular replicon from each of the two arcs.[23] If one of the two origins is defective for termination (but not initiation) and the other is defective for initiation (but not termination), only one of the two replicons is produced. Such constructs have been used to create miniature phage nanoparticles that are only about 50 nm long and have only about 95 pVIII subunits.[24]

In many research contexts, it is advantageous to reduce the DNA copy number by disabling the *minus* strand origin.[17] The *minus* strand is still synthesized without the origin, but much less efficiently. Virion yield drops only about twofold to 5×10^{11} virions ml^{-1} under ordinary culture conditions, but cannot be improved significantly by vigorous aeration. Infectivity drops about fivefold to $\sim 10\%$, because the rate at which *minus*-strand synthesis is initiated on the infecting single-stranded viral DNA is much lower than when the *minus* strand origin is present. Plaque size is greatly reduced. Generally, replication-defective phage carry an antibiotic resistance gene and are detected and quantified as colony-forming units after spreading infected cells on nutrient agar containing the antibiotic. The inconveniences of low infectivity are in many cases more than compensated for by the near elimination of cell killing and of more subtle selection against slight functional defects (see Section 1.5). Even a short peptide displayed on a few hundred copies of pVIII can be lost after a few rounds of non-selective propagation in wild-type phage, whereas the same peptide displayed in the same way on replication-defective phage is fully retained under the same conditions.[16] The first large-scale type 8 library (Figure 1.4), displaying short random peptides on every copy of pVIII, could only be constructed successfully in a replication-defective vector.[25] Use of replication-defective vectors will also probably greatly reduce a problem that can vex affinity selections from phage-display libraries in non-replication-defective vectors: dominance of 'target-unrelated' clones. These are clones that are favored, not because they display a peptide with high affinity for the immobilized target selector, but rather because they harbor a mutation that gives them a growth advantage.[26] In short, the use of replication-defective vectors gives the engineer much more flexibility in the creation of new types of phage nanoparticles, even if ultimately the replication defect must be removed in order to produce the final nanoparticle in sufficient quantity and purity for the intended end use.

Acknowledgements

This work was supported by US NIH grants P50 CA103130 to Wynn A. Volkert and R21 CA127339 to the author.

References

1. R. Webster, in *Phage Display: a Laboratory Manual*, ed. C. F. Barbas III, D. R. Burton, J. K. Scott and G. J. Silverman, Cold Spring Harbor Laboratory Press, Cold Spring Harbor, NY, 2001, pp. 1.1–1.37.
2. D. A. Marvin, L. C. Welsh, M. F. Symmons, W. R. Scott and S. K. Straus, *J. Mol. Biol.*, 2006, **355**, 294–309.
3. S. K. Straus, W. R. Scott, M. F. Symmons and D. A. Marvin, *Eur. Biophys. J.*, 2008, **37**, 521–527.
4. J. Greenwood, G. J. Hunter and R. N. Perham, *J. Mol. Biol.*, 1991, **217**, 223–227.
5. G. J. Hunter, D. H. Rowitch and R. N. Perham, *Nature*, 1987, **327**, 252–254.
6. D. H. Rowitch, G. J. Hunter and R. N. Perham, *J. Mol. Biol.*, 1988, **204**, 663–674.
7. M. F. Symmons, L. C. Welsh, C. Nave, D. A. Marvin and R. N. Perham, *J. Mol. Biol.*, 1995, **245**, 86–91.
8. K. R. Yamamoto, B. M. Alberts, R. Benzinger, L. Lawhorne and G. Treiber, *Virology*, 1970, **40**, 734–744.
9. M. Y. Zakharova, A. V. Kozyr, A. N. Ignatova, I. A. Vinnikov, I. G. Shemyakin and A. V. Kolesnikov, *Biotechniques*, 2005, **38**, 194, 196, 198.
10. G. P. Smith and T. R. Gingrich, *Biotechniques*, 2005, **39**, 879–884.
11. S. F. Parmley and G. P. Smith, *Gene*, 1988, **73**, 305–318.
12. N. D. Zinder and K. Horiuchi, *Microbiol. Rev.*, 1985, **49**, 101–106.
13. A. S. Kang, C. F. Barbas, K. D. Janda, S. J. Benkovic and R. A. Lerner, *Proc. Natl. Acad. Sci. USA*, 1991, **88**, 4363–4366.

14. V. A. Petrenko, G. P. Smith, X. Gong and T. Quinn, *Protein Eng.*, 1996, **9**, 797–801.
15. D. Pratt, H. Tzagoloff and W. S. Erdahl, *Virology*, 1966, **30**, 397–410.
16. G. P. Smith and A. M. Fernandez, *Biotechniques*, 2004, **36**, 610–614, 616, 618.
17. G. P. Smith, *Virology*, 1988, **167**, 156–165.
18. J. Armstrong, J. A. Hewitt and R. N. Perham, *EMBO J.*, 1983, **2**, 1641–1646.
19. X. Jin, J. R. Newton, S. Montgomery-Smith and G. P. Smith, *Biotechniques*, 2009, **46**, 175–182.
20. J. K. Scott and C. F. Barbas III, in *Phage Display a Laboratory Manual*, ed. C. F. Barbas III, D. R. Burton, J. K. Scott and G. J. Silverman, Cold Spring Harbor Laboratory Press, Cold Spring Harbor, NY, 2001, pp. 2.1–2.19.
21. G. P. Dotto and N. D. Zinder, *Proc. Natl. Acad. Sci. USA*, 1984, **81**, 1136–1340.
22. J. D. Boeke, P. Model and N. D. Zinder, *Mol. Gen. Genet.*, 1982, **186**, 185–192.
23. K. Horiuchi, *Proc. Natl. Acad. Sci. USA*, 1980, **77**, 5226–5229.
24. L. Specthrie, E. Bullitt, K. Horiuchi, P. Model, M. Russel and L. Makowski, *J. Mol. Biol.*, 1992, **228**, 720–724.
25. G. P. Dotto and N. D. Zinder, *Nature*, 1984, **311**, 279–280.
26. L. A. Brammer, B. Bolduc, J. L. Kass, K. M. Felice, C. J. Noren and M. F. Hall, *Anal. Biochem.*, 2008, **373**, 88–98.

CHAPTER 2

The Roles of Structure, Dynamics and Assembly in the Display of Peptides on Filamentous Bacteriophage

STANLEY J. OPELLA

Department of Chemistry and Biochemistry, University of California, San Diego, La Jolla, CA 92035, USA

2.1 Molecular and Structural Biology of Filamentous Bacteriophage

Filamentous bacteriophage (inoviruses) are rod-shaped nucleoprotein particles that infect bacteria without killing them.[1-3] Filamentous bacteriophage consist of about 90% by weight protein, most of which is the major coat protein subunits arrayed on the exterior of the particle and 10% is DNA that is encased within the protein coat on the interior. Although the viruses themselves do not have a membrane, their lifecycle is closely associated with that of the host cell's membranes,[4] especially the structure, dynamics and conformational transitions of the major coat protein. These viruses have about 10 genes encoded in single-stranded circular DNA that is encased within a cylindrical coat that consists almost entirely of several thousand copies of the major coat protein.

There are many isolates of these bacteriophage, but they are principally divided into two classes. The best-studied class, referred to as Class I (fd, M13,

RSC Nanoscience & Nanotechnology No. 17
Phage Nanobiotechnology
Edited by Valery A. Petrenko and George P. Smith
© Royal Society of Chemistry 2011
Published by the Royal Society of Chemistry, www.rsc.org

f1), infects *Escherichia coli*. The major coat protein of M13 differs from that of fd in only a single amino acid, hence for all practical purposes they are mutants of the same virus and are used interchangeably in biochemical and biophysical studies. They utilize the F pilus of the bacteria during infection and have a complex lifecycle; the viral DNA is replicated through a duplex intermediate and the newly synthesized DNA is covered with a single-stranded DNA-binding protein in the cell. Pf1 infects *Pseudomonas aeruginosa* and is a Class II filamentous bacteriophage and its comparisons with Class I virions such as fd provide many insights into the structure, dynamics and assembly of all filamentous bacteriophage.[5,6]

Filamentous bacteriophage have been studied extensively as examples of a wide range of prokaryotic molecular and structural biology. They are very well characterized experimental systems for studying the insertion and processing of membrane proteins. However, they are perhaps best known as tools for many laboratory procedures of molecular biology and biotechnology,[7] including the cloning and sequencing of DNA and the generation and screening of peptide libraries, where peptides can be selected from phage display libraries.[8-11] They are also increasingly being utilized in nanotechnology applications.[12]

In phage display of peptides, there are three types of bacteriophage particles that need to be considered in order to study a range of peptides. Short sections of each type of bacteriophage are displayed schematically in Figure 2.1. The genome of filamentous bacteriophage fd has been engineered to allow peptide epitopes to be displayed on the surface of the virus particles as part of the N-terminal region of the major coat protein.[13,14] The phage particles in Figure 2.1A have wild-type coat proteins, those in Figure 2.1B have recombinant coat proteins with 6–8 inserted residues and those in Figure 2.1C have larger peptides that must be diluted with wild-type coat proteins in order to assemble into virus particles. The coat protein subunits are packed around the DNA, which is not visible in these illustrations. The wild-type coat protein subunits have a few mobile residues at the N-terminus that are exposed to the aqueous solution and they are shown in blue while the rest of the residues are in the gray in Figure 2.1A. Figure 2.1B and C illustrate two types of coat proteins displaying added peptide sequences at their N-termini. It is possible to prepare

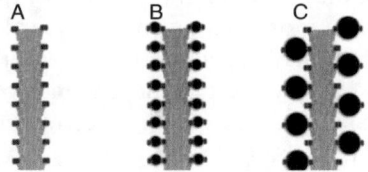

Figure 2.1   Cartoons illustrating the types categories of filamentous bacteriophage coat proteins used in the research. (A) Wild-type coat proteins. (B) Recombinant coat proteins with small (6-residue) peptide inserts. (C) Hybrid coats where some have larger (12–20 residue) peptide inserts and some are wild-type coat proteins. Both (B) and (C) represent phage particles used to generate peptide libraries.

recombinant coat proteins, where every coat protein subunit contains the altered sequence, when there are only about six residues added to the protein. This is illustrated in Figure 2.1B. Even longer peptide sequences can be added to the coat proteins, but then it is necessary to generate hybrid bacteriophage particles where the majority of coat proteins have the wild-type sequence and only a fraction have the larger, bulkier polypeptides exposed to the solution. This is shown, roughly to scale for a 20-residue peptide, in Figure 2.1C.

2.2   Packaging of the Genome into Filamentous Bacteriophage

As suggested by Figure 2.1, the primary focus of attention is on the major coat protein subunits of the filamentous bacteriophage in fundamental studies of their biology and also for practical applications to phage display and biotechnology. However, the DNA must also be taken into account in understanding the structures and lifecycles of filamentous bacteriophage in addition to the facility for integrating extra nucleotide sequences into the genome, as done for large-scale procedures such as cloning entire proteins or more subtle changes such as adding a few residues to the N-terminal portions of the major coat proteins for phage display of peptides. An overriding advantage of working with the filamentous bacteriophage is the ability to link the genomic change in the DNA to the phenotypic change in the modified coat protein subunits, because this enables large quantities to be made and for the experiments to reproduced.

   NMR spectroscopy is particularly well suited for studying the properties of DNA in these particles because of its ability to obtain separate signals from the DNA and the protein, something that is not possible with diffraction or other spectroscopic experiments. NMR shows that the DNA is packed very differently in Class I (fd) and Class II (Pf1) bacteriophage,[15,16] and this provides some of the strongest evidence that these are entirely different viruses, rather than variants of the same organism. The only phosphorus atoms in the virus particles are in the phosphodiester backbone of the DNA. Consequently, the ^{31}P NMR spectra reflect the structure and packing of the DNA backbone. The chemical shift is one of the most fundamental physical interactions that affect NMR spectra. In samples that are immobile on the time-scale of the NMR experiments, about 10 kHz, the angular dependence of the spin interactions, like the chemical shift, are manifested in the spectra and are extremely reliable indicators of structure and dynamics. The properties of the ^{31}P interactions in nucleic acids are well characterized. The simulated non-axially symmetric chemical shift powder pattern for a rigid phosphodiester group as found in the backbone of DNA is shown in Figure 2.2.

   For fd, a Class I bacteriophage, both the unaligned (powder) and aligned samples yield similar spectra, which appear to be powder patterns; this indicates that the DNA is packed randomly inside the virus particles. This is

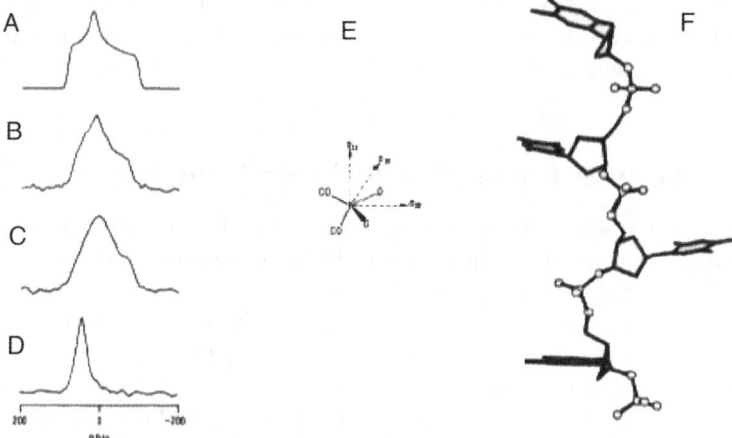

Figure 2.2   ^{31}P NMR spectra of filamentous bacteriophage. (A) Simulated powder pattern for a phosphodiester group. (B) Spectrum of an unoriented sample of fd. (C) Spectrum of an aligned sample of fd. (D) Spectrum of an aligned sample of Pf1. (E) Orientation of the ^{31}P chemical shift tensor in the backbone of DNA. (F) A structure of the DNA in Pf1 consistent with the data in (D). The phosphates are oriented as determined by the experimental chemical shift in the context of the shield tensor in (E) with an extended conformation based on the length of the particle and the number of nucleotides.[15,16]

remarkable because the protein subunits that surround the DNA and form the tubular coat are nearly perfectly ordered and aligned.

In contrast, Pf1, a Class II bacteriophage, gives very different ^{31}P NMR results that demonstrate that the phosphodiester linkages of its DNA are uniformly oriented in the virus particles. A detailed analysis shows that the chemical shift frequency is approximately the same as that of the downfield principal value of the chemical shift tensor. This demonstrates unequivocally that each of the phosphates as the same structural properties and orientation in the virus particle. Furthermore, the orientationally dependent frequency observed for the chemical shift imposes very strong structural constraints for the modeling of the DNA in Pf1. With the incorporation of geometric restraints for the DNA in Pf1,[17] a structure consistent with the NMR data for one of the strands of the DNA is shown in Figure 2.2.

Although they differ from some interpretations of diffraction results,[18] the NMR data are unambiguous and are very straightforward to interpret, since essentially no intervening theory is needed. Further, the chemical composition of the virus particles strongly supports the NMR findings. The ratios of nucleotides to coat protein subunits are strikingly different at 2.4:1 for the Class I fd and 1:1 for the Class II Pf1. The relatively large and non-integral ratio for fd is consistent with the lack of symmetry in its packing, whereas the 1:1 ratio is fully consistent with the precise alignment and necessary packing with the coat protein subunits in Pf1. This suggests that it will be easier to modify the nucleotide composition in fd than in Pf1 because of its physically less stringent packing requirements.

2.3   Structural Form of the Major Coat Protein

There are two stable, biologically relevant forms of the filamentous bacteriophage major coat protein: the membrane-bound form and the structural form that constitutes the protective coat. The protein is stored in the cell membrane prior to assembly, where it becomes the most abundant membrane protein. Assembly occurs at the cell membrane as the newly formed virus particles are extruded with the coat protein subunits in their structural form encasing the viral DNA. A major structural rearrangement is required during the assembly process for the coat protein subunits to go from their membrane-bound conformation to the structural form in the bacteriophage particles. Fiber diffraction studies of filamentous bacteriophage particles[2,5,19] have provided essential structural information, in particular the overall architecture of the virus particles and packing of the coat protein subunits about the DNA. Supplemented by spectroscopic studies,[6,20–22] the fiber diffraction studies demonstrate that the major coat protein subunits are highly helical, the C-terminal residues are buried and interact with the DNA and the N-terminal residues are exposed on the surface. Cryo-electron microscope studies have contributed additional insights into the structure of the particles.[23] Both magic angle sample spinning solid-state NMR[24,25] and oriented sample (OS) solid-state NMR[3,26–29] have provided atomic resolution structural details about the coat proteins in the filamentous bacteriophage particles.

The principal approach that we have used to determine the three-dimensional structures of the major coat protein subunits in the filamentous bacteriophage particles utilizes OS solid-state NMR spectroscopy because each of the coat protein subunits is aligned as part of the intact virus particles by the magnetic field.[26] The high degree of uniaxial alignment results from the cumulative diamagnetic anisotropy of the peptide groups arranged in helices along the long axis of the particles combined with the liquid crystalline character of concentrated solutions of the filaments. Importantly, since all of the protein subunits are identical and are arranged symmetrically, differing only by rotation and translation, each ^{15}N-labeled amide site contributes a narrow single-line resonance to the NMR spectrum.

In spite of this being, at least in principle, a very favorable situation for structural studies by OS solid-state NMR, the one-dimensional NMR spectrum of a uniformly ^{15}N-labeled sample of a filamentous bacteriophage is an essentially unresolved lump due to the overlap among the 40 or so resonances from rigid, helical residues aligned approximately along the direction of the magnetic field, even though each individual resonance is relatively narrow because of the high degree of alignment. The overlap problem can be addressed through multidimensional NMR experiments and isotopic labeling. For example, the

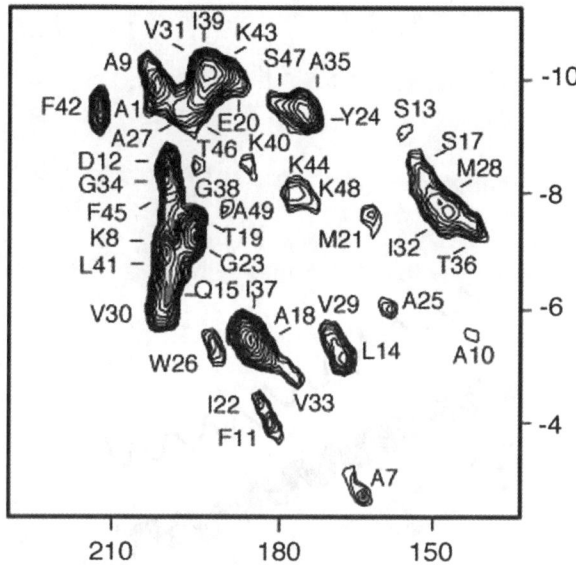

Figure 2.3 Experimental two-dimensional $^1H/^{15}N$ solid-state NMR spectrum of uniformly ^{15}N-labeled fd coat protein in filamentous bacteriophage particles aligned in the magnetic field of the NMR spectrometer.[27] The sequential assignments of the amide resonances are indicated. Each resonance is characterized by the orientationally-dependent frequencies associated with the ^1H–^{15}N heteronuclear dipolar coupling and ^{15}N chemical shift interactions These two frequencies provide sufficient angular constraints to calculate the atomic-resolution structure of the coat protein in the virus particles shown in Figure 2.4.

two-dimensional separated local field NMR spectrum of uniformly ^{15}N-labeled fd coat protein in aligned bacteriophage particles is shown in Figure 2.3.[27] In this experiment, the frequencies that characterize the dipole–dipole couplings between the directly bonded ^1H and ^{15}N at each amide site in the backbone of the protein are correlated with the ^{15}N chemical shift for that site. Both the dipole–dipole and chemical shift interactions are anisotropic, that is, their spectral frequencies are determined by the angles between the interactions as located in the molecule and the direction of orientation in the magnetic field.[30] Complementary data from selectively isotopically labeled samples enabled the resonance assignments to be made, as marked in Figure 2.3, and provided assurances about which resonances contribute to partially overlapped peaks in the spectrum. The thorough analysis of the spectroscopic data permitted a detailed structural analysis.[27]

　　Figure 2.4 presents a graphical analysis of the orientationally dependent frequencies measured from the signals in the two-dimensional spectrum in Figure 2.3. For clarity, the protein sequence is divided into three parts, shown in red, blue and green to provide a correlative color scheme for this presentation. The left column (Figure 2.4A) represents the experimental frequencies as

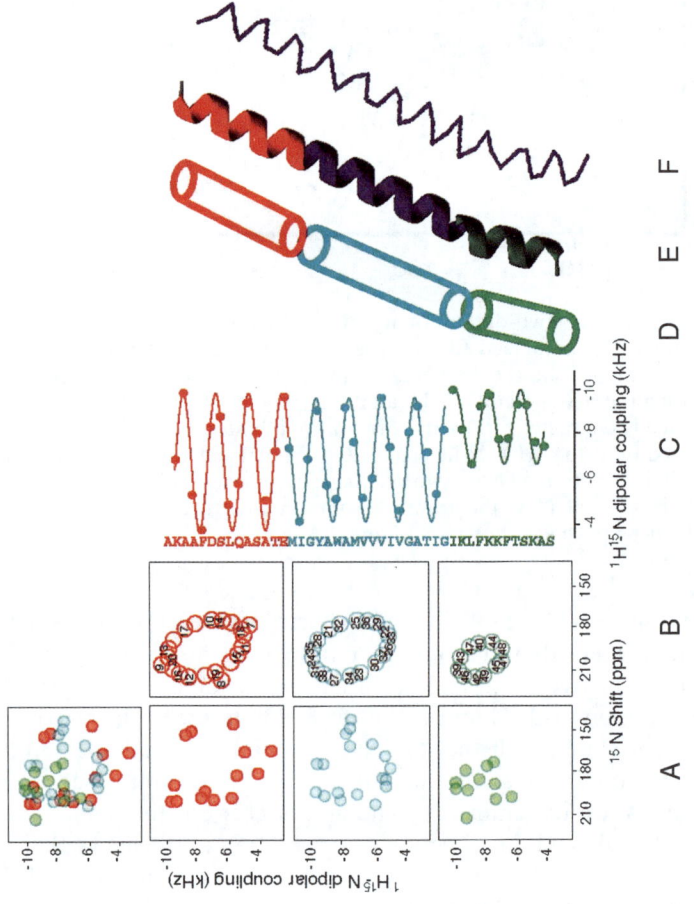

Figure 2.4 (A) Representations of the experimental NMR spectrum of fd coat protein in aligned bacteriophage particles. (B) PISA wheels corresponding to the experimental data in (A) and the dipolar waves is (C). (D) Tube representation of fd coat protein in the bacteriophage particles. (E) Ribbon representation of fd coat protein in the bacteriophage particles. (F) Structure of fd coat protein in bacteriophage particles determined by solid-state NMR.[27]

dots identified by color for comparison with ideal, calculated PISA wheels[31,32] in Figure 2.4B. Perfect matches between the experimental frequencies and those calculated for ideal helices are not expected because there is site-to-site variation among the chemical shift interactions at each amide nitrogen site. This variation is small enough that it does not interfere with the identification or interpretation of the pair of frequencies associated with each resonance. Residues 7–20 are the amphipathic helix that lies on the surface of bilayers in the membrane-bound form of the protein, as it is stored prior to assembly, and a break in the helix was anticipated near residue 21 at the 'hinge' between the surface amphipathic helix and the transmembrane hydrophobic helix. Remarkably, both the PISA wheels and dipolar waves,[33] which are plots of the dipolar coupling frequencies as a function of residue number, show that residues 7–39 are in a continuous helix without evidence of a kink or distortion near residue 21. On the other hand, there is a distinct kink near residue 39. This was unexpected. For clarity, the major coat protein is represented as tubes in Figure 2.4D and ribbons in Figure 2.4E. These representations are fairly close to that obtained by a rigorous mathematical structural fitting to the spectroscopic data shown in Figure 2.4F. For a fully assigned spectrum, structural fitting is equivalent to the direct calculation of the structure from the individual frequencies.

The structure of fd coat protein determined by solid-state NMR spectroscopy shown in Figure 2.4F is substantially different from that derived from X-ray fiber diffraction, which characterizes the entire protein as 'a single gently curving a helix'.[19] However, differences in the resolution of the two structures may account for most of the differences. Thorough comparisons between X-ray crystal structures and simulations demonstrate that well-fitted dipolar waves, as in Figure 2.4C, correspond to resolution better than that found in a 1 Å crystal structure. In contrast, the fiber diffraction structure results from model fitting with a resolution of about 7 Å.[2,19] However, the fiber diffraction data provide valuable complementary information in the form of the inter-subunit spacing and symmetry that enable the coat protein subunit structures to be assembled into virus particles, as shown in Figure 2.5. The most notable features are the open volume in the center where the DNA is packaged and the N-terminal residues of the coat protein subunits that extend out into the aqueous solution and are where the additional residues are added for phage display of peptides.

Comparisons between Class I (fd) and Class II (Pf1) bacteriophage are instructive in understanding their architectures[15] and lifecycles. One of the major differences is in how the DNA is packaged, as demonstrated by the data in Figure 2.2. Even though both of their coat proteins are predominantly helical, there are substantial differences in detail, most notably that the structural form of the Pf1 coat protein contains strong residual evidence of the hinge involved in the transition from the membrane-bound during the assembly of the virus particles.

A distinctive property of the three-dimensional structures of Pf1 coat protein determined by solid-state NMR spectroscopy is that the N-terminal segment of the protein forms a six-residue 'double hook', which can be seen in both forms

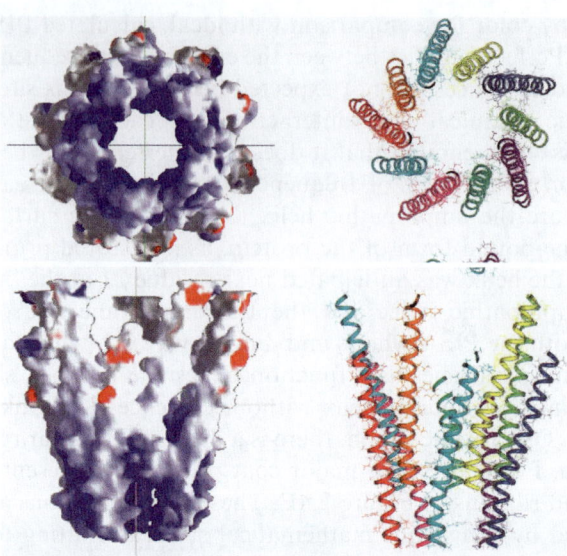

Figure 2.5   Representation of the viral coat based on the structure of coat protein in
Figure 2.4F and the organization of the viral particles from fiber diffraction.[2,19] The unit cell is a pentamer and the helical assembly has a twofold
screw axis with a 1.6 nm pitch. The positive charges line the interior for
interactions with the DNA, with the single negative charge exposed to the
solvent.

of the protein displayed in Figure 2.6. For comparison, previous descriptions of
these N-terminal residues in the structures based on X-ray fiber diffraction have
included helical[34] and extended, unstructured[35] conformations.

   Interestingly, Pf1 differs from fd in that it undergoes a temperature-dependent
structural transition.[36,37] At low (0 °C) and relatively high (>20 °C) temperatures, Pf1 coat protein has the same basic secondary structure and topology,
which consists predominantly of three distinct helical segments, in addition to
the distinctive N-terminal segment. Each helical segment has a slightly different
tilt angle relative to the filament axis; there are kinks near Q16 and A29. Solid-
state NMR spectra of aligned samples are sensitive monitors of protein structure
because the resonance frequencies reflect orientations with respect to the filament axis and magnetic field; therefore, the rather dramatic spectroscopic
changes that occur as a function of temperature reflect fairly subtle changes in
protein structure.[29] This structural transition of the coat proteins may reflect
the balancing of packing interactions; perhaps it is advantageous for the survival
of a long, thin, filamentous bacteriophage particle under the wide range of
temperatures and conditions encountered under widely varying environmental
situations in Nature. When the central helical segments of the two structures are
overlaid in Figure 2.6 using only uniaxial rotations and translations consistent
with the experimental NMR data, the C-terminal segments have different tilt
angles. Notably, there are differences in the first five N-terminal residues at the

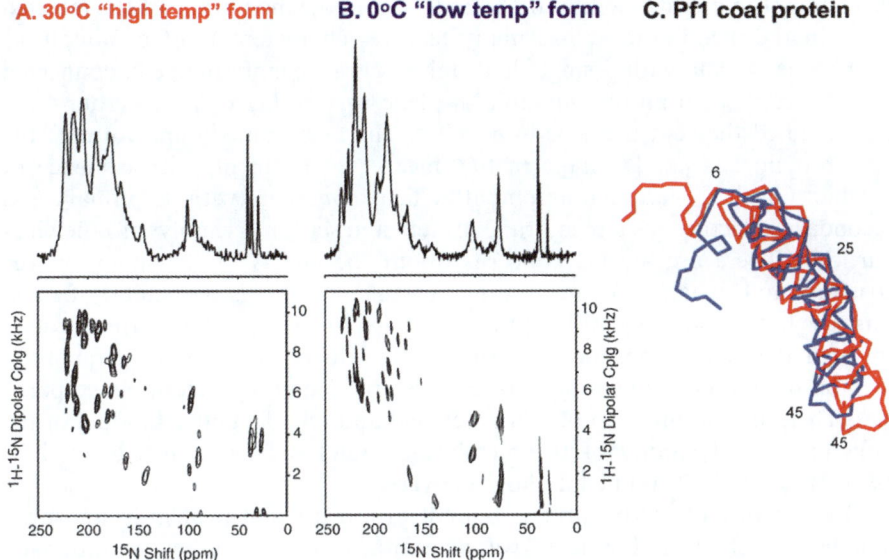

Figure 2.6   Comparison of the spectra and structures of the high- and low-temperature forms of Pf1 coat protein in the bacteriophage. (A) High-temperature
form (30 °C). (B) Low-temperature form (0 °C). (C) The two structures are
superimposed with the high-temperature form in red and the low-temperature form in blue.[29]

two temperatures examined, although they both appear to have distinctive
'double hook' conformations.

In considering phage display of peptides, the structure, dynamics and solvent
exposure of the first few residues at the N-terminus are of crucial importance.
In Pf1, they are structured with the already noted 'double hook' arrangement
that appears to bring the N-terminus in towards the body of the particle and
allowing little flexibility for the addition or insertion of amino acid residues.
In contrast, in fd coat protein, the first five residues are mobile and unstructured and do not have measurable dipolar couplings, and residue 6 is a proline
that terminates the N-terminal helix. Hence Class I bacteriophage such as
fd and M13 appear to be very well suited for display of added peptides at their
N-terminus.

2.4   Membrane-bound Form of Filamentous Bacteriophage Coat Proteins

In vivo, the coat protein of fd is expressed as the procoat protein with a
23-residue leader sequence. The procoat protein is inserted into the cell membrane through the intervention of the translocase YidC.[38,39] After processing,
which consists of cleaving off the leader sequence, the bacteriophage major coat

protein is stored in its membrane-bound form in the membrane of the infected cells. In the membrane, it has many features characteristic of a monotopic membrane protein with a single hydrophobic transmembrane helix connected by a short loop to an amphipathic in-plane surface helix. Many groups[40–44] have studied the coat proteins in micelles. The membrane-bound form of the coat protein has an 'L' shape in the micelle environment, with evidence of flexibility in the N-terminal amphipathic helix consistent with its primary and secondary structures as a membrane protein and set for assembly into the virus particles. There are always concerns about the effects of the lipids on the structure and dynamics of membrane proteins and this is particularly true in this case because the assembly of the virus particles occurs as they are extruded through the cell membrane with the coat proteins transitioning from their membrane-bound form to their structural form found in bacteriophage particles. Therefore, in order to obtain a detailed and reliable understanding of the structure and dynamics of the membrane-bound form of the protein, it is essential to study it in phospholipid bilayers.

The structure of the membrane-bound form of fd coat protein in phospholipid bilayers shown in Figure 2.7 was determined solely from the orientationally dependent frequencies measured from solid-state NMR spectra,[45] similar to

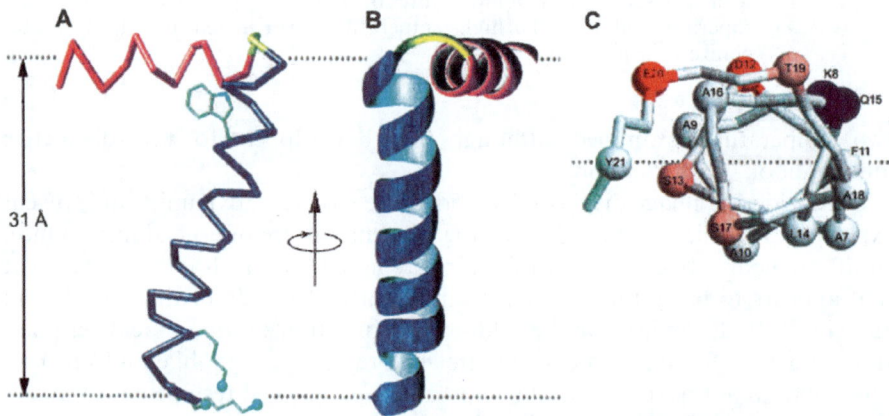

Figure 2.7   Representations of the membrane-bound form of fd coat protein in phospholipid bilayers determined by NMR spectroscopy.[45] The surface amphipathic helix is shown in magenta, the hydrophobic transmembrane helix in blue and the short inter-helical connecting loop in yellow. The flexible N- and C-terminal residues are not shown. The dotted lines mark the lipid–water boundary. (A) Side view showing the 26° tilt of the transmembrane helix. The Trp 26 side-chain is shown in its experimentally determined orientation. The arrow shows the direction of the applied magnetic field, which is collinear with the normal to the bilayers. The side-chains for Lys40, Lys43 and Lys44 side-chains are modeled using the program MolMol and face the cytoplasmic side of the membrane. (B) Front view. (C) View of the amphipathic surface helix from the C-terminus. The α-carbons are shown as spheres.

those shown in Figures 2.3 and 2.6. Because the samples consisted of the protein in phospholipid bilayers, the results are more definitive than those obtained on samples of the coat protein in micelles. These results demonstrate unambiguously that the membrane-bound form of fd coat protein has two distinct α-helical segments. The in-plane helix is amphipathic and rests on the membrane surface with the boundary separating the polar and hydrophobic residues parallel to the bilayer surface and the hydrophobic residues facing the hydrocarbon core of the lipid bilayer. The aromatic residues F11 in the in-plane helix and Y21 in the transmembrane helix are near this boundary region. The transmembrane helix crosses the membrane at an angle of 26° to near residue 40, where the helix tilt changes to 16°. The helices are connected by a short turn (Thr19 and Glu20), which differs from the substantially longer loop (residues 17–26) found for the same protein in micelles. In addition to providing information about the hinge that is key to the structural rearrangement that accompanies assembly, this is among the earliest examples of showing differences between proteins in micelles and bilayers, reinforcing the view that membrane proteins have to be evaluated in their native phospholipid bilayer environment.

The differences between Class I and Class II bacteriophage are also manifested in the properties of the membrane-bound forms of their coat proteins, in addition to the structural difference in the bacteriophage particles. As seen with fd coat protein, solid-state NMR signals from the N-terminal segment of Pf1 coat protein definitively show that the helix lies on the bilayer surface. Solution NMR experiments demonstrated that these residues form an amphipathic α-helix in micelles, and recent solution NMR experiments that take advantage of the ability to prepare samples of membrane proteins in DHPC micelles with two different types of alignment have enabled us to determine the structure of the protein in micelles.[46] NMR data obtained on DHPC micelle samples also demonstrate that Pf1 coat protein forms an α-helix in the N-terminal region. The protein has a rigid transmembrane helix and the mobile interhelical loop and N-terminal helix (Figure 2.8). The structure of the membrane-bound form

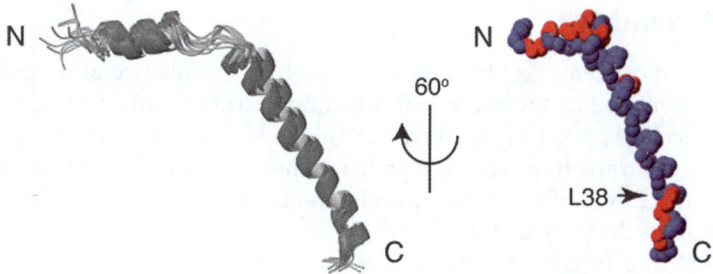

Figure 2.8 Structure of the membrane-bound form of Pf1 coat protein in micelles.[46] Left: superimposition of 15 calculated backbone structures of Pf1 coat protein. Right: 60° rotation of one structure to the vertical axis. A slight kink at residue 38 is indicated. The hydrophilic residues are colored red and the hydrophobic residues blue.

of Pf1 coat protein was calculated using all of the available restraints measured from solution NMR experiments in micelles and solid-state NMR experiments in bicelles and bilayers. Excellent correlations were obtained between the experimental and back-calculated RDCs, DCs and CSAs, when all were used as restraints in simulated annealing calculations.

The structure of the membrane-bound form of Pf1 coat protein is shown in Figure 2.8. Comparison with the previously determined structure of the coat protein incorporated in the bacteriophage particle (Figure 2.6) shows that the tilt of the hydrophobic helix relative to the axis of alignment (*i.e.* the membrane normal and the bacteriophage long axis) differs by $\sim 8°$ in the two forms ($\sim 30°$ in the membrane and $\sim 22°$ in the bacteriophage), as expected based on the NMR data.

In bilayers, the coat protein's transmembrane α-helix spans residues 23–42 and is tilted 30° from the lipid bilayer normal. In virus particles, the structural form of the protein consists of a long α-helix that spans residues 7–46 that is interrupted by several kinks. Interestingly, the kink at A29 observed in the structural form is also found in bicelles, as indicated by the deviation of the dipolar coupling measured for residue 29 from the sinusoidal fitting function. We have previously shown that the secondary structure of the Pf1 coat protein is preserved in lipid environments.[47] However, its dipolar wave plot has a different phase, which indicates that the helix rotation differs by 160° in the two forms of the protein, as shown in the corresponding helical wheel projections.

Notably, the 30° tilt of the transmembrane helix of membrane-associated Pf1 coat protein is very similar to that found for the membrane-associated fd bacteriophage coat protein, whose structure has been determined in glass-aligned planar lipid bilayers by solid-state NMR.[45] The two aromatic residues, Y25 and Y40, in Pf1 are situated near the periplasmic (Y40) and cytoplasmic (Y25) faces of the bacterial membrane and are likely to play a role in determining transmembrane helix tilt and orientation, similar to Y21, Y24, W26 and F45 in the fd coat protein.

2.5   Assembly

Among the most remarkable features of the filamentous bacteriophage is that they are assembled at the cell membrane and extruded out of the cell without lysing the cell. The principal player in this is the major coat protein, which undergoes the transition from a membrane protein to the structural coat surrounding the DNA of the virus particles. This involves major structural rearrangements of the polypeptides.[3]

Both Pf1 and fd transmembrane helices are rotated so that the C-terminal positively charged (R44 and K45 in Pf1; K40, K43 and K44 in fd) and polar (S41 in Pf1; T36 in fd) residues face the cytoplasmic side of the membrane. This topology may have functional significance for the bacteriophage assembly process, during which the single-stranded phage DNA is extruded through the inner bacterial membrane after being enclosed within a tube of coat protein subunits. The orientation of the coat protein transmembrane helix enables the charged amino groups of the lysine and arginine side-chains to emerge from the membrane interior into the cytosol and thus become available at the membrane surface for DNA binding during bacteriophage extrusion.

In contrast, small hydrophobic residues (Gly, Ala) line the transmembrane helix side facing the extracellular leaflet of the membrane. Some of these residues (G24, A36) are conserved in the sequences of the coat proteins of Pf1, fd and M13 bacteriophage, suggesting that they could play a role in mediating helix–helix interactions during viral assembly and in the assembled viral particles. Indeed, the small-xxx-small motif found in the G24xxxG28 sequence of Pf1 is conserved in the sequences of other filamentous bacteriophage coat proteins (*e.g.* G23xxxA27 in fd and M13) and was found to be important for stabilizing the interaction between the coat protein subunits in the bacteriophage capsid.

In the Pf1 bacteriophage particle, each coat protein subunit at a given position is closely packed with its neighbors at positions $i \pm 5$, $i \pm 6$ and $i \pm 11$, resulting in extensive hydrophobic interactions, which as in the case of fd contribute to the stability of the capsid. The side-chain of R44 is directed inwards, as expected for a side-chain that interacts with DNA and the charged and polar residues at the N-terminus face the exterior of the phage particle to confer solubility in water. Inspection of the phage-associated form of the protein indicates that the side-chains of K20 in subunit $i + 11$ and of D4 in subunit i could be sufficiently close to interact *via* a salt bridge.

During phage assembly, as the DNA–protein complex is extruded through the inner bacterial membrane, the unpaired positive charges of K20 on subunits i, $i + 5$ and $i + 6$ could direct the assembly of the next subunit ($i + 11$) and its transformation from the membrane-bound to the phage-bound form, by interacting with acidic residues in the N-terminal membrane-associated helix (Figure 2.9). This electrostatic interaction between membrane-bound coat protein and phage capsid, together with the electrostatic interaction between basic residues in the protein's C-terminus and the viral DNA, could then position the coat protein for further hydrophobic interactions with other capsid-incorporated coat protein subunits and thus drive its conformational change, whereby the helix-connecting loop itself adopts a helical conformation straightening the coat protein (Figure 2.9).

2.6   Phage Display

Peptides are pervasive throughout biology and, as a result, there has been a large amount of research devoted to understanding their functions and structures and adapting them for applications in medicine and biotechnology. Their many roles in biochemistry and biotechnology are largely a consequence of their binding to proteins with high specificity. They function in signaling as hormones and other types of effector molecules, hence they are ripe for roles in

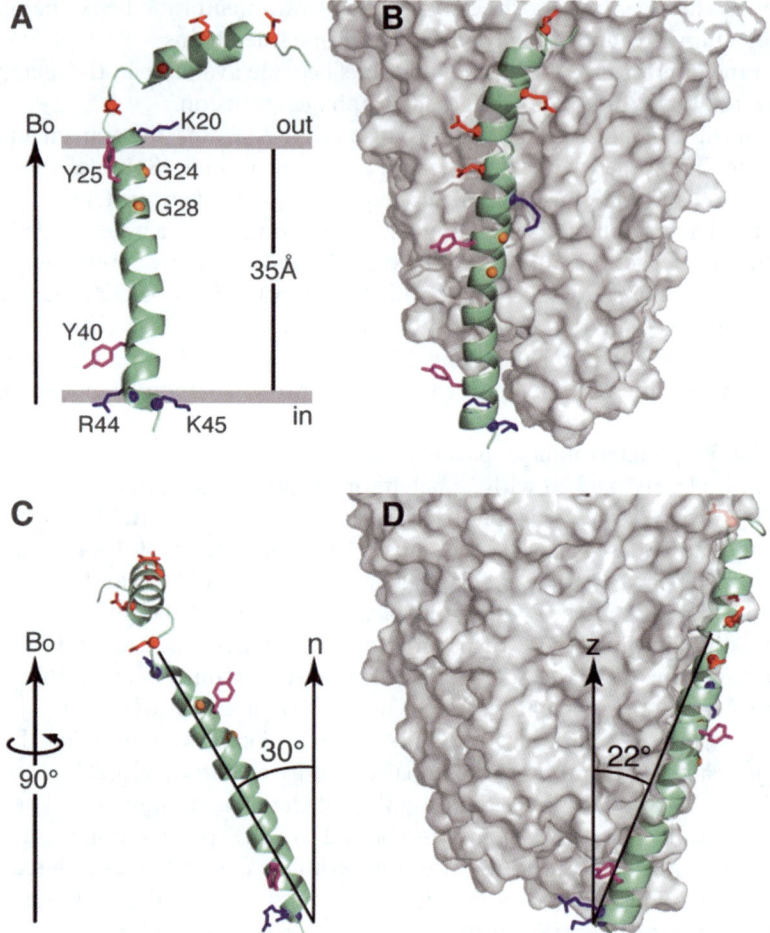

Figure 2.9   (A, C) Structures of the Pf1 coat protein in the membrane. (B, D) Structures of the Pf1 coat protein in the intact bacteriophage particles as determined by NMR spectroscopy. The protein's N-terminus is exposed to the bacterial periplasmic space (out) and its C-terminus is in the cytoplasm (in). The orientation of the magnetic field (B_0) is parallel to the membrane normal (n) and to the long axis of Pf1 bacteriophage (z). Acidic residues (Asp, Glu) are shown in red, glycines in the transmembrane helix in yellow, basic residues (Arg, Lys) in blue and interfacial tyrosines in pink. In the Pf1 phage particle, R44 and K45 face the DNA interior.

drug discovery. They are also epitopes when they bind to antibodies and this has consequences for the development of vaccines.

There are two main sources of peptide sequences for drug design or vaccine development. Peptides can be selected from libraries, which can be prepared by biological methods, in particular phage display,[8] or by a variety of synthetic strategies. The purpose of the library is to generate as many peptide sequences as possible within certain constraints, for example, the total number of residues or the presence of disulfide linkages. The next step is to select those sequences of greatest interest based on their binding affinities for target receptors.[9,10] It is also possible to design peptide sequences from first principles of protein structure or starting with sequences found in biologically active peptides or proteins.

Few structures of linear peptides have been determined because of the difficulties they present for the standard experimental approaches to structure determination. Peptides rarely crystallize in forms suitable for X-ray diffraction and when they do there are concerns about the biological relevance of their conformations. Peptides are typically too flexible in solution to be analyzed definitively by NMR spectroscopy and these studies also suffer from uncertainties about the averaging among multiple conformations. Solvent effects are also of concern.

The filamentous bacteriophage provide opportunities to apply the methods of structural biology to peptides in a novel setting. Of course, this is of particular relevance because these same bacteriophage are the most widely used systems for the generation of peptide libraries. The N-terminal region of the major coat protein of Class I bacteriophage provides an environment conducive to stabilizing peptide conformation, as shown by the increased biological activity of amino acid sequences inserted into the coat protein as compared with the peptide,[48] and also the results from our own NMR studies.[49,50] The residues in the larger inserted peptide sequence are completely immobilized in a uniquely folded conformation when the coat proteins are in the intact virus particles.[50] Further, the peptide sequence of interest is aligned along with the virus particles by the magnetic field of the NMR spectrometer. The same filamentous bacteriophage coat protein that is used to generate the peptide sequences also provides an ideal platform for determining the structures of the peptides by the solid-state NMR methods that we are developing. During our fruitful collaboration with Richard Perham at Cambridge University, we determined the structures of three different peptide epitopes displayed on the N-terminus of fd coat protein by NMR spectroscopy.

The principal neutralizing determinant of many strains of HIV-1 is a polypeptide loop, designated V3, in the third hypervariable region of the envelope glycoprotein gp120, with the conserved sequence GPGRAF at its tip. We were able to determine the structure of the six-residue sequence GPGRAF inserted in the N-terminal region of the major coat protein with solution NMR experiments on samples of the recombinant protein in micelles.[49] The two-dimensional spectrum of the 56 residue recombinant fd coat protein containing the GPGRAF insert displays extra resonances from five of the inserted residues (the Pro imino nitrogen is not observed). The observation of a large number of inter-residue backbone NOEs in a three-dimensional NOESY spectrum was remarkable and indicative of a highly populated stable conformation for the six inserted residues. Nearly 60 short- and medium-range backbone NOEs were detected for residues 2–10 of the recombinant protein. No NOE peaks were observed between the six residues of the insert and the remainder of the coat

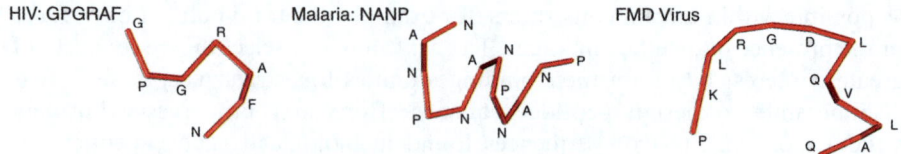

Figure 2.10   Structures of peptides determined by NMR as part of the fd coat protein. The structures represent the three epitopes labeled in the figure that were inserted near the N-terminus of fd coat protein.[49,50]

protein and the resonances for residues 12–50 are unaffected by the presence of the six added residues, indicating that the bulk of the recombinant protein has the same structure as the wild-type protein and does not interact with the six inserted residues. The three-dimensional structure of the GPGRAF residues shown in Figure 2.10 was calculated from the NMR data. The GPGRAF epitope adopts a well-defined double-bend, S-shaped conformation similar to the structure determined in the antibody–epitope complex.[51] The similarity between the conformations of the unbound and antibody-bound epitope is significant and identifies the role of the GPGPRAF double-turn structure in antibody–antigen recognition. The compact structure of turn regions can maximize the contact area between the antibody and the peptide, which might result in stronger binding.[52] All of the evidence points to the GPGRAF sequence in the contest of the N-terminal region of the coat protein adopting a double-turn structure similar to that in the native gp120 protein.

Much of the immunology of malaria is focused on the circumsporozoite (CS) protein. In the case of *Plasmodium falciparum*, the sequence NANP is repeated 41 times in the protein. Considerable effort has gone into developing vaccines for malaria utilizing this repeated peptide sequence.[53] The conformation of the repeating peptide (NANP)n is thus of interest for the design of vaccines and understanding the CS protein.[54] Several proteins containing several copies of the NANP repeating peptide unit have been successfully used to raise a high titer of antibodies. When the epitope is displayed on the surface of the filamentous bacteriophage fd, an even higher immunogenic response is observed. We studied three repeats (NANP)3 of this sequence inserted between residues 3 and 4 of fd coat protein.[13] Hybrid phage are used to display this peptide because the recombinant coat protein has a total of 62 residues, which is too large to form particles with 100% coverage with the peptide. The extra resonances from the inserted (NANP)3 residues are smaller than the others in the solution NMR spectrum because only about 30% of the proteins carry the insert and the other peaks represent sites present in both the 50-residue wild-type protein and the rest of the recombinant protein. The many NOEs measured from these samples permitted the calculation of the structure in Figure 2.10.[50]

The foot and mouth disease (FMD) virus is a single-stranded RNA virus.[55] The icosahedral shell consists of 60 copies of each of four structural proteins VP1–4. The VP1–3 proteins are partly exposed on the surface whereas VP4 is internal and has an N-terminal myristic acid. The major antigenic site has been localized to a region known as the G–H loop (residues 134–160) in the VP1 protein[56] and synthetic peptides corresponding to this region were found to elicit neutralizing antibody responses. In addition, vaccination with these peptides protects animals against subsequent challenges by the virus. A 12-residue sequence from this G–H loop region of the VP1 protein was inserted between residues 3 and 4 of fd coat protein by Perham and co-workers.[48] This construct enabled us to express and isotopically label the peptide containing coat protein. This is an interesting example, because direct comparisons can be made with the structure of the same residues in the intact protein characterized by X-ray diffraction.[55] The structure determined from the NOE data is shown in Figure 2.10.

2.7   Conclusion

It is not a coincidence that one of the simplest organisms is turning out to be among the most useful for practical applications in biotechnology and nano-technology. The small coat protein provides opportunity for manipulation of its sequence, especially near the exposed, flexible N-terminus of the fd coat protein. Also, the DNA code for the altered coat protein is linked to the expression of the phenotype through the intimate packaging of the DNA by the protein coat. This also provides the mechanism for production of large quantities of particles that display interesting properties as starting points for drugs, vaccines, electrical circuits and other nanotechnology applications.

The structure determinations of both the membrane-bound and structural forms of the coat proteins of a Class I (fd) and a Class II (Pf1) bacteriophage described in this chapter contribute to defining the boundaries of what can be assembled and displayed in the bacteriophage particles that are essential to assay the properties of the modified proteins. As described in this volume, the field is maturing; with more sophisticated medical applications and nanodevices in the offing. The information about the structure, dynamics and assembly of the coat protein subunits will increase in importance in their refinement. As that process goes forward, it will be necessary to revisit many of the issues discussed in this chapter to interrogate the synergy among the various components and to improve the structural resolution, especially for the unusual features of turns, kinks, interactions and other features that were surprising to find in what appeared to be simple helical peptides when the filamentous bacteriophage were first discovered.[57]

Acknowledgements

The research summarized in this chapter was performed in many laboratories throughout the world, attesting to the importance of the filamentous bacterio-phage in biology and medicine. The research from the author's laboratory benefited from long-term collaborations with Lee Makowski and Richard Perham and discussions with many of the other researchers in the field. It also benefited from the contributions of many students and postdoctoral researchers in the group, as recognized as authors of the cited references. This research was supported by grants from the National Institutes of Health. It utilized the Biomedical Technology Resource for NMR Molecular Imaging of Proteins at the University of California, San Diego, which is supported by grant P41EB002031.

References

1. R. Webster, Filamentous phage biology, in *Phage Display: a Laboratory Manual*, ed. C. F. Barbas, D. R. Burton, J. K. Scott and G. J. Silverman, Cold Spring Harbor Laboratory Press, Cold Spring Harbor, NY, 2001, pp. 1–37.
2. D. A Marvin, *Curr. Opin. Struct. Biol.*, 1998, **8**, 150–158.
3. S. J. Opella, A. C. Zeri and S. H. Park, *Annu. Rev. Phys. Chem.*, 2008, **59**, 635–657.
4. M Russel, *Trends Microbiol.*, 1995, **3**, 223–28.
5. R. Nambudripad, W. Stark and L. Makowski, *J. Mol. Biol.*, 1991, **220**, 359–379.
6. G. J. Thomas, B. Prescott and L. A. Day, *J. Mol. Biol.*, 1983, **165**, 321–356.
7. J. Sambrook, E. F. Fritsch, T. Maniatis, N. Irwin and T. Maniatis, *Molecular Cloning: a Laboratory Manual*, Cold Spring Harbor Laboratory Press, Cold Spring Harbor, NY, 1989.
8. G. P. Smith, *Science*, 1985, **228**, 1315–1317.
9. J. K. Scott and G. P. Smith, *Science*, 1990, **249**, 386–390.
10. V. Petrenko, *Expert Opin. Drug Deliv.*, 2008, **5**, 825–836.
11. D. J. Rodi, L. Makowski and B. K. Kay, *Curr. Opin. Chem. Biol.*, 2001, **6**, 92–96.
12. C. Mao, D. J. Solis, B. D. Reiss, S. T. Kottmann, R. Y. Sweeney, A. Hayhurst, G. Georgiou, B. Iverson and A. M. Belcher, *Science*, 2004, **303**, 213–217.
13. J. Greenwood, A. E. Willis and R. N. Perham, *J. Mol. Biol.*, 1991, **220**, 821–827.
14. P. Malik and R. N. Perham, *Gene*, 1996, **171**, 49–51.
15. T. A. Cross, P. Tsang and S. J. Opella, *Biochemistry*, 1983, **22**, 721–726.
16. P. Tsang and S. J. Opella, *Biopolymers*, 1986, **25**, 1859–1864.
17. L. A. Day, R. L. Wiseman and C. J. Marzec, *Nucleic Acids Res.*, 1979, **7**, 1393–1403.
18. M. F. Symmons, L. C. Welsh, C. Nave, D. A. Marvin and R. N. Perham, *J. Mol. Biol.*, 1995, **245**, 86–91.
19. M. J. Glucksman, S. Bhattacharijee and L. Makowski, *J. Mol. Biol.*, 1992, **226**, 455–470.
20. B. A. Clack and D. M. Gray, *Biopolymers*, 1989, **28**, 1861–1873.
21. G. E. Arnold, L. A. Day and A. K. Dunker, *Biochemistry*, 1992, **31**, 7948–7956.

22. S. A. Overman, D. M. Kristensen, P. Bondre, B. Hewitt and G. J. Thomas, Jr., *Biochemistry*, 2004, **43**, 13129–13136.
23. Y. A. Wang, X. Yu, S. Overman, M. Tsuboi, G. J. Thomas Jr and E. H. Egelman, *J. Mol. Biol.*, 2006, **361**, 209–215.
24. S. J. Opella, M. H. Frey and T. A. Cross, *J. Am. Chem. Soc.*, 1979, **101**, 5856–5857.
25. A. Goldbourt, B. J. Gross, L. A. Day and A. E. McDermott, *J. Am. Chem. Soc.*, 2007, **129**, 2338–2344.
26. T. A. Cross and S. J. Opella, *J. Am. Chem. Soc.*, 1983, **105**, 306–308.
27. A. C. Zeri, M. F. Mesleh, A. A. Nevzorov and S. J. Opella, *Proc. Natl. Acad. Sci. USA*, 2003, **100**, 6458–6463.
28. D. S. Thiriot, A. A. Nevzorov, L. Zagyanskiy, C. H. Wu and S. J. Opella, *J. Mol. Biol.*, 2004, **341**, 869–879.
29. D. S. Thiriot, A. A. Nevzorov and S. J. Opella, *Protein Sci.*, 2005, **14**, 1064–1070.
30. S. J. Opella, P. L. Stewart and K. G. Valentine, *Q. Rev. Biophys.*, 1987, **19**, 7–49.
31. F. M. Marassi and S. J. Opella, *J. Magn. Reson.*, 2000, **144**, 150–155.
32. J. Wang, J. Denny, C. Tian, S. Kim and Y. Mo, *et al., J. Magn. Reson.*, 2000, **144**, 162–167.
33. M. F. Mesleh and S. J. Opella, *J. Magn. Reson.*, 2003, **163**, 288–299.
34. D. A. Marvin, *Int. J. Biol. Macromol.*, 1990, **12**, 125–138.
35. A. Gonzalez, C. Nave and D. A. Marvin, *Acta Crystallogr., Sect. D*, 1995, **51**, 792–804.
36. E. J. Wachtel, R. J. Marvin and D. A. Marvin, *J. Mol. Biol.*, 1976, **107**, 379–383.
37. L. C. Welsh, M. F. Symmons and D. A. Marvin, *Acta Crystallogr., Sect. D*, 2000, **56**, 137–150.
38. J. C. Samuelson, M. Chen, F. Jiang, I. Moller and M. Wiedmann, *et al.*, *Nature*, 2000, **406**, 637–641.
39. M. Chen, K. Xie, J. Yuan, L. Yi and S. J. Facey *et al.*, *Biochemistry*, 2005, **44**, 10741–10749.
40. T. A. Cross and S. J. Opella, *Biochem. Biophys. Res. Commun.*, 1980, **92**, 478–484.
41. F. C. Almeida and S. J. Opella, *J. Mol. Biol.*, 1997, **270**, 481–495.
42. G. D. Henry and B. D. Sykes, *Biochemistry*, 1992, **31**, 5284.
43. F. J. Van de Ven, J. W. Van Os, J. M. Aelen, S. S. Wymenga and M. L. Remerowski *et al.*, *Biochemistry*, 1993, **32**, 8322–8328.
44. K. A. Williams, N. A. Farrow, C. M. Deber and L. E. Kay, *Biochemistry*, 1996, **35**, 5145–5157.
45. F. M. Marassi and S. J. Opella, *Protein Sci.*, 1996, **12**, 403–411.
46. S. H. Park, W. S. Son, R. Mukhopadhyay, H. Valafar and S. J. Opella, *J. Am. Chem. Soc.*, 2009, **131**, 14140–14141.
47. R. A. Schiksnis, M. J. Bogusky and S. J. Opella, *J. Mol. Biol.*, 1988, **200**, 741–743.

48. F. D. M. Veronese, A. E. Willis, C. Boyer-Thompson, E. Appella and R. N. Perham, *J. Mol. Biol.*, 1994, **243**, 167–172.

49. R. Jelinek, T. D. Terry, J. J. Gesell, P. Malik, R. N. Perham and S. J. Opella, *J. Mol. Biol.*, 1997, **266**, 649–655.

50. M. Monette, S. J. Opella, J. Greenwood, A. E. Willis and R. N. Perham, *Protein Sci.*, 2001, **10**, 1150–1159.

51. J. B. Ghiara, E. A. Stura, R. L. Stanfield, A. T. Profy and I. A. Wilson, *Science*, 1994, **264**, 82–85.

52. T. Scherf, R. Hiller, F. Naider, M. Levitt and J. Anglister, *Biochemistry*, 1992, **33**, 6884–6897.

53. V. Nussenzweig and R. S. Nussenzweig, *Adv. Immol.*, 1989, **45**, 283–334.

54. H. J. Dyson, A. C. Satterthwait, R. A. Lerner and P. E. Wright, *Biochemistry*, 1990, **29**, 7828–7837.

55. R. Acharya, E. Fry, D. Stuart, G. Fox, D. Rowlands and F. Brown, *Nature*, 1989, **337**, 706–716.

56. J. L. Bittle, R. A. Houghten, H. Alexander, T. M. Shinnick, J. Sutcliffe, J. Gregor and R. A. Lerner, *Nature*, 1982, **298**, 30–33.

57. D. A. Marvin and E. J. Wachtel, *Nature*, 1975, **253**, 19–23.

CHAPTER 3

Quantitative Analysis of Peptide Libraries

LEE MAKOWSKI

Department of Electrical and Computer Engineering and Chemistry and Chemical Biology, Dana Research Center, Northeastern University, 360 Huntington Avenue, Boston, MA 02115, USA

3.1 Introduction

Affinity screening of phage-displayed combinatorial peptide libraries is carried out in order to identify a peptide sequence or sequence motif that binds tightly to a particular molecular target. The result of the experiment is a population of phage particles that represent a subset of the original library that has been enriched for binding to the molecular target. The nucleic acid sequences of the inserts in these phage particles are used to determine the sequences of the displayed peptides whose binding properties presumably led to their selection. In some cases, a sequence motif common to some (or many) of the derived sequences is immediately apparent from a visual inspection of the selected sequences. Frequently, no readily discernible motif can be identified from a visual inspection. This does not necessarily mean that the experiment failed. There are numerous examples in which weak motifs have been identified only after detailed computational analysis of the sequences. This chapter reviews the existing computational methods available for analyzing populations of peptides and identification of motifs within those populations. We consider five issues that can be effectively addressed using existing informatic tools.

The first issue considered is quantitative analysis of the quality of a peptide library. Experimentally, the quality of a peptide library is often taken to be

RSC Nanoscience & Nanotechnology No. 17
Phage Nanobiotechnology
Edited by Valery A. Petrenko and George P. Smith
Published by the Royal Society of Chemistry, www.rsc.org

equivalent to the number of clones extant within the library post-transfection. This measure does not take into account the possibility that different sequences may be present in different abundances. These libraries are produced within the context of a complex virus–host system. Each sequence in the library is subject to selective pressures with some, inevitably, more favorable for viral replication than others, leading to bias in their relative abundances. In order to take these biases into account, we have developed a quantitative measure of the diversity of a library that can be estimated from the sequences of about 100 peptides. This metric provides an intuitively reasonable measure of the relative utility of different libraries for affinity screening based on the degree of sequence bias within the library.

The second issue considered is the quality or effectiveness of an affinity screen. Experimentally, this can be measured by the 'amplification' associated with each round of affinity purification, or the input phage titer divided by the titer of phage eluted. Computationally, it can be quantitated as a change in diversity of the library. We will see that affinity selection has less effect on the diversity of a peptide library than might be expected. This is because, in many cases, slight variants in sequence are not excluded by the selection process and because the binding motif is usually shorter than the displayed peptide. This means that (a) there is a selective pressure on only a portion of the displayed peptide and (b) the motif may appear at a number of different positions within the peptide. Taking these factors into account, the concept of diversity provides an informative measure of the success of an affinity selection experiment.

A complementary measure of the quality of an affinity screen can be constructed by considering the likelihood of a particular peptide being observed within the library. This likelihood is expressed in terms of a measure of the 'information' associated with each peptide; information being greater for less abundant peptides. For instance, if I were to tell you it was 80 °F and sunny in Los Angeles, that might not be considered significant news. However, if it were snowing there – a relatively unlikely event – the statement associated with that fact would have much higher information content. Similarly, the observation of a peptide with a relatively low probability of occurrence has a higher level of associated information than the observation of a peptide that has a much higher expected abundance. Considering the relative abundances of rare – 'high-information' – peptides can provide insight into the effectiveness of an affinity selection.

The third issue is identification of sequence motifs. Recognition of one (or more) binding motifs from an affinity-selected population can be a substantial challenge depending on the strength, manner and specificity of binding. Standard informatic tools such as BLAST and FASTA have limited utility for the identification of sequence similarities in these peptides. BLAST uses a similarity matrix, most frequently a BLOSUM or PAM matrix, optimized on the basis of amino acid replacement frequencies during evolution. The use of this similarity matrix for detecting sequence motifs in very short peptide sequences (typically 12 or fewer amino acids in phage display experiments) is generally unsuccessful. For instance, the weighting of tryptophan in BLOSSUM is so great that it may dominate the comparison of short peptide segments, excluding all but trypto-phan-containing motifs. To correct this bias, a new similarity matrix has been developed here for use in the analysis of populations of peptides where short sequence motifs are being sought or compared.

A fourth issue involves phage display experiments that go beyond the identification of a peptide motif that binds a particular molecular target. Many groups have attempted to use the selected peptides to identify binding sites on a protein that may display binding properties similar to the peptides. In these experiments, the investigator must compare the set of affinity-selected peptides with the sequence of a protein (one known or hypothesized to bind to the relevant molecular target) and use the level of similarity between the selected peptides and the protein to predict which portions of the protein will be most likely to display binding to the molecular target. The informatic challenge here is similar to that of identifying a binding motif in a set of peptides and we will find that similarity matrices that are successful for one of these applications will be comparably useful for the other.

Finally, in a world dominated by '-omics', it is worth asking whether the sequences of peptides selected for affinity to a particular molecular target could also be used for the identification of proteins that bind to that particular target within an entire genome. The chapter concludes with a consideration of this possibility.

3.2 Assessing the Quality of a Phage-displayed Library

The success of an affinity selection experiment using a phage-displayed com-binatorial library is highly dependent on the complexity of the library – the more peptide sequences contained within the library, the more likely it is that a sequence exhibiting binding affinity to the target will be present. The experi-mental conditions routinely used to construct a phage display library effectively limit the primary unamplified library to $\sim 10^9–10^{10}$ clones.[1] Since it is physi-cally impractical to carry out an affinity selection experiment with samples containing more than $\sim 10^{11}$ phage,[2,3] an experiment may utilize an average of 10–100 copies of each sequence present in the primary library.[1] The quality of the library will be highly dependent on the relative abundances of those sequences which are, in turn, dependent on the biological processes involved in the creation of the library.[4,5]

3.2.1 Peptide Sequence Censorship

Because phage-displayed libraries are created and amplified within an *Escheri-chia coli* host, the disruption of a biological process essential to phage replica-tion by a displayed peptide will result in censorship of that peptide from the library and a resultant decrease in library diversity.[4,5] Any peptide that is toxic – or even metabolically challenging – to the host cell may be censored from a phage library or be present at very low abundance. This censorship

extends to the presence of rare codons which, because of the paucity of corresponding tRNA isoacceptors in *E. coli*, occur at decreased abundance in
some phage libraries.[6] Libraries that are constructed in such a way as to
minimize the censorship enforced by the biological system will exhibit increased
diversity and potentially enhanced utility.[7,8] A key issue in the quantification of
censorship is the choice of a metric for the diversity of these libraries.

3.2.2 Experimental Measures

Many experimental methods have been used to assess the quality of a peptide
library. A number of groups have investigated biochemical diversity using
restriction digestion pattern analysis of small numbers of group members and
colony hybridization with primers.[1,9-11] Experimentally, the titer of a library –
the total number of infectious particles in a library – is readily measured.
However, titer, even in an uncensored library, is only useful as a relative
measure. Scott and Smith[2] constructed a library of 1.3×10^{14} phage and estimated that this library would include 69% of the 6.4×10^7 possible hexapeptides
assuming Poisson statistics and equal probabilities for all sequences. However,
this estimate assumes equal probability of occurrence for all peptides, which
has been ruled out. Noren and Noren[1] suggested that the best assay for
sequence diversity in a library is to pan against a readily obtainable target and
look for known consensus sequences. For instance, affinity selection using
streptavidin as a target typically yields the biotin mimotope sequence HPQ.

3.2.3 Conceptual Measures

The utility of combinatorial libraries is directly dependent on the sequence
diversity of the library. But what is meant by 'diversity'? We require a definition
of 'diversity' that will help in evaluating the utility of a library for use in affinity
selection experiments. One possible definition is 'completeness' – corresponding
to the proportion of possible sequences in the library that are present in at least
one copy. As argued by Noren and Noren,[1] the more sequences are present in
the library, the greater is the chance that any one peptide that will bind to the
target is present. A problem with this metric is that it ignores the relative
abundances of different peptides in the library. In the context of an affinity
selection experiment, very rare sequences may not be detected and may be
overwhelmed by weaker binding by highly abundant sequences that could
dominate the experimental results. There are two possible conceptual
approaches to the definition of diversity – 'technical diversity' or completeness,
corresponding to the percentage of possible members of a population that exist
at any copy number within a population, and 'functional diversity', which takes
into account the copy number of each distinct member of the population.

   In the latter case, if the copy numbers of the members present in the population are dramatically different, the diversity is intrinsically lower. Experiments that utilize limited sequence information to estimate the technical

diversity cannot provide accurate estimates of completeness since very rare members of the population will inevitably be under-sampled. However, limited sequence information is capable of estimating the 'functional diversity' of a peptide library. Since an affinity screen usually involves the selection (and sequencing) of a limited number of peptides, this 'functional diversity' is the relevant measure of peptide population diversity.

Makowski and Soares[12] introduced a measure of 'functional diversity', which takes into account the copy number of each clone. They reasoned that virtually every step of an affinity selection experiment involves a random – or biased – selection of multiple peptides from the library. Maximizing diversity should minimize the chance that the same peptide is selected twice in any one of these steps. For instance, if a few highly abundant sequences dominate the library, the utility of the library will be limited and the probability of selecting two identical sequences from the library will be high. If, on the other hand, all sequences are present at equal abundance, the utility of the library will be high and the probability of selecting the same sequence twice will be relatively low. They went on to develop a measure of diversity that is both consistent with this argument and intuitively reasonable. Although distinctly different libraries may have the same diversity by this measure, the probability of selecting the same peptide twice is the same for all libraries of a given diversity value.

3.2.4 Quantitative Measures

3.2.4.1 Probability and Information

An important first step in the quantitative analysis of peptide libraries is calculation of the probability of occurrence of a particular peptide. The expected abundance of the peptide is proportional to that probability. In general, the probability of a peptide, p_k, being observed is calculated as the product of the probabilities, p_{ik}, of each individual amino acid, i, occurring in the peptide:

$$p_k = \prod_i p_{ik} \tag{3.1}$$

This makes the implicit assumption that the probability of each amino acid is independent of the identity of the other amino acids in the peptide. Although this is only approximately true, it is, in most cases, a reliable approximation. The exceptions involve specific motifs that may be over- or under-represented within the peptide population, leading to some interdependence of probabilities among adjacent sites. Scott and Smith[2] pointed out that because of the way in which their library was constructed, amino acid sequences coded for by Bgl 1 sites would be excluded from the library. Nevertheless, the absence of these sequences will generate only very weak correlations. Similarly, we have shown recently that sequences displaying a high affinity for ATP are preferentially excluded from some libraries (manuscript in preparation). Detection of the interdependences among amino acids occupying adjacent sites in the peptides (if they occur at all) would require analysis of the sequences of perhaps 1000 peptides and, in some cases, as many as 10 000.

The choice of probabilities for individual amino acids, p_{ik}, requires careful consideration and may not be straightforward in this context: Consider the problem of calculating the probability of a peptide occurring in a phage-displayed library. One approach might be to compute the probability of each amino acid occurring at each position based on the number of codons that code for that amino acid and then multiplying these probabilities together. Using this approach will lead to a predicted set of abundances for peptides that does not agree with observation for any real phage-displayed library. For reasons that may or may not be related to biological censorship,[4] some amino acids are overexpressed and some are underexpressed in these libraries relative to the abundances expected on the basis of codon number. It has also been observed that the relative abundances of amino acids are dependent on the position within the displayed peptide.[4,8] Comparison of the calculated and observed amino acid abundances in these libraries has provided some insight into the biological processes that censor library content.[4] From the relevant amino acid abundances observed, we can calculate the probability of observing any given peptide using equation (3.1). The probabilities computed in this way are frequently very small, being of the order of 0.05^{12} ($=2.4\times10^{-16}$) for a dodecameric (12-mer) peptide – reflecting the fact that there are 20^{12} ($=\sim4\times10^{15}$) different possible 12-mers. The 'information', I_k, associated with peptide k is defined as

$$I_k = -\ln(p_k) \tag{3.2}$$

and provides a more convenient measure of the likelihood of a peptide occurring than the probability itself. For instance, for a dodecamer, typical values of information are in the range of $-\ln(0.05^{12}) = 35.9$, with smaller values corresponding to peptides that are more likely and larger values less likely.

3.2.4.2 Diversity

Makowski and Soares[12] recognized that, regardless of the form of the library, the relevant measure of quality depended on the probability of selecting the same peptide sequence twice in independent selections. The lower that probability, the greater is the functional diversity of the library. They went on to demonstrate that diversity, d, defined as

$$d = 1/(N \sum_k p_k^2) \tag{3.3}$$

where N is the total number of peptides possible and p_k is the probability of peptide k, provides an intuitively reasonable measure of library quality. On the basis of this definition, a library that has all possible members present at equal abundance has a diversity of 1 (the highest possible value for diversity as defined here) and a population composed entirely of a single peptide has a diversity of $(1/N)$ – the lowest possible value for diversity as defined in equation (3.3). Furthermore, those libraries in which the abundances of amino acids are more nearly equal will have the greater diversity.

Figure 3.1 demonstrates the intuitive nature of this definition for diversity. In the diagrams, libraries are represented as plots in which the peptides are arranged from left to right in descending order of abundance. Figure 3.1a plots three hypothetical populations: one with equal abundances of all peptides, one with a linear distribution of abundances and one in which a relatively small number of peptides dominate the population. Figure 3.2 shows distributions for which the diversity [equation (3.3)] is readily computed.

Using the definition in equation (3.3), Krumpe and co-workers[7,8] demonstrated that phage libraries displayed on T7 exhibit greater diversity than those displayed on M13 and libraries constructed from trinucleotide cassettes have even greater diversity. The greater diversity of the T7 libraries compared with M13 is due, presumably, to the restrictions intrinsic to the membrane-associated assembly mechanism of M13,[4,5] whereas the use of trinucleotide cassettes allows for the construction of a library in which each amino acid is coded for by a single codon resulting in nearly equal abundances for all amino acids in the library.

Diversity can be calculated for each position of a library. Figure 3.2 is a plot of the diversity for each position of two 12-mer libraries – one an M13 library and the other a T7 library. Plots of this form can be used to identify positions in libraries that are under substantial selective pressures from the phage–host system.[4] In the cases shown, the diversity of the M13 library is substantially lower than that of the T7 library at every position, a probable consequence of the restrictions imposed on the M13 library by the obligate translocation of the insert across the inner membrane and/or the membrane-associated assembly process.[7,8]

The diversity of protein sequences in a variety of genomes and peptide libraries as computed from the definition of diversity in equation (3.3) is shown in Table 3.1. The diversity of proteomes is calculated per amino acid. The human proteome is more diverse than that of *E. coli* because the abundances of amino acids in human proteins are more uniformly distributed than for *E. coli* proteins. The diversity of computationally generated libraries has been computed for genetic codes that use 20 codons; 32 codons and 61 codons assuming equal use of each codon. For these virtual libraries, the diversity is quoted for 12-mers (on the left) and also the diversity per amino acid (on the right). The diversities of experimentally generated phage-displayed peptide libraries (including 12-, 7- and 6-mer libraries) are also tabulated here. Where the diversity quoted is for a 6- or 7-mer, that fact is noted, because diversity is highly dependent on the length of the peptide. For instance, if only half of the possible amino acids are present in a library its diversity is 0.5 if it is a single amino acid in length, 0.25 if it is two amino acids in length, 0.125 if it is three amino acids in length, and so on. Consequently, the comparison of libraries of peptides of different length is best carried out in terms of their diversity per residue.

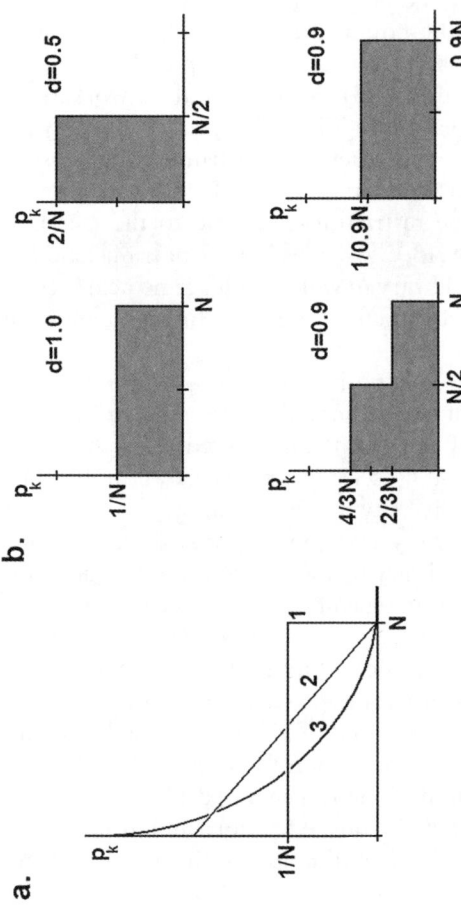

Figure 3.1   Hypothetical populations of peptides chosen to demonstrate the properties of the measure of diversity defined in equation (3.3). (a) Plots of abundance of peptides in three hypothetical populations of peptides where the peptides are arranged in order of decreasing abundance. All three populations include all possible peptides. In population 1, all peptides have equal abundance, in population 2, there is a linear decrease in abundance throughout the population, and in population 3, a relatively small number of peptides are present at relatively high levels. (b) Plots of abundance for four additional populations of peptides. In the first, all peptides are present at equal abundance and the diversity, as defined by equation (3.3), is 1.0. In the second population, half of the possible peptides are present and all have equal abundance. The diversity of this population is 0.5. In the third population, all possible peptides are present, but half are present at twice the abundance of the other half. From equation (3.3), the diversity of this population is 0.9, which is the same diversity as that of the fourth population which contains only 90% of possible peptides, but all at the same abundance.

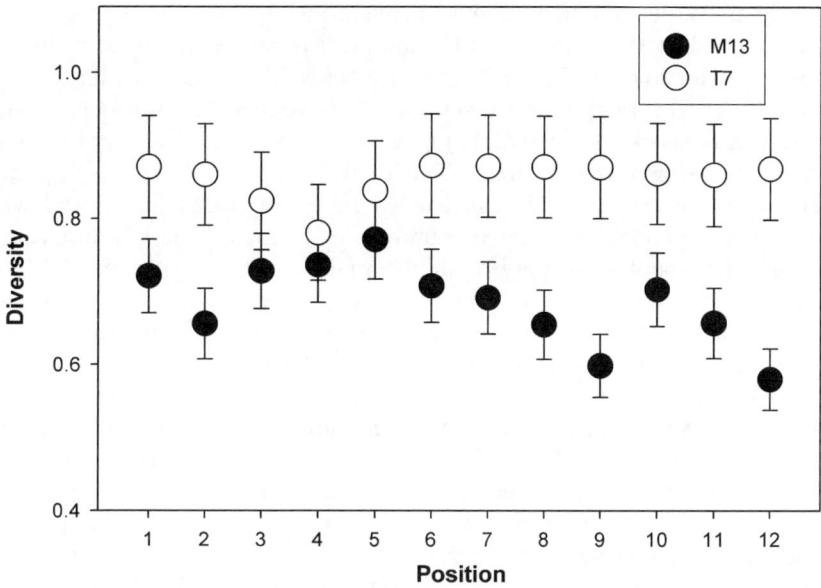

Figure 3.2 Diversity of phage-displayed peptides as a function of position in the peptide. The diversity of a population of peptides displayed on T7 is compared with that of a population displayed on M13. Both libraries were constructed using a 32 codon genetic code.[1,7]

Table 3.1 Diversity of libraries and genomes.

	12-/7-mer	*Per aa*
Diversity of proteomes		
Human	–	0.85
E. coli	–	0.82
Diversity of computationally generated peptide libraries		
20 codon	1.000	1.00
32 codon	0.107	0.83
61 codon	0.059	0.79
Diversity of parent libraries		
Ph.D.-12[1]	0.043	0.77
Ph.D.-C7C[1]	0.079 (7-mer)	0.70
T7 Trinuc[8]	0.320	0.91
T7 NNK[7]	0.142 (7-mer)	0.85
p8[13]	0.075 (6-mer)	0.65
Diversity after selection for binding to TonB[14]		
Ph.D.-12	0.004	0.63
Ph.D.-C7C	0.066 (7-mer)	0.68
Diversity after selection for binding to Taxol[15]		
Ph.D.-12	0.011	0.69
Ph.D.-C7C	0.065 (7-mer)	0.68

The diversity per amino acid may provide insight into the degree of censorship applied to the library by the biological processes involved in its generation.[4] For instance, in Table 3.1, the T7 NNK library has greater diversity than the Ph.D. libraries which are peptides displayed on the p3 protein of M13. The membrane-associated assembly process appears to act as a filter to remove some sequences from the library.[4,7] The T7 trinucleotide library displays even greater diversity because it is constructed to avoid biases introduced when amino acids are coded by different numbers of codons.[8] The p8 library, constituting a hexapeptide displayed on the major coat protein, p8 of M13,[13] exhibits lower diversity than the p3 display libraries because the very large number of copies of p8 per phage particle has the effect of amplifying the effect of any perturbation an insert may have on the function of p8.[4]

It should be noted that in several estimates of the diversity of the Ph.D.-12 library made using peptides that have been isolated over a period of several years, the diversity has appeared to vary well beyond the error bars of the diversity calculation. It is possible that the diversity of a library may be sensitive to the details of its generation, including the metabolic state of the host cells during viral replication and the number of replicative cycles that it undergoes during amplification. These issues, although noted anecdotally, have never been systematically explored.

3.3   Assessing the Quality of an Affinity Screen Experiment

A phage display experiment usually involves three to four affinity screening steps alternated with phage amplification steps. The screening steps serve to select mainly those peptides that bind to the molecular target. Peptides exhibiting non-specific binding or binding to the container in which the molecular target is immobilized can contaminate the selected population, complicating the analysis and obscuring potential binding motifs. The amplification steps carried out during a phage display experiment are designed to increase all selected sequences equally in order to facilitate further purification steps. However, each displayed peptide may have a different impact on the speed of production, biasing the results of the amplification step and altering the relative abundances of the selected peptides. This is one reason why the number of selection/amplification cycles is so important. In spite of these potential sources of distortion and contamination, the identification of motifs within a population of affinity-selected peptides is frequently straightforward, with the motif being readily identified through visual inspection of the sequences. In other cases, a visual inspection does not result in the identification of a motif and the computational methods outlined below must be employed.

Prior to attempting motif identification within a population of affinity-selected peptides, it is useful to have a means of gauging the success of the screen. One method is to calculate the degree to which diversity of the population is decreased by the selection. A second is to track the distribution of 'rare' peptides within the population. If there is no detectable decrease in the diversity of the population after the selection, then there is reason to believe that the affinity screen failed. If there is a detectable change, then observation of the change in distribution of peptides with different associated information may provide guidance as to which peptides contain relevant binding motifs.

3.3.1 Change in Diversity

A successful affinity selection experiment should lead to a decrease in peptide sequence diversity. As reflected in the numbers in Table 3.1, selection for binding to TonB[14] or Taxol[15] led to a significant decrease in the diversity of the population when a 12-mer library was used and to a more modest decrease in diversity when a constrained 7-mer library was used. The Ph.D.-12 and Ph.D. C7C libraries are commercially available (New England BioLabs) and the T7 trinuc and T7 NNK libraries are those described by Krumpe and co-worlers.[7,8]

The change in diversity between an unselected population and a subset generated by an affinity selection experiment can be relatively modest. In the case of selection for Taxol binding peptides,[15] the diversity of the primary M13 dodecamer library was 0.043 and that of the selected peptide population was 0.011. At first glance, it would seem that this fourfold decrease in diversity would be too small to reflect a successful screen. In fact, there are multiple factors in the experiment that limit the decrease in diversity observed. In addition to those factors listed above – all of which add to the 'noise' in the sequence data generated – there is the simple fact that even in a perfect experiment, there are a great many peptides that will bind to any molecular target. Consider an ideal experiment using a population of 12-mer peptides with a diversity of 1.0 selected for binding to a molecular target in which five amino acids participate in binding in such a way that every selected peptide has the motif incorporated. If the binding motif is limited to the N-terminal position of this peptide, this population would contain 20^7 ($= 1.28 \times 10^9$) peptides that would bind to the target. Add to that the possibility that the motif can occupy any of eight positions in the dodecamer, and the total number of possible peptides that bind increases to $\sim 10^{10}$. Although a relatively large number, it is still a decrease by a factor of 400 000 from the total number of dodecapeptides possible. The diversity of this subset of peptides containing the binding motif is still considerable, as every amino acid is allowed at every position in the peptide. In this ideal experiment, the diversity, d, decreases from 1.0 prior to selection to 0.212 after selection. Comparison of this decrease in diversity with the experimental values in Table 3.1 provides a measure of the noise level inherent in this type of experiment.

3.3.2 Change in Information

An affinity screen will result in a change in the relative abundances of all the peptides in a population. Prior to the screen, the relative abundances will be governed largely by the methods used to synthesize the library (*e.g.* number of codons coding for each amino acid) and the relative selective pressures applied by the different peptides to the biological requirements of the system. After affinity selection, the relative abundances will be due to these factors *and* the selectivity of the screen. The affinity screening will enrich a population for peptides exhibiting binding properties independent of the abundance of these peptides in the original population. This may or may not involve peptides that are rare in the original population. Tracking the change in the relative abundances of peptides provides a clue to the identification of peptides in the selected population that have been enriched because of affinity to the molecular target.

In Figure 3.3, the distribution of information associated with each peptide in a population is plotted for an unselected population and for a population enriched for peptides that exhibit affinity for the anti-cancer drug Taxol. As can be seen, the number of peptides with relatively high information content increases due to affinity selection. The majority of these 'unexpected' peptides (peptides with low *a priori* chance of occurring) were found to contain at least a portion of the Taxol-binding motif.[15] Of course, there is no guarantee that peptides containing a binding motif will necessarily be of high information – they may contain sequences that are highly likely in the library used (*i.e.* low information). On the other hand, if a peptide that is unlikely to occur in the

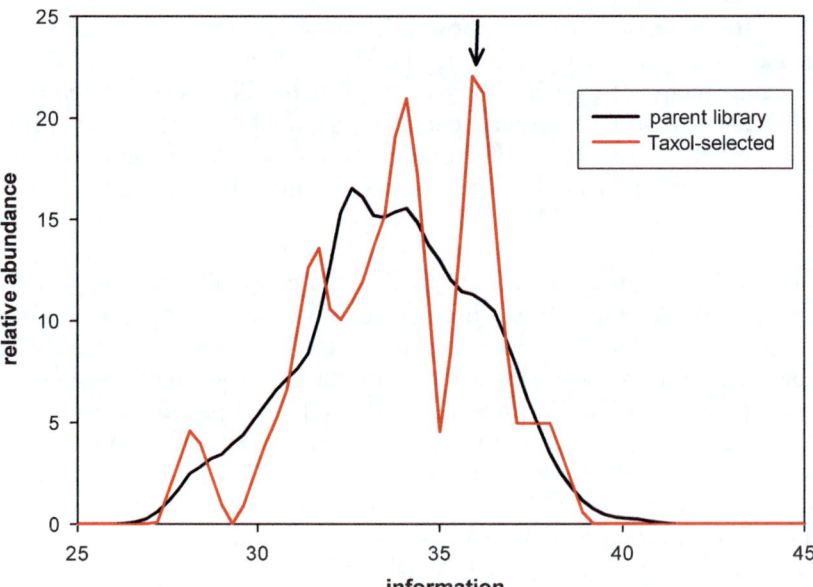

Figure 3.3   Histogram of the relative abundances of peptides with different information levels. The distribution of information in the parent library (black) is shifted by selection for binding to Taxol (red) by a decrease in the number of low-information peptides and an increase in the number of high-information peptides.

unscreened library is observed after screening, its presence is very likely due to its binding properties.

3.4 Identification of Motifs in a Peptide Population

The identification of binding motifs in a set of peptide sequences obtained from affinity selection involves a process akin to the search for protein sequence similarities using an informatic tool such as BLAST,[16,17] the Smith–Waterman algorithm,[18] FASTA[19] and ParAlign.[20] However, the search for relatively short sequence motifs among a relatively large number ($\sim$100) of peptide sequences requires a process that is different in subtle ways from the alignment of entire protein sequences. The most important difference is the adjustment of the similarity matrix for use in comparison of very short sequence segments. Also, the treatment of gaps must be altered since a gap in a sequence motif that may be only five amino acids long may have a much greater functional consequence than a gap in a protein sequence 250 amino acids in length.

Motifs that are completely conserved or reflect only conservative amino acid substitutions can be identified using relatively unsophisticated search algorithms. Three such search tools are available on the RELIC server.[21] MOTIF1 searches for user-specified contiguous amino acid sequence motifs within a peptide population. Experience suggests that alignment of short (*e.g.* 3–4) amino acid motifs may aid in the identification of weaker consensus sequences that extend to either side of this anchor sequence. MOTIF2 searches for patterns of three amino acid identities allowing for gaps of identical length in each peptide. MOTIF3 searches for motifs containing four amino acids in which at least three of the four amino acids are identical. Each of these three software tools may be useful as an initial method for identifying potential binding motifs that occur at statistically significant levels in the selected peptide population. Incorporation of similarity score and gap penalties may lead to identification of motifs not identified with the MOTIF programs. For instance, Table 3.2 shows sequence motifs mined from a set of peptides selected for affinity to Taxol. For this calculation, the similarity threshold was set at 17 with a gap penalty of 6. Occurrence of identical or similar amino acids within the motif pairs recognized are indicated in Table 3.2 by shading.

3.5 Similarity Matrices

The most important change in similarity matrices required for identification of motifs in a peptide population is in the diagonal elements. Because similarities are being calculated over very short segments, the differences in the magnitude of diagonal elements in the BLOSUM matrices lead to huge bias in the amino acid content of the motifs identified. For instance, as can be seen in Table 3.3a, the diagonal element in BLOSUM62 for alanine is 4, for cysteine is 9 and for tryptophan is 11. In searching for motifs, it would take nearly three alanine–alanine identities to achieve a similarity score comparable to a single tryptophan–tryptophan identity. However, a single amino acid identity provides no information on the

Table 3.2   Motifs identified within a Taxol-selected population.

12 21.00	p--shpst	sdlntdpshpst
19 21.00	pisshpst	lkpisshpstva
12 19.00	ntdpshps	sdlntdpshpst
34 19.00	nsppsfps	mrtansppsfps
15 20.00	htphp	htphpmnlvaea
60 20.00	htphp	htphpdasiqgv
21 17.00	li-stnaatl	alistnaatltp
43 17.00	liest-aagi	rpliestaagim
37 17.00	--sqpqvlsts	fydsqpqvlstt
23 23.00	stsqptvstlr	stsqptvstlrp
39 23.00	shhsptvtvlr	shhsptvtvlrs
24 17.00	tppdeap	hmmdtppdeapv
68 17.00	tppmqsp	gtppmqspsate
34 17.00	tanspps	mrtansppsfps
28 18.00	ssppsk	syssppskavre
64 18.00	tsppak	tsppakifsmva
31 17.00	lmptarssms-	mplmptarssms
34 17.00	-mrtansppst	mrtansppsfps
47 19.00	slsrpypl	slsrpyplnmhf
51 19.00	slttpyll	slttpyllwntl
48 17.00	ptwr-sms	ptwrsmsviapp
31 17.00	ptarssms	mplmptarssms
56 18.00	tttlslvplrp	tttlslvplrpn
69 18.00	ttshpivyhrp	nttshpivyhrp
64 18.00	tsppak	tsppakifsmva
28 18.00	ssppsk	syssppskavre

presence of similar sequence motifs. Consequently, the BLOSUM62 matrix was transformed to a matrix, BINARY62-4, in which all diagonal elements are equal to 4.0 and the off-diagonal elements are proportional to those in BLOSUM62 as weighted by the factors required to transform the diagonal elements. Each off-diagonal element was then rounded to the nearest integer. This matrix was found, empirically, to be successful in the identification of motifs within populations of peptides and to significantly out perform BLOSUM62 in this particular application. Table 3.3 includes similarity matrices (a) BLOSUM62 and (b) BINARY62-4. In this form of the matrices, 'x' is used to indicate an unknown/ unidentified amino acid (next to last row; next to last column).

Table 3.3 BLOSUM62 matrix and BINARY 62-4 similarity matrix.

(a) BLOSUM62 matrix

	a	c	d	e	f	g	h	i	k	l	m	n	p	q	r	s	t	v	w	x	y
a	4	0	-2	-1	-2	0	-2	-1	-1	-1	-1	-2	-1	-1	-1	1	0	0	-3	0	-2
c	0	9	-3	-4	-2	-3	-3	-1	-3	-1	-1	-3	-3	-3	-3	-1	-1	-1	-2	0	-2
d	-2	-3	6	2	-3	-1	-1	-3	-1	-4	-3	1	-1	0	-2	0	-1	-3	-4	0	-3
e	-1	-4	2	5	-3	-2	0	-3	1	-3	-2	0	-1	2	0	0	-1	-2	-3	0	-2
f	-2	-2	-3	-3	6	-3	-1	0	-3	0	0	-3	-4	-3	-3	-2	-2	-1	1	0	3
g	0	-3	-1	-2	-3	6	-2	-4	-2	-4	-3	0	-2	-2	-2	0	-2	-3	-2	0	-3
h	-2	-3	-1	0	-1	-2	8	-3	-1	-3	-2	1	-2	0	0	-1	-2	-3	-2	0	2
i	-1	-1	-3	-3	0	-4	-3	4	-3	2	1	-3	-3	-3	-3	-2	-1	3	-3	0	-1
k	-1	-3	-1	1	-3	-2	-1	-3	5	-2	-1	0	-1	1	2	0	-1	-2	-3	0	-2
l	-1	-1	-4	-3	0	-4	-3	2	-2	4	2	-3	-3	-2	-2	-2	-1	1	-2	0	-1
m	-1	-1	-3	-2	0	-3	-2	1	-1	2	5	-2	-2	0	-1	-1	-1	1	-1	0	-1
n	-2	-3	1	0	-3	0	1	-3	0	-3	-2	6	-2	0	0	1	0	-3	-4	0	-2
p	-1	-3	-1	-1	-4	-2	-2	-3	-1	-3	-2	-2	7	-1	-2	-1	-1	-2	-4	0	-3
q	-1	-3	0	2	-3	-2	0	-3	1	-2	0	0	-1	5	1	0	-1	-2	-2	0	-1
r	-1	-3	-2	0	-3	-2	0	-3	2	-2	-1	0	-2	1	5	-1	-1	-3	-3	0	-2
s	1	-1	0	0	-2	0	-1	-2	0	-2	-1	1	-1	0	-1	4	1	-2	-3	0	-2
t	0	-1	-1	-1	-2	-2	-2	-1	-1	-1	-1	0	-1	-1	-1	1	5	0	-2	0	-2
v	0	-1	-3	-2	-1	-3	-3	3	-2	1	1	-3	-2	-2	-3	-2	0	4	-3	0	-1
w	-3	-2	-4	-3	1	-2	-2	-3	-3	-2	-1	-4	-4	-2	-3	-3	-2	-3	11	0	2
x	0	0	0	0	0	0	0	0	0	0	0	0	0	0	0	0	0	0	0	0	0
y	-2	-2	-3	-2	3	-3	2	-1	-2	-1	-1	-2	-3	-1	-2	-2	-2	-1	2	0	7

(b) BINARY 62-4 similarity matrix

	a	c	d	e	f	g	h	i	k	l	m	n	p	q	r	s	t	v	w	x	y
a	4	0	-1	-1	-1	0	-1	-1	-1	-1	-1	-1	-1	-1	-1	1	0	0	-1	-1	-1
c	0	4	-1	-1	-1	-1	-1	-1	-1	-1	-1	-1	-1	-1	-1	-1	-1	-1	-1	-1	-1
d	-1	-1	4	1	-1	-1	-1	-1	-1	-1	-1	1	-1	0	-1	0	-1	-1	-1	-1	-1
e	-1	-1	1	4	-1	-1	0	-1	1	-1	-1	0	-1	1	0	0	-1	-1	-1	-1	-1
f	-1	-1	-1	-1	4	-1	-1	0	-1	0	0	-1	-1	-1	-1	-1	-1	-1	1	-1	1
g	0	-1	-1	-1	-1	4	-1	-1	-1	-1	-1	0	-1	-1	-1	0	-1	-1	-1	-1	-1
h	-1	-1	-1	-1	0	-1	4	-1	-1	-1	-1	1	-1	0	0	-1	-1	-1	-1	-1	1
i	-1	-1	-1	-1	0	-1	-1	4	-1	1	1	-1	-1	-1	-1	-1	-1	1	-1	-1	-1
k	-1	-1	-1	1	-1	-1	-1	-1	4	-1	-1	0	-1	1	1	0	-1	-1	-1	-1	-1
l	-1	-1	-1	-1	0	-1	-1	1	-1	4	1	-1	-1	-1	-1	-1	-1	1	-1	-1	-1
m	-1	-1	-1	-1	0	-1	-1	1	-1	1	4	-1	-1	0	-1	-1	-1	1	-1	-1	-1
n	-1	-1	1	0	-1	0	1	-1	0	-1	-1	4	-1	0	0	1	0	-1	-1	-1	-1
p	-1	-1	-1	-1	-1	-1	-1	-1	-1	-1	-1	-1	4	-1	-1	-1	-1	-1	-1	-1	-1
q	-1	-1	0	1	-1	-1	0	-1	1	-1	0	0	-1	4	1	0	-1	-1	-1	-1	-1
r	-1	-1	-1	0	-1	-1	0	-1	1	-1	-1	0	-1	1	4	-1	-1	-1	-1	-1	-1
s	1	-1	0	0	-1	0	-1	-1	0	-1	-1	1	-1	0	-1	4	1	-1	-1	-1	-1
t	0	-1	-1	-1	-1	-1	-1	-1	-1	-1	-1	0	-1	-1	-1	1	4	0	-1	-1	-1
v	0	-1	-1	-1	-1	-1	-1	1	-1	1	1	-1	-1	-1	-1	-1	0	4	-1	-1	-1
w	-1	-1	-1	-1	1	-1	-1	-1	-1	-1	-1	-1	-1	-1	-1	-1	-1	-1	4	-1	1
x	-1	-1	-1	-1	-1	-1	-1	-1	-1	-1	-1	-1	-1	-1	-1	-1	-1	-1	-1	-1	-1
Y	-1	-1	-1	-1	1	-1	1	-1	-1	-1	1	-1	-1	-1	-1	-1	-1	-1	1	-1	4

3.6 Identification of Binding Sites in Proteins

Phage display experiments are designed to identify peptide sequences that exhibit affinity to a particular molecular target. In some cases, there is a further need to determine the location on a particular protein of sequences that display that affinity. For instance, Carter *et al.*[14] demonstrated the use of phage display to identify peptide sequences that interacted specifically with the *E. coli* protein TonB and went on to identify, through a comparison of these sequences with that of FhuA, the region of FhuA that binds to TonB. In order to carry out this identification, it was necessary to compare the sequences of all the selected peptides with that of FhuA. Experience has indicated that even in the absence of a clearly identifiable consensus sequence motif, weaker, conserved sequence patterns may be embedded within the data. These sequences may not be exact matches for the protein sequence but, when aligned with the entire protein sequence, may cluster together providing evidence for a sequence similarity that may reflect an underlying binding affinity.

In the comparison of a population of peptides with the sequence of a protein, matches are made using a similarity matrix such as BINARY62-4. To perform this calculation, one has to choose the length of peptide sequence compared, the handling of gaps and the threshold below which similarity scores are not included in the final calculation. A threshold is required because we have found that if all matches, no matter how weak, are allowed to contribute to the final similarity score, the noise in the calculation overwhelms any potentially useful signal. We have found that a five amino acid stretch within which no gaps are allowed is the most useful length of window for these calculations. Within this window, a threshold of 13 is usually used for identification (when using BIN-ARY62-4). This translates into requiring three identities and one similarity out of five amino acids for the peptide to be identified as having a significant similarity with a particular protein sequence. Because of biases in the amino acid compositions of peptide populations and variations in the amino acid usages among proteins, these parameters will occasionally result in either too few or too many matches to make possible a preliminary identification of a potential binding site. In these cases, it may be useful to set the bar lower or higher, respectively, to produce potentially useful patterns of peptide matches.

Figure 3.4 shows the results of a calculation carried out using the program MATCH on the RELIC server.[21] In this case, peptides selected for affinity to ATP were compared with the sequence of an ATP-binding protein, phosphoenolpyruvate kinase. In Figure 3.4a the selected sequences have been aligned with that of the protein and the amino acids in each peptide that are identical with or similar to those of the protein are highlighted. In this diagram, amino acids within 10 Å of the ATP are marked with an 'x'. The region of highest similarity to the ATP-selected peptides is at position 257, where over a dozen ATP-selected peptides align. Figure 3.4b is a plot of the similarity score as a function of position along the protein sequence and Figure 3.4c is a three-dimensional rendering created with a file generated from HETEROalign. HETEROalign utilizes the same algorithm as MATCH but also inputs a .pdb file with the atomic coordinates of the protein and generates a new pdb file in which the temperature factors from the input file have been replaced by the similarity scores from the MATCH alignment algorithm. Many graphics routines used for displaying the three-dimensional structure of a protein allow the amino acids to be colored according to temperature factor. By replacing the temperature factor with similarity, it is possible to produce three-dimensional renditions of a protein in which the colors reflect the level of similarity to the selected peptides as in this rendering of phosphoenolpyruvate kinase.

(a)

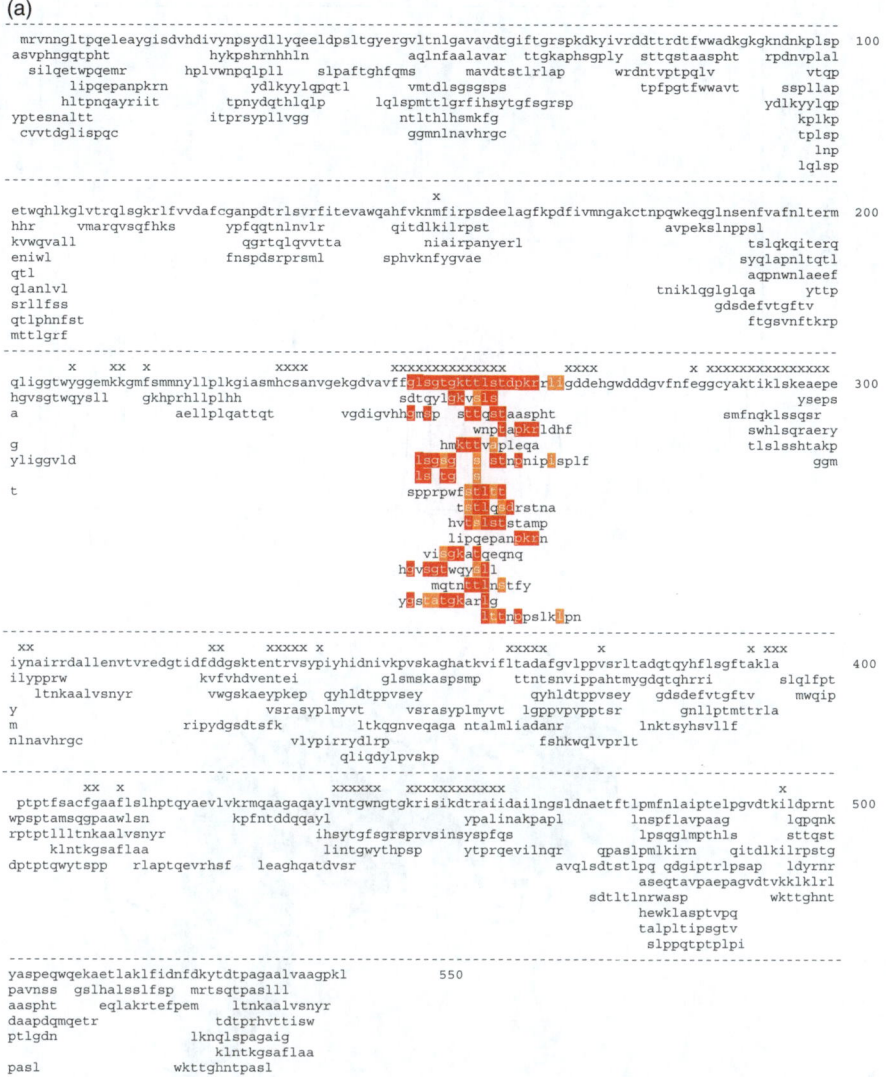

Figure 3.4 Comparison of a set of ATP-selected peptides with the sequence of an ATP-binding protein. Part (a) shows the alignment of ATP-selected peptides with the sequence of the protein, demonstrating the clustering of those peptides along the protein sequence. Amino acids identical or similar to the protein sequence are highlighted in yellow. (b) Similarity score as a function of position along the protein. (c) A rendering of the three-dimensional structure of the protein with colors chosen to reflect the similarity scores plotted in (b). Here, red corresponds to high similarity scores and blue to low similarity scores. The region of highest similarity is immediately adjacent to the bound ATP.

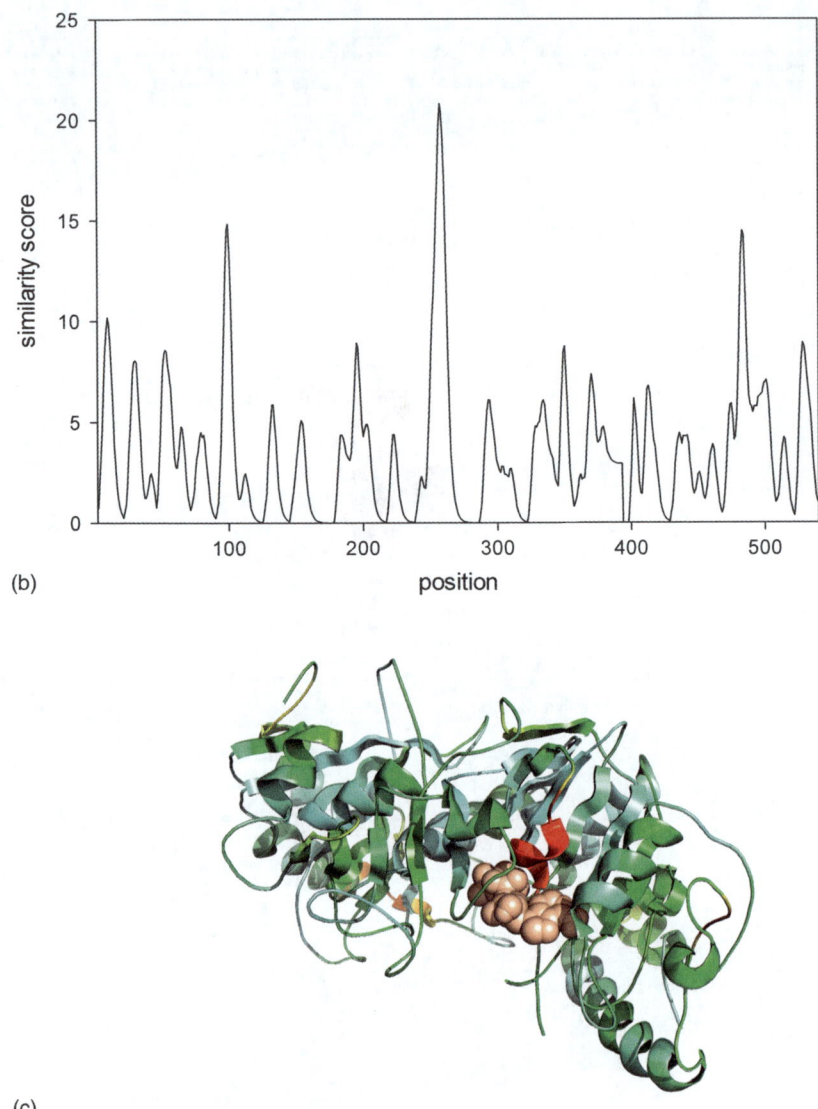

(b)

(c)

Figure 3.4   Continued.

3.7   Identification of Binding Proteins in a Proteome

In principle, it should be possible to utilize the same algorithms as used for
identifying binding sites on a single protein for analysis of all the proteins coded
for by a genome. If successful, this could provide an important tool for
annotation of proteins of unknown function. In practice, the degree to which

this approach will be successful is still unclear. The RELIC server includes an application, FASTAskan that is capable of this calculation. The input for the program is a set of peptide sequences and the protein sequences for an entire proteome. The output is a ranking of the proteins in descending order by peak similarity score. This output is designed so that the proteins most likely to bind to the ligand used for affinity screening will be clustered at the top. Applying this approach to the *E. coli* genome using peptides selected for binding to ATP results in the list presented in Table 3.4 (adapted from D. J. Rodi *et al.*, in preparation). This list includes the 50 *E. coli* proteins exhibiting the greatest similarity to a set of peptides selected for affinity to ATP (as judged by the highest similarity score in the protein). In this table, proteins in red font are ATP binders. Orange font indicates the protein binds to an ATP-like target. Blue font indicates no known binding to ATP and green font indicates no information available. Of the top 50 hits, 36 are known to bind to ATP or a molecule exhibiting at least a portion of the structural motifs present on ATP. Only eight are demonstrably false positives. This result suggests that the approach has substantial merit for identification within a proteome of proteins that bind to specific molecular targets.

3.8 RELIC

A suite of informatic tools that carry out all of the computations reviewed in this chapter is available at the RELIC bioinformatics server (http://relic.bio. anl.gov).[21] This suite includes programs for calculating the frequency of amino acids observed in a peptide population and the diversity of a peptide population, for identification of sequence motifs that are present multiple times in a peptide population and for the alignment of peptides with a protein sequence. The tools in RELIC allow the calculation of diversity for each position in a peptide population, making possible generation of plots such as those in Figure 3.2, for calculating information associated with each peptide and generating the data required to produce Figure 3.3, for the alignment of peptides with a protein sequence making possible the generation of diagrams such as Figure 3.4a and for the determination of the similarity score as a function of position along a protein sequence, as required in the generation of Figure 3.4b. When the three-dimensional structure of the protein is available, the similarity score (such as that plotted in Figure 3.4b) as a function of position in the protein sequence, can be transferred to the 'temperature factor' column of a pdb file, making it possible to visualize, in three dimensions, the positions of high-similarity regions in the protein. This can be carried out using almost any software package (*e.g.* Rasmol[22]) for producing three-dimensional images of a protein, since virtually all of them allow the protein to be drawn with a color scheme coded to the temperature factor (which in the pdb files output from HETEROalign are replaced by similarity scores that have been scaled to values typical for temperature factors in proteins).

Table 3.4 List of *E. coli* proteins exhibiting the highest level of similarity to ATP-binding peptide.

22.17 > gi|1787942|gb|AAC74725.1| member of ATP-dependent helicase superfamily II
21.73 > gi|2367341|gb|AAC77025.1| diadenosine tetraphosphatase
21.43 > gi|1787197|gb|AAC74049.1| methylglyoxal synthase
21.13 > gi|1787702|gb|AAC74514.1| putative virulence protein
21.13 > gi|1786345|gb|AAC73262.1| ATP-binding component of hydroximate-dependent iron transport
20.98 > gi|1787668|gb|AAC74483.1| split orf
20.24 > gi|1787121|gb|AAC73980.1| anaerobic dimethyl sulfoxide reductase subunit A
20.24 > gi|1788092|gb|AAC74861.1| putative amino acid/amine transport protein
20.09 > gi|1787885|gb|AAC74673.1| putative transport protein
20.09 > gi|1788863|gb|AAC75568.1| orf, hypothetical protein
19.35 > gi|1788112|gb|AAC74880.1| orf, hypothetical protein
19.35 > gi|1787051|gb|AAC73916.1| putative ATP-binding component of a transport system
19.20 > gi|1790393|gb|AAC76938.1| phosphoenolpyruvate carboxylase
18.90 > gi|1789971|gb|AAC76573.1| 3-methyladenine DNA glycosylase I, constitutive
18.90 > gi|1789962|gb|AAC76565.1| putative ATP-binding component of dipeptide transport system
18.90 > gi|1789453|gb|AAC76107.1| aerotaxis sensor receptor, flavoprotein
18.90 > gi|1789230|gb|AAC75904.1| orf, hypothetical protein
18.75 > gi|1788472|gb|AAC75210.1| ATP-binding component of methyl galactoside transport and galac
18.75 > gi|1786804|gb|AAC73690.1| ferric enterobactin transport protein
18.30 > gi|1790736|gb|AAC77240.1| IS30 transposase
18.30 > gi|1788909|gb|AAC75610.1| phosphoribosylformylglycinamide synthetase = FGAM synthetase
18.15 > gi|1788344|gb|AAC75093.1| putative glucose transferase
18.01 > gi|1789286|gb|AAC75956.1| putative enzyme
18.01 > gi|1788151|gb|AAC74916.1| orf, hypothetical protein
17.86 > gi|1790375|gb|AAC76921.1| cystathionine gamma-synthase
17.86 > gi|1787318|gb|AAC74162.1| flagellar biosynthesis, cell-distal portion of basal-body rod
17.86 > gi|1790611|gb|AAC77126.1| *N*-acetylmuramoyl-l-alanine amidase II; a murein hydrolase
17.86 > gi|1789350|gb|AAC76014.1| glycolate oxidase iron-sulfur subunit
17.71 > gi|1786336|gb|AAC73254.1| poly(A) polymerase I
17.56 > gi|1790857|gb|AAC77349.1| right origin-binding protein
17.56 > gi|1786801|gb|AAC73687.1| ATP-dependent serine-activating enzyme
17.41 > gi|1790472|gb|AAC77009.1| chorismate lyase
17.41 > gi|1786521|gb|AAC73430.1| orf, hypothetical protein
17.26 > gi|1788767|gb|AAC75480.1| orf, hypothetical protein
17.26 > gi|1788524|gb|AAC75256.1| cytochrome *c*-type biogenesis protein
17.26 > gi|1787547|gb|AAC74372.1| putative ATP-binding protein of peptide transport system
17.26 > gi|1786852|gb|AAC73734.1| a minor lipoprotein
17.26 > gi|1786549|gb|AAC73456.1| putative transport protein
17.11 > gi|1789345|gb|AAC76009.1| orf, hypothetical protein
17.11 > gi|1787529|gb|AAC74356.1| DNA topoisomerase type I, omega protein
16.96 > gi|1789336|gb|AAC76001.1| transport of nucleosides, permease protein
16.96 > gi|1787906|gb|AAC74692.1| repressor of malX and Y genes
16.82 > gi|2367260|gb|AAC76690.1| regulator of uhpT
16.67 > gi|1790745|gb|AAC77248.1| regulator for fec operon, periplasmic
16.67 > gi|1788557|gb|AAC75286.1| orf, hypothetical protein
16.67 > gi|1787909|gb|AAC74694.1| enzyme that may degrade or block biosynthesis of endogenous mal
16.67 > gi|1787771|gb|AAC74568.1| orf, hypothetical protein
16.52 > gi|1790835|gb|AAC77328.1| peptide chain release factor RF-3
16.52 > gi|1790828|gb|AAC77322.1| 2-component transcriptional regulator
16.52 > gi|1790354|gb|AAC76902.1| orf, hypothetical protein

3.9 Discussion

The screening of a combinatorial phage-displayed peptide library for members that exhibit affinity to a molecular target results in the selection of a subset of the original library with properties that can provide substantial insight into the nature of the interaction of peptides with the target. In many cases, visual inspection of the sequences of the selected peptides will result in the identification of binding motifs. In other cases, binding motifs or sequence patterns may not be readily apparent. In most cases, quantitative analysis of the sequences of the parent library and the affinity-selected subset can provide substantial insight into the quality of the original library, the quality of the affinity selection and the identity of peptides that exhibit affinity for the target. In this chapter, we have reviewed a number of these methods and discussed briefly the principles on which these methods are based.

As in most fields of experimental science, the results of these analyses will be limited by the noise level in the data. In analysis of combinatorial libraries of peptides, the 'noise' in the data is generated by biological processes that result in censorship of libraries and biases that lead to differences in the relative abundances of library members. In the search for consensus motifs within a selected population, the 'noise' is generated both by the processes that lead to biases in the parent libraries and by the imperfections of the selection process – imperfections that lead to the inclusion of peptides in a selected population that are present for reasons other than affinity to the molecular target. In most cases, 'noise' in data can be overcome by the collection of larger volumes of data. In the analysis of peptide populations, analysis of additional sequences will almost always enhance the probability of a favorable outcome – whether that outcome involves the identification of a consensus binding motif, or location of a binding site on a protein. Experimental approaches to minimizing 'noise' should be pursued, including the production of more highly diverse libraries, added attention to the preparation of surfaces and the standardization of conditions for removal of unbound/bound phage particles.

Phage display experiments have a great deal to teach us about the way in which molecules interact in addition to acting as an important tool in drug development and molecular engineering and design. Methods such as those discussed in this chapter provide ways of extracting as much information as possible from these experiments.

References

1. K. A. Noren and C. J. Noren, *Methods*, 2001, **23**, 169–178.
2. J. K. Scott and G. P. Smith, *Science*, 1990, **249**, 386–390.
3. G. P. Smith and V. A. Petrenko, *Chem. Rev.*, 1997, **97**, 391–410.
4. D. J. Rodi, A. Soares and L. Makowski, *J. Mol. Biol.*, 2002, **322**, 1039–1052.
5. D. J. Rodi, S. Mandava and L. Makowski, *Filamentous bacteriophage structure and biology*, in *Phage Display in Biotechnology and Drug Discovery*, ed. S. S. Sidhu, Marcel Dekker, New York, 2005, pp. 1–62.

6. D. J. Rodi and L. Makowski, Transfer RNA isoacceptor availability contributes to sequence censorship in a library of phage displayed peptides, in *Proceedings of the 22nd Tanaguchi International Symposium, 18–21 November 1996*, Matsushita, Ltd., Kei Han Na Science City, Japan, 1997.
7. L. R. H. Krumpe, A. J. Atkinson, G. W. Smythers, A. Kandel, K. M. Schumacher, J. B. McMahon, L. Makowski and T. Mori, *Proteomics*, 2006, **6**, 4210–4222.
8. L. R. H. Krumpe, K. M. Schumacher, J. B. McMahon, L. Makowski and T. Mori, *BMC Biotechnol.*, 2007, **7**, 65.
9. S. E. Cwirla, E. A. Peters, R. W. Barrett and W. J. Dower, *Proc. Natl. Acad. Sci. USA*, 1990, **87**, 6378–6382.
10. J. D. Marks, H. R. Hoogenboom, T. P. Bonnert, J. McCafferty, A. D. Griffiths and G. Winter, *J. Mol. Biol.*, 1991, **222**, 581–597.
11. M. E. DeGraaf, R. M. Miceli, J. E. Mott and H. D. Fischer, *Gene*, 1993, **128**, 13–17.
12. L. Makowski and A. Soares, *Bioinformatics*, 2003, **19**, 483–489.
13. V. A. Petrenko, G. P. Smith, X. Gong and T. Quinn, *Protein Eng.*, 1996, **9**, 797–811.
14. D. M. Carter, J.-N. Gagnon, M. Damlaj, S. Mandava, L. Makowski, D. J. Rodi, P. D. Pawlek and J. W. Coulton, *J. Mol. Biol.*, 2006, **357**, 236–251.
15. D. J. Rodi, R. W. Janes, H. J. Sanganee, R. Holton, B. A. Wallace and L. Makowski, *J. Mol. Biol.*, 1999, **285**, 197–204.
16. S. F. Altschul, W. Gish, W. Miller, E. W. Myers and D. J. Lipman, *J. Mol. Biol.*, 1990, **215**, 403–410.
17. S. F. Altschul, T. L. Madden, A. A. Schäffer, J. Zhang, Z. Zhang, W. Miller and D. J. Lipman, *Nucleic Acids Res.*, 1997, **25**, 3389–3402.
18. T. F. Smith and M. S. Waterman, *J. Mol. Biol.*, 1981, **147**, 195–197.
19. W. R. Pearson and D. J. Lipman, *Proc. Natl Acad. Sci. USA*, 1988, **85**, 2444–2448.
20. T. Rognesa, *Nucleic Acids Res.*, 2001, **29**, 1647–1652.
21. S. Mandava, L. Makowski, J. Uzubell, S. Devarapalli and D. J. Rodi, *Proteomics*, 2004, **4**, 1439–1460.
22. R. A. Sayle and E. J. Milner-White, *Trends Biochem. Sci.*, 1995, **20**, 374–376.

CHAPTER 4
Phage-mediated Drug Delivery

VALERY A. PETRENKO AND PRASHANTH K. JAYANNA*

Department of Pathobiology, Auburn University, Auburn, AL 36849, USA

4.1 Introduction

'To the biochemical chemotherapist, it is not only a matter of faith, but an obvious fact, that every cell type must have a characteristic biochemical pattern and therefore be susceptible to attack at some locus or loci critical for its survival and replication.' By these words, the concept of targeted therapeutics was formulated by Nobel Laureate George H. Hitchings in one of his lectures. This idea, although extremely intuitive, was not the first of its ilk. Another Nobel Laureate, Paul Ehrlich, had several decades earlier envisioned a similar scenario, proposing that a suitable reprieve from diseases could be achieved by chemical targeting. In his landmark Nobel Lecture in 1908, Ehrlich conjectured that the use of optical aids such as microscopy as a cellular investigative tool had reached its limits and further inquiries into the processes of cellular life would require the conception of the chemical nature of the cell as a living and evolving entity ('chemism'). He proposed that a true understanding of the life processes and also the cornerstones for a rational pharmacotherapy could be obtained by a conceptual breakdown of the cell into a number of specific functions, in a process similar to extracting individual wheels and cogs of a complex mechanical device such as a watch and studying them individually.

---

*Present address: Department of Chemistry, Sam Houston State University, Huntsville, TX 77340, USA.

RSC Nanoscience & Nanotechnology No. 17
Phage Nanobiotechnology
Edited by Valery A. Petrenko and George P. Smith
© Royal Society of Chemistry 2011
Published by the Royal Society of Chemistry, www.rsc.org

Appealing to results of his experiments proving the existence of such chemical groupings, which he termed 'receptors', Ehrlich proposed that differences in receptors and their modulation are responsible for the differential toxicities of many medications. He concluded that systematic study of such receptors would ultimately yield a compound that could achieve a complete cure with a single dose, the *therapia sterilisans magna*.

Ehrlich's vision adumbrated modern chemotherapy. Most importantly, his analysis of immunological specificity formed the basis for his 'magic bullet' concept, which in turn was the progenitor of the present-day theory of targeted therapy in which a non-specific molecule is specifically targeted to a pathogen by conjugation to a pathogen-specific antibody or other targeting ligand (Figure 4.1). A case in point would be the humanized anti-CD74 monoclonal antibody (hLL1 milatuzumab or IMMU-115). This antibody has been used as an anticancer drug by itself, as a carrier for cytotoxic drugs to increase their target localization and also as a navigating ligand for liposomes loaded with doxorubicin thus forming a component of a drug delivery system (DDS) (Figure 4.2).[1]

A recent review generalized the concept of targeting ligands beyond antibodies to so-called hunter–killer peptides (HKPs).[2] These are chimeric molecules consisting of two functional domains, one serving as a navigating moiety and the other possessing cytotoxic properties. The navigating moiety delivers the cytotoxic payload to specific target areas where the proapoptotic domain

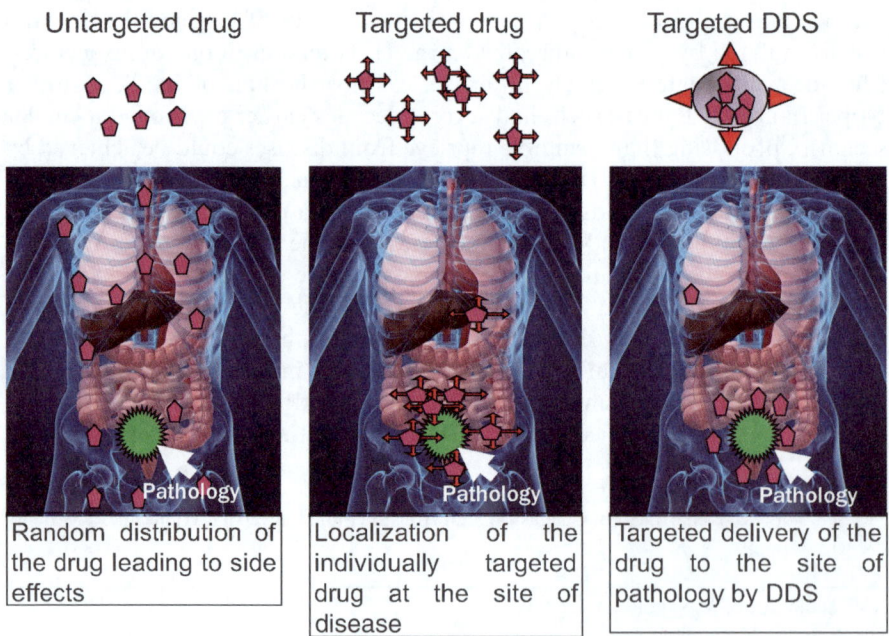

Untargeted drug	Targeted drug	Targeted DDS
Random distribution of the drug leading to side effects	Localization of the individually targeted drug at the site of disease	Targeted delivery of the drug to the site of pathology by DDS

Figure 4.1   Illustration demonstrating the concept of targeting of drugs and drug delivery systems.

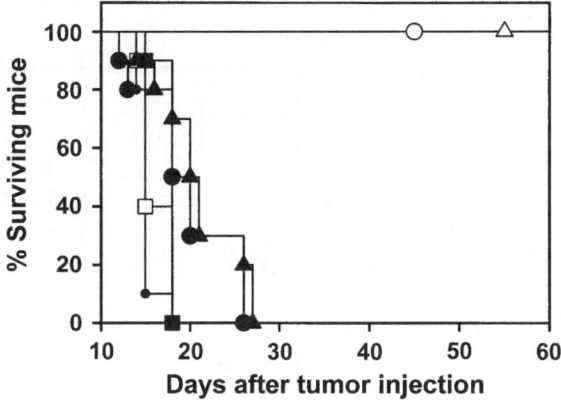

Figure 4.2 Free doxorubicin (dox), alone or in combination with unconjugated human LL1 (hLL1), had no therapeutic effect. Groups of mice were treated at day 5 with single 350 µg doses of dox conjugates of hLL1 (open circles) and murine LL1 (open triangles), in comparison with control groups including animals given the same dose of free dox (solid circles), unconjugated hLL1 alone (solid squares) or a mixture of free dox + unconjugated hLL1 (solid triangles). Other control groups were untreated (small solid squares), or injected with a non-reactive conjugate, dox–human MN-14 (open squares). No additional deaths occurred to the end of the experiment, at day 112. Both dox-LL1 groups were significantly different from all of the control groups ($p < 0.001$). Adapted from G. L. Griffiths *et al.*[1]

kills the cells by the disruption of mitochondrial membranes. One of the applications of this bioconjugate was in cancer therapeutics.[3] A tumor vasculature targeting cyclic peptide, CNGRC, was coupled to the proapoptotic peptide KLAKLAKKLAKLAK using a glycine linker. Its therapeutic efficiency was tested in nude mice bearing MDA-MB-435-derived human breast carcinoma xenografts. The authors observed a dramatic improvement in survival and a reduction in tumor growth and metastasis in the treated mice compared with control mice (Figure 4.3). Following this pioneering study, other peptide–drug combinations were synthesized and utilized as putative therapeutics for cancer,[4,5] arthritis,[6] and obesity.[7] Although the magic bullet hypothesis was realized first with monoclonal antibodies and peptides as exemplified above, advances in our understanding of physiology and breakthroughs in biomolecular technology have allowed the use not only of chemical compounds, but also of physical forces such as magnetic fields and heat to realize targeted localization of drugs or drug carriers.[8,9]

An efficacious cure is achieved when a drug appropriate for the pathology reaches the site of its action at an optimal concentration, which directly depends on various pharmacokinetic factors such as solubility, absorption kinetics and degradation/inactivation in the body. These factors very often diminish the effective concentration of the drug at its site of action and preclude a higher doses due to the toxic effects of most drugs towards normal tissues.

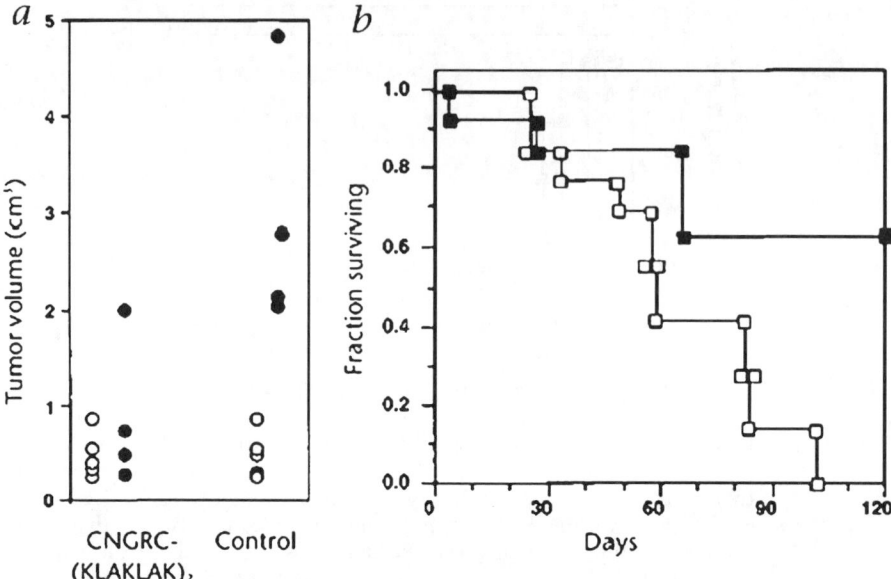

Figure 4.3 Treatment of nude mice bearing MDA-MB-435-derived human breast carcinoma xenografts with CNGRC-GG-D(KLAKLAK)$_2$. (a) Tumors treated with CNGRC-GG-D(KLAKLAK)$_2$ are smaller than control tumors treated with CARAC-GG-D(KLAKLAK)$_2$, as shown by differences in tumor volumes between day 1 (open circles) and day 50 (solid circles). $p = 0.027$, t-test. One mouse in the control group died before the end of the experiment. (b) Mice treated with CNGRC-GG-D(KLAK-LAK)$_2$ (solid squares) survived longer than control mice treated with an equimolar mixture of D(KLAKLAK)$_2$ and CNGRC (open squares), as shown by a Kaplan–Meier survival plot ($n = 13$ animals per group). $p < 0.05$, log-rank test. Adapted from Ellerby *et al.*[3]

This issue becomes significant in malignancies and viral infections where the disease either originates from host tissue or is deeply integrated into host cellular machinery.

A new challenge in disease-targeted therapy is also the resistance to available drugs, which again requires higher than therapeutic concentrations to overcome the resistance. Hence the problem basically is the achievement and maintenance of a concentration differential between diseased and normal tissues. Circumventing this problem involves either creating drugs that are more target specific or modifying existing drugs so as to result in preferential localization in target tissue.

4.2 Targeting of Drugs/Drug Carrier Systems

Creation of specific drugs is complicated by the lack of unique targets and in cases where a specific target is available, by the capacity of the pathology to initiate resistance by creating bypasses to blockades in cellular machinery. Therefore, modification of existing drugs by their coupling with targeting moieties or their encapsulation into targeted vehicles remains a logical choice to achieve higher concentrations of the drug at desired locations in the body, thereby improving their therapeutic index.

The choice of targeting strategy, which may involve the modification of a drug itself or a DDS depends on various factors, such as the nature of the target, drug and targeting moiety, and also the vehicular system used. It is commonly believed that targeting of DDS is more efficient as the stoichiometric ratios of drug and targeting ligand needed to achieve cytotoxic concentrations are much less than in directly targeted drugs. For example, hundreds of drug-conjugated ligands may be required to deliver the desired therapeutic concentration of drugs to the target cells. On the other hand, by targeting a DDS, a few targeting moieties are used to decorate the surface of a nanocarrier (*e.g.* liposomes) containing thousands of drug molecules. A DDS sometimes provides additional therapeutic advantages. For example, most DDSs alter the drug's pharmacokinetics, which can reduce its toxicity toward normal tissues. The enclosed drug mainly demonstrates the pharmacokinetic profile of the vehicle, thereby acquiring enhanced therapeutic efficacy by virtue of a better bioavailability and a decline in non-specific effects. Furthermore, drugs carried within the DDS are protected from degradation and clearance, contributing further to the delivery of an efficacious and stable payload to the site of disease. Also, in the case of colloidal DDSs, passive targeting is observed wherein a higher concentration of the drug accumulates in the tumor due to histological and physiological characteristics peculiar to the tumor.[10,11] In summary, DDSs such as liposomes can combine the benefits of the DDSs with selective homing to pathological tissues.

4.3   Targeting Ligands

A plethora of targeting ligands have been proposed, including antibodies or their fragments, peptides, growth factors, glycoproteins, carbohydrates and receptor ligands. Although most studies have used antibodies, these therapies show limited efficacy in solid tumors, mainly because their large size ($\sim$160 kDa) precludes efficient penetration, a problem aggravated by the elevated interstitial pressure in many tumors.[12] Furthermore, optimal tumor penetration may be hindered by a high-affinity interaction between the antibody and the first few antigen molecules with which it comes in contact, thereby preventing deep penetration into the tumor mass, a phenomenon first described as 'binding site barrier'[13] and later demonstrated experimentally.[14] Also, immunogenicity and non-specific uptake by the reticuloendothelial system (RES) contribute to a less than favorable clinical outcome. Advances in antibody engineering have partly alleviated these problems by creating functional fragments of target-specific antibodies. Examples include single-chain Fv fragments ($\sim$25 kDa),[15] VHH domains and nanobodies ($\sim$15 kDa),[16] which are better able to meet the need for increased tumor penetration while decreasing immunogenicity and non-specific removal. Research indicating that pharmacokinetic properties are improved with the use of smaller ligands underscores the need for smaller molecules for tumor directed therapies.[17] Peptides, being in the range of 1–2 kDa and demonstrating acceptable affinity and specificity to their targets, represent an attractive alternative to antibodies. Many problems associated with larger sized antibodies appear to be resolved with peptides. They possess impressive tumor-penetrating capacity, are generally not recognized by the RES and are less likely to induce immune responses. In addition, they exhibit a higher activity per unit mass and greater stability.[18] Their main disadvantages, rapid proteolysis and relatively poor affinity, can be resolved by derivatization of peptidomimetics by N- and/or C-terminal blocking, cyclization and incorporation of D-amino acids.[19–21] Furthermore, the availability of combinatorial peptide libraries from which potential therapeutics can be extracted adds another availing dimension to the concept of peptidic biopharmaceuticals.

4.4 Phage-displayed Libraries as a Source of Peptide Targeting Ligands

Phage display[22] has emerged as an increasingly powerful tool in molecular biology, with widespread applications and far-reaching implications. Its power lies in the self-assembly and self-perpetuation of the query molecules on the viral surface through the maintenance of the viral life cycle. Additional features are the stability and the longevity of the phage particles, which are largely shared by the peptides that they bear,[23] Selection from very large random peptide libraries (10 billion peptides or more) provides a generic system to discover ligands for a great diversity of molecular and tissue targets, yielding, in favorable cases, phage probes with high affinity, specificity and selectivity. To date, phage libraries have been the most common combinatorial technology used to screen for cancer-specific ligands.[24] In the affinity selection process, alternatively called 'biopanning', a ligand library displayed on the surface of phage particles is panned (screened) against the target molecule or tissue serving as the 'selector'. After washing away phage particles that do not bind the selector (the overwhelming majority in the initial library), particles that remain bound to the selector are eluted with acid or alkali, enzymatic cleavage or cell lysis with mild detergents, and amplified by infecting fresh bacterial host cells, yielding a sub-library that is enriched for selector-binding phage. The amplified sub-library is then used in subsequent rounds of selection to identify and isolate ligands with highest affinity for the target. A single round of affinity selection will enrich the target-specific clones by many orders of magnitude and three or four rounds can suffice to select guest peptides with high affinity for the selector from a multibillion clone library. The amino acid sequences of the selected ligands are easily determined by sequencing their coding sequences in the phage

DNA. The ligand displayed on the surface of the phage can range from large antibodies to short peptides, thus permitting the harnessing of this technology to a wide variety of applications from epitope mapping to the development of anti-cancer therapeutics.[25–32] Indeed, a number of reviews have expatiated on the utility of phage-displayed libraries and also listed a number of examples of peptide ligands obtained using phage display technology.[19,33–36] Furthermore, the success of *in vivo* affinity selection, in which the selector is tissue in a living animal, makes phage display an even more attractive platform for discovery of target-specific ligands.[37,38]

Within the new vista of bionanotechnological applications of phage display, landscape phage stand out. Landscape phage are constructs in which the foreign peptide is displayed on all copies of the major coat protein pVIII. They are produced by the in-frame splicing of degenerate synthetic oligonucleotides into phage gene *VIII*, so that the peptide encoded by a particular oligonucleotide is fused to the N-terminus of each pVIII subunit. Since the capsid length is altered to match the size of the enclosed DNA by adding proportionally fewer or more pVIII protein subunits during phage assembly, landscape phage with a 9.2-kilobase genome have ~4000 copies of pVIII, compared with ~2700 copies in the wild-type phage.[39] The thousands of guest peptides are arranged in a dense, geometrically reiterative pattern on the particle surface, subtending a substantial fraction of the total surface area (Figure 4.4). Phage with different guest peptides can have dramatically different physicochemical properties while

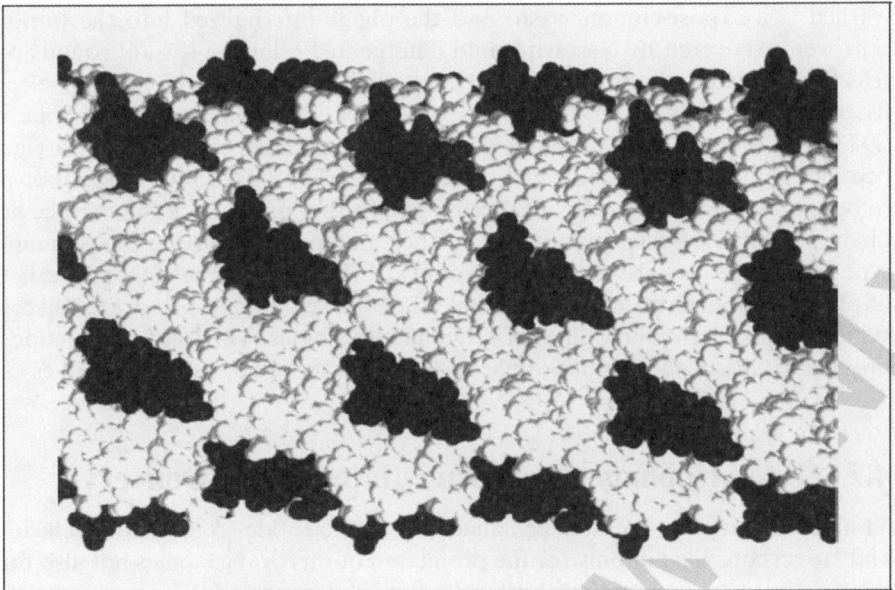

Figure 4.4 Segment of landscape phage. Foreign peptides are pictured with dark atoms; their arrangement corresponds to the model of Marvin *et al.*[79] Adapted from Petrenko *et al.*[95]

retaining the ability to be propagated in bacterial host cells; even more than other phage constructs, they fit the conventional notion of a nanoparticle, nanomaterial or nanotube. A large collection of such phage, displaying up to a billion different guest peptides, is called a 'landscape library'. By many orders of magnitude, more distinct nanomaterials have been created in this way than in the entire history of conventional, non-biogenic nanomaterial science. This concept was first demonstrated by a study by Petrenko and Smith,[40] in which phage selected from a landscape library were shown to functionally mimic antibodies in their ability to bind protein antigens. Since the affinity-selected phage is used as a framework for the binding moiety, this technology bypasses one of the most troublesome steps of phage–antibody technology: re-engineering of the selected antibody genes to express them at a high level.

Landscape phage probes have been identified for a variety of targets belonging to both the organic and inorganic realms encompassing a wide spectrum of complexity.[41] As a pertinent example, highly specific and selective landscape phage probes were selected for the PC3 prostate cancer cell line.[42] Before positive selection, the input landscape library was progressively depleted of clones that bound plastic, serum proteins and normal fibroblast cells. Only then was the depleted library positively selected with the target PC3 cells. After reaction with the depleted library, the PC3 cells were washed 20 times to eliminate unbound phage and also weak binders. Following incubation, phage associated with the cells were recovered using two different approaches as shown in Figure 4.5. Phage bound to tumor cell surface were eluted using an acidic elution buffer of pH 2.1. Following the acid elution, the cells were washed twice (post-elution wash) and the phage internalized into the tumor cells were recovered by lysis with mild detergent. Following several rounds of selection, isolated phage clones were evaluated for their specificity and selectivity. Two clones from a landscape library displaying random 8-mers (DTDSHVNL and DTPYDLTG, designated by a sequence of a fused foreign peptide) and a single clone from a 9-mer library (DVVYALSDD) were shown to be specific and selective for PC3 cells. The further use of such tumor-specific phage for targeted therapy requires further manipulation of the phage genotype/phenotype and also an intensive study of the physico-chemical properties of phage components that would enable them to be harnessed as drug delivery vehicles or as essential modules of DDS like liposomes. The following sections provide details of the various applications of bacteriophage in drug delivery.

4.5   Bacteriophage Capsid-mediated Drug Delivery

Viral capsids represent extremely efficient polynucleotide-transporting vehicles and are responsible not only for the protection of the viral genomes but also for their delivery to appropriate host cells for viral procreation. They are self-assembling cage composed of repeating units of the coat protein(s), which are involved in numerous interactions with relevant biological molecules *in vivo* to achieve successfully the transfer of viral genomes.

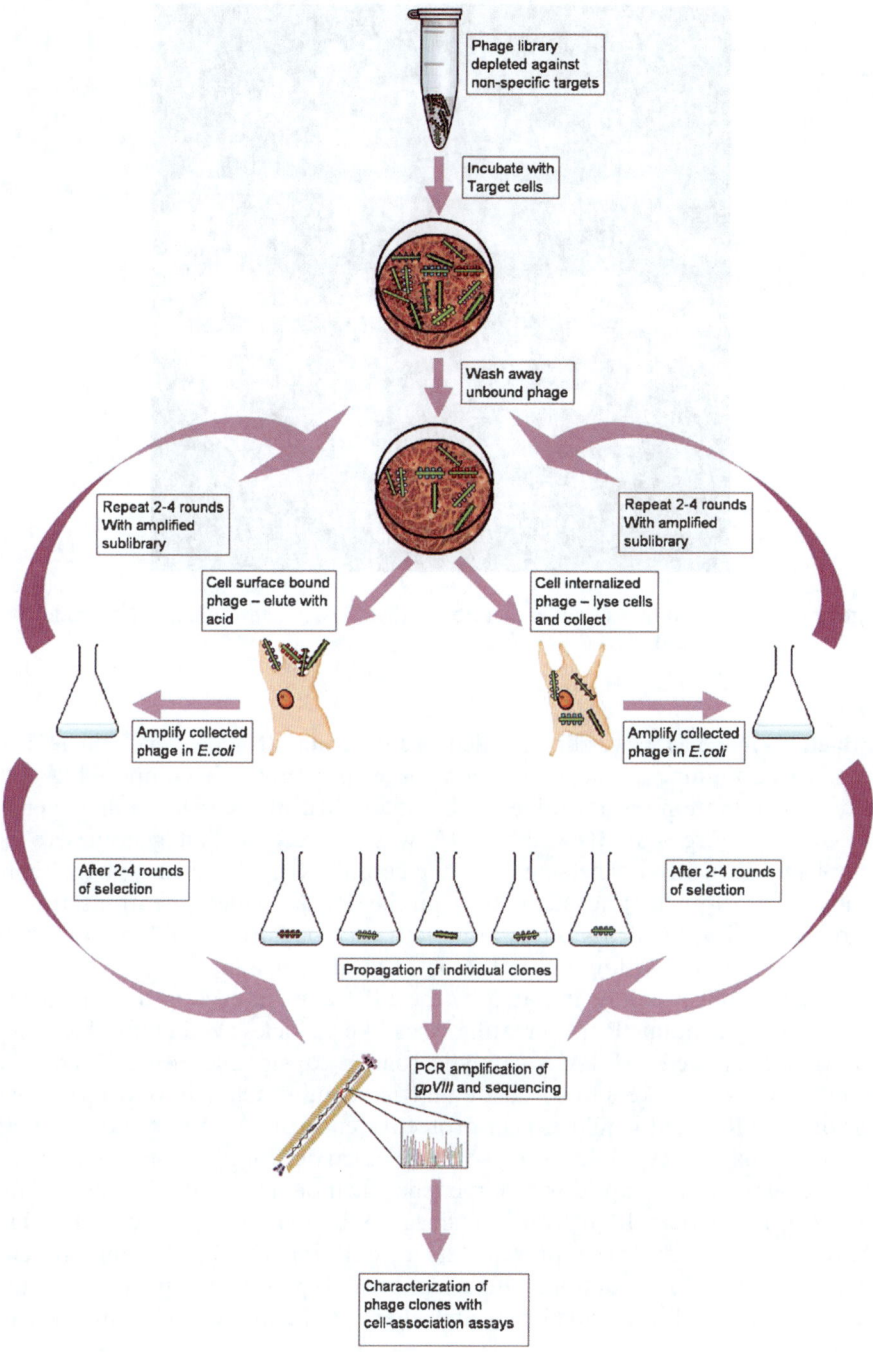

Figure 4.5   Scheme of selection of landscape phage probes for tumor cells. Adapted from Jayanna *et al.*[42]

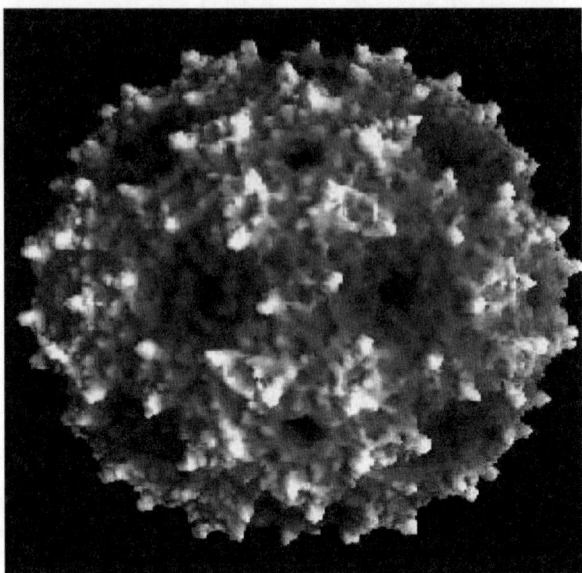

Figure 4.6 Three-dimensional structure of the MS2 bacteriophage. Adapted from Valegard *et al.*[51]

Bacteriophage MS2 was first studied as a potential drug delivery vehicle. It is an RNA-containing *Escherichia coli* virus whose capsid is composed of 180 coat protein monomers assembled into an icosahedral shell with a diameter of 27–34 nm (Figure 4.6). Its small RNA was one of the first genomes to be completely sequenced. The signal for the capsid assembly during phage replication is a 19-nucleotide stem–loop in the RNA, called the transcription repression (TR) signal. Its interaction with the coat protein subunits triggers termination of translation and initiation of viral packaging.[43]

Capsid assembly can be initiated by the TR fragment even in the absence of the remaining genomic RNA, creating virus-like particles (VLPs); the TR signal was thus presumed to play a catalytic role in capsid nucleation.[44] This led Stockley and co-workers to speculate that covalent attachment of drugs to one end of the TR signal would trigger protective encapsidation of the drug along with the signal, effectively creating a vehicle or carrier for the drugs. Decoration of the exterior of the capsid with target-specific moieties would then enable the viral capsid carriers to deliver the drug to the intended target cells. The encapsidation of the drug prevents its recognition by the immune system, protects it from degradation, affords improved pharmacokinetic properties especially for small-molecule drugs and offers opportunities for combinatorial drug therapy.

In their initial study in 1995, Stockley's group encapsidated the ricin A chain in MS2 capsids by coupling it covalently to the TR signal and incubating the conjugate with previously acid-disassembled MS2 coat protein subunits.[45] The

MS2-TR-RAC particles complexed with anti-MS2 antibodies were taken up by Pu518 cells and produced a toxic effect *via* receptor-mediated endocytosis of the immune complex, followed by escape of the conjugated RAC from the endosome due to instability of viral capsids at the endosomal pH. Further, in order to demonstrate targeted toxicity, the MS2-TR-RAC particles were decorated with transferrin molecules as the targeting moiety and its cytotoxicity was assayed in HeLa cells and HL-60 human leukemia cells. The results indicated a cell-specific delivery and toxicity with the targeted cytotoxic viral capsids, providing the basis for the concept of MS2 capsid-mediated drug delivery, which the authors called synthetic virion (SV) technology.

In a modification of this concept, the same group developed MS2 virus-like particles to deliver specifically and selectively cytotoxic antisense oligodeoxynucleotides (ODNs) to tumor cells. The authors targeted the proliferation-associated nucleolar antigen p120, which has been shown to be highly expressed in malignant tumors.[46] Previously, it had been shown that antisense ODN-mediated blockade of this mRNA resulted in cytocidal effects in human tumor cells with nuclear and nucleolar aberrations characteristic of apoptosis.[47,48] This capsid-mediated delivery overcomes a severe obstacle to therapeutic use of naked ODNs: rapid degradation by nucleases during transit to the target.

Wu *et al.* introduced an innovation that greatly simplifies the use of MS2 VLPs for delivery of ODNs.[49] The selected ODN was synthesized as an extension of the TR RNA and the TR-ODN hybrids were introduced directly into pre-formed VLPs by 'soaking', which is based on the observations that overexpression of a recombinant coat protein vector results in the self-assembly of excess coat proteins into RNA-free VLPs and that those VLPs are permeant to RNAs containing the TR RNA stem–loop.[50,51] The RNA fragments enter the protein cage through solvent channels in the crystal lattice and also via the pores at the capsid symmetry axes and make stable complexes with every coat protein dimer constituting the capsid architecture. Thus, delivery vehicles for antisense ODNs could be produced simply by co-incubation of empty VLPs and TR-ODNs. Covalent ligation of transferrin molecules to the surface of the MS2-TR-ODN particles allowed navigating the cytotoxic payload to tumor cells overexpressing the transferrin receptor. Encapsidation of ODNs not only increased their internalization by target cells but also prevented their efflux resulting in high intracellular concentrations. This was reflected in a cytotoxic effect that was nearly eightfold higher than that observed with free ODNs and that was sustained over a period of several days.

Despite these promising results *in vitro*, translating the VLP approach to drug delivery *in vivo* presents some hurdles. One is the immunogenicity of the phage capsid, which can lead to rapid clearance of the phage vehicles and to other specific effects involving both humoral and cell-mediated immunity.[52,53] In fact, these effects were exploited later to apply tumor-specific phage in a targeted tumor immunotherapy approach.[54] PEGylation of the VLPs to create stealth counterparts can alleviate this problem, but would require further manipulation of the surface of the phage carrier. The stability of the formulation both on the shelf and *in vivo* after such extensive modifications will need to be investigated thoroughly. An MS2 VLP's payload is limited to the number of TR RNA conjugates accommodated within the protein lattice (90 molecules each binding to 90 coat protein dimers unless the drug molecules are concatenated). The quantity of drug reaching its target may be therapeutically insufficient to gain a significant clinical advantage.

One way to surmount this problem would be to modify the interior of the coat protein units with genetically fused or covalently coupled payload molecules. In an elegant use of orthogonal conjugation chemistry, Francis and co-workers modified the internal surface of empty MS2 VLPs to allow specific covalent coupling of extraneous molecules.[55] These internal modification reactions are possible because of the presence of 32 pores per capsid, each approximately 1.8 nm in diameter, which allow reagents and moderately sized molecules ready access to the inner milieu of the viral particles. MS2 bacteriophage were freed of their RNA by alkaline hydrolysis and the empty capsids were shown to resist disassembly for 12 h over a range of pH. Diazonium-coupling reactions can then be used to introduce reactive sites on the amino acid residue Tyrosine 85 located on the inner surface of the capsid to which imaging agents, drugs or other functional molecules can be linked (Figure 4.7a). In proof of concept experiments, fluorescein dye molecules were conjugated to the reactive tyrosine 85.

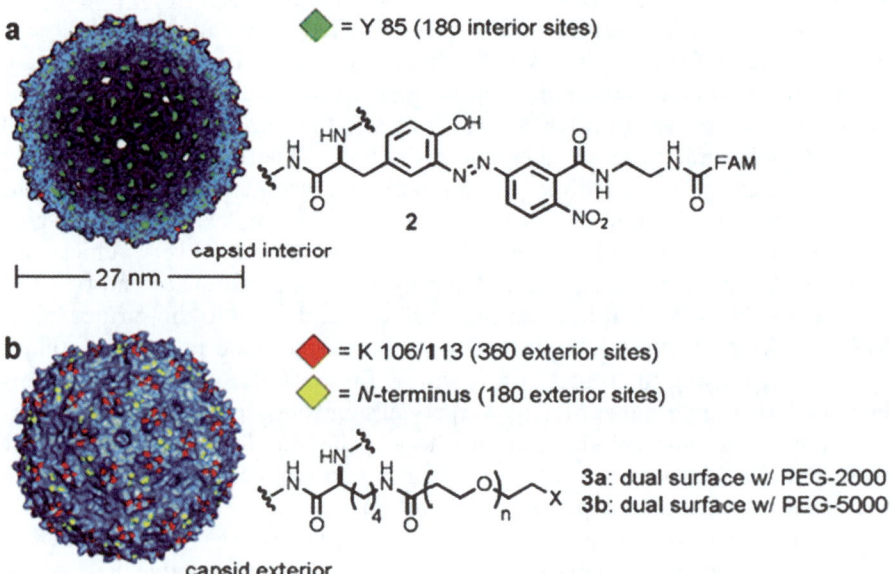

Figure 4.7 Dual surface modification of MS2 capsids. (a) Tyrosine 85 of the interior capsid surface undergoes rapid diazonium coupling with *p*-nitroaniline derivatives, including large dye conjugates. (b) Up to 360 accessible amino groups (lysines 106, 113 and the *N*-terminus) on the capsid exterior are readily modified with PEG-NHS esters. Adapted from Kovacs *et al.*[55]

In an effort to overcome the immunosurveillance instigated by the viral particles, the external surface of the empty capsids was then adorned with poly(ethylene glycol) (PEG) chains. PEGylation of the capsids reduced their recognition by rabbit anti-MS2 polyclonal antibodies by 65–90% (Figure 4.7b). In order to achieve targeting of the VLPs, the authors developed a modular strategy for the attachment of a wide variety of molecules at the ends of the PEG chains through oxime formation. Targeting was demonstrated by modifying the surface of the MS2 capsid surface with PEG-conjugated biotin followed by incubation with avidin beads. The mixture was centrifuged to remove the avidin-bound MS2 particles from the solution and the supernatant was analyzed for the presence of MS2 capsids. More than 90% of the PEGylated avidin-targeted VLPs were removed from the solution, indicating a successful interaction between the ligand biotin and target avidin.

The outlined data clearly demonstrate that VLPs can serve as efficient vehicles for biologically and medically relevant payloads by various modification and functionalization strategies. Successful removal of the genetic material without compromising capsid architecture gives the researcher access to a generic, stable, pre-formed scaffold for a great diversity of internal drug payloads and external targeting moieties. This combinatorial construction bypasses the need to fabricate an entirely new nanoparticle for each application.

4.6 Drug-bearing Filamentous Phage as Targeted Chemotherapeutics

One of the most serious emerging medical problems is bacterial resistance to traditional antibiotics. Bacterial strains resistant to new antibiotics often emerge soon after the drugs are introduced into clinical practice, a dilemma phenomenon with alarming epidemiological implications.

Chemical modification of existing antibacterial agents to upgrade their potency or spectrum has long been a mainstay approach to overcoming bacterial resistance. At the same time, application of new bioinformatics strategies in the post-genomic era have spurred the discovery of entirely novel agents. Nevertheless, lack of selectivity of both chemically modified and newly discovered agents remains a looming problem. An added complication is that very high concentrations of the antibiotic are often required to overcome the bacteria at the site of infection.

A novel approach based on targeting antibiotics to specific bacteria promises a reprieve from both dilemmas, provided that suitable targeting moieties and drug carrier can be discovered and effectively combined. The unique architecture and biological function of bacteriophage especially commend them for this purpose.

Benhar's group has pioneered the use of filamentous phage as a targeted drug carrier (Figure 4.8). Their overall goal was to achieve a high antibiotic concentration in the immediate vicinity of the targeted bacteria. At the same time, by using a payload prodrug conjugate that can be activated only in the presence

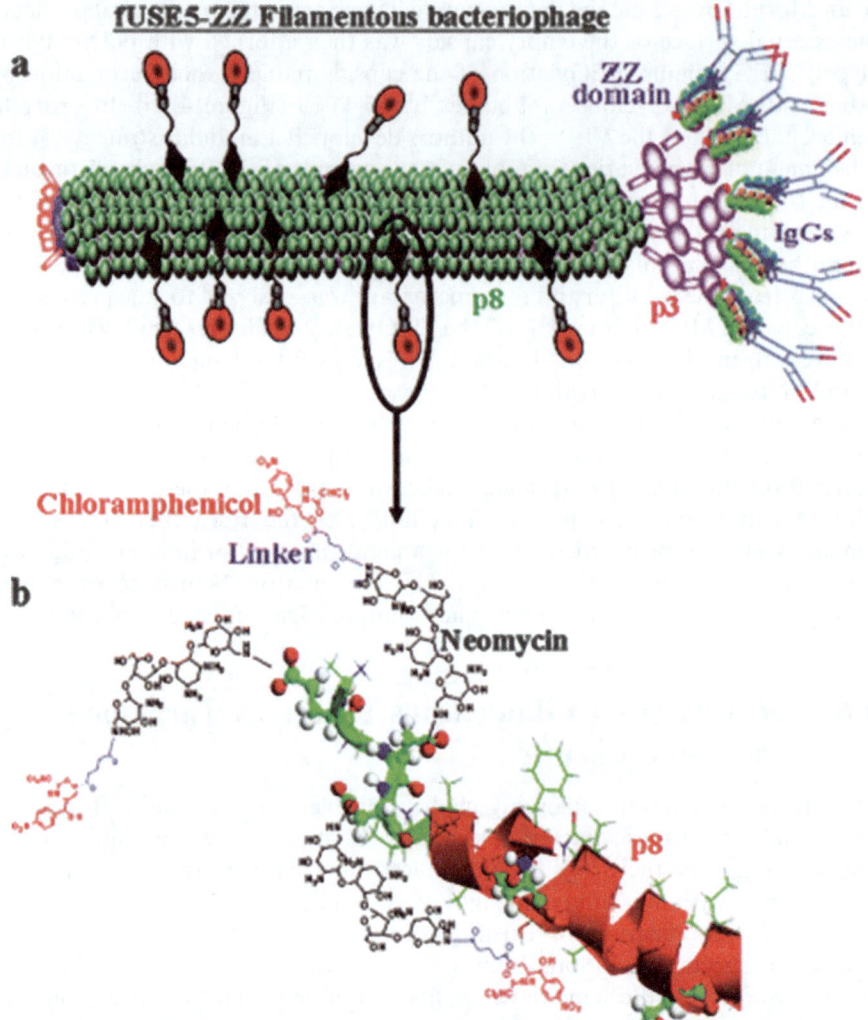

Figure 4.8   Schematic representation of drug-carrying bacteriophage. (a) Drawing of
a single fUSE5 ZZ-displaying bacteriophage. Small turquoise spheres
represent major coat protein p8 monomers. Purple sphere and sticks
represent the five copies of minor coat protein p3, which is fused to a
three-color helix representing the IgG binding ZZ domain. The Y-shaped
structure represents complexed IgG. (b) The red helix (represented by
ribbons) represents a partial structure of a major coat protein p8 monomer, conjugated through three N-terminal carboxyl side-chains (aspartyl
residues, represented by balls and sticks) to three molecules of neomycin
(black). Each neomycin molecule is conjugated through a labile ester bond
linker (blue) to a molecule of chloramphenicol (red). This represents
conjugation at minimal capacity (not using other free amine groups of
neomycin). Adapted from Yacoby *et al.*[57]

of the bacteria, they hoped to reduce or eliminate systemic toxicity. They adopted two approaches to targeting prodrug-bearing phage.[56] One was to use target-specific peptides displayed on the major coat protein pVIII in mosaic phage. The other relied on antibacterial antibodies (IgG) attached to the phage via an IgG-binding ZZ domain displayed on the minor phage coat proteins pIII. The *N*-hydroxysuccinimide (NHS)-modified chloramphenicol prodrug was conjugated to surface-exposed amines on the phage pVIII major coat protein subunits. Up to 3000 prodrug molecules could be coupled to the phage particle in this way without causing precipitation. From a large random peptide library, they affinity selected phage that bound specifically to live *Staphylococcus aureus* cells. Some of the binding phage displayed a guest peptide, VHMVAGPGREPT, with high affinity towards the target bacteria that was not affected by conjugation with the chloramphenicol prodrug. The drug–phage conjugate demonstrated good stability, only 15% of the prodrug being released after 1 h in serum (source of enzymes able to cleave the link between drug and linker), with linear kinetics. *S. aureus* treated with the drug-decorated phage in the presence of serum showed a retarded growth rate when the overall concentration of chloramphenicol released from the phage carriers was below the minimum inhibitory concentration.

In another targeting modality, phage was modified to display the immunoglobulin-binding ZZ domain on its minor coat protein pIII. Specific antibodies on the target bacteria would enable drug-carrying phage to bind the bacteria and deliver their antibacterial payload. The targeted phage inhibited the growth of *S. aureus* 10 times more effectively than free chloramphenicol. The authors postulated that the bacteriostatic nature and low potency of chloramphenicol, combined with a limited conjugation capacity, were the cause of the limited retardation of bacterial growth. Despite this limitation, the authors demonstrated that phage could be used as targeted vehicles for drugs.

Many issues must be addressed before the technique can become clinically feasible. One of them is the low number of the payload molecules on each phage. In a subsequent publication, the same group reported a novel conjugation chemistry utilizing a hydrophilic aminoglycoside antibiotic.[57] They cross-linked chloramphenicol–aminoglycoside adducts to the carboxyl groups on the phage major coat protein with a water-soluble carbodiimide, achieving a payload density of $\sim 10\,000$ molecules per phage, compared with ~ 3000 with the previous technique. Another advantage of carbodiimide chemistry is that it also allows for the cross-linking of the targeting moiety (*S. aureus* specific antibodies) to the phage minor coat protein units. The authors calculated a potency improvement factor of about 20 000 for targeted drug in comparison with the free drug with this system of targeting. The dramatic improvement was attributed to an overall improvement in solubility of the drug-bearing platform, and also to increased stability of the antibody targeting moieties achieved by covalently coupling them to the phage.

In an extension of the concept, Benhar and co-workers demonstrated the utility of the bacteriophage platform for targeted delivery of anticancer therapeutics. They designed phage nanovehicles that carry covalently coupled hygromycin or doxorubicin tethered to the major coat protein through a cathepsin B-cleavable peptide.[58] The nanovehicles were targeted using antibodies that had previously been shown to be effective in delivering cytotoxic payloads to target cells and tumor models: Trastuzumab and chFRP5, which target ErbB2; and Cetuximab, which targets EGFR.[59-61] Different approaches were employed for the attachment of the drugs to the phage carriers. One exploited carbodiimide chemistry to attach drugs directly to the carboxyl groups on the phage major coat protein units. Another relied on tethering the drug to the major coat protein through the cathepsin B-cleavable peptide mentioned above. Both approaches appeared to be feasible *a priori*, but differed in their efficacy depending the drug used. Covalently linked hygromycin was effective in inhibiting tumor cell growth, whereas a similar experiment with doxorubicin showed no effect. In contrast, doxorubicin tethered through the cathepsin B-cleavable peptide showed potency in impeding the growth of tumor cells, whereas hygromycin tethered in the same way was inactive. Using fluorescence confocal microscopy, the authors demonstrated that the targeted phage nanoparticles are internalized whether conjugated with the drug or not. Phage particles possessing hygromycin as payload were able to inhibit cell growth almost 1000 times better than free drug, whereas phage particles possessing doxorubicin attached *via* an enzyme sensitive linker showed anti-tumor effects, albeit non-specific. The authors postulated that the presence of a large number of hydrophobic molecules on the surface of the phage particles causes them to bind non-specifically to bacteria and cells. Despite the lack of specificity, this technique provides a critical proof of principle for the application of bacteriophage as targeted nanocarriers for anticancer chemotherapeutics. The findings that filamentous nanoparticles exhibit longer circulation times than their spherical counterparts as a result of various nanomechanical factors further invigorates the idea of using filamentous phage as targeted bioselective vehicles.[62]

4.7   Phage Fusion Proteins as Targeting Ligands for Nanomedicines

As was mentioned above, among other proteinaceous targeting ligands developed for drug-delivery and imaging vehicles, peptides possess considerable potential owing largely to their small size. Discovery of tumor-specific peptides was routinized recently using phage display technology.[33,63-67] The development of *in vivo* selection techniques has further provided an impetus in the broad spectrum application of phage display methods.[38,68]

The successful use of peptides as targeting ligands has been demonstrated hitherto in different systems.[69-74] In particular, the development of PEGylated stealth liposomes, including constructs bearing end-functionalized PEG derivatives, make liposomal formulations a very attractive platform to which target-specific peptides can be coupled. Several examples of peptide-interfaced targeted liposomes can be found.[69,71,73,74] The technology for incorporating targeting peptides into liposomes has evolved. Numerous covalent coupling techniques can be used, including disulfide bond formation, cross-linking between primary amines, reactions between a carboxylic acid and primary amine, maleimide-mediated coupling to thiols and reactions between primary amines or hydrazide and aldehydes,[75] Conjugation procedures adapted from the standard armamentarium of synthetic organic chemistry are fairly efficient for preparation of targeted liposomes on a scale sufficient for preliminary laboratory and clinical study. However, the cost and reproducibility of these derivatives in quality and quantity sufficient for pharmaceutical applications become prohibitive problems. Also, preparative conditions for the addressed vesicles differ idiosyncratically from one targeted particle to another, creating a bottleneck in the development of targeted drugs.

These considerations led us to evaluate the potential of recombinant phage coat proteins as the targeting probes for drug-loaded liposomes and micelles. In our approach, pVIII major coat protein molecules bearing disease-specific peptides are prepared by solubilizing landscape phage particles selected to be specific for a target organ, tissue or cell. The solubilized proteins can be directly inserted into drug-loaded liposomes and micelles by exploiting the intrinsic amphiphilic properties of the pVIII coat protein (Figure 4.9). Since the targeting peptide ligand is a natural physical extension of the protein's membrane-spanning anchor, there is no need for conjugation. Incorporation of the protein into the lipid bilayer is dominated by the highly evolved properties of the natural coat protein, not by idiosyncratic properties of the fused guest peptide. One of the most troublesome and labor-intensive steps of the conjugation

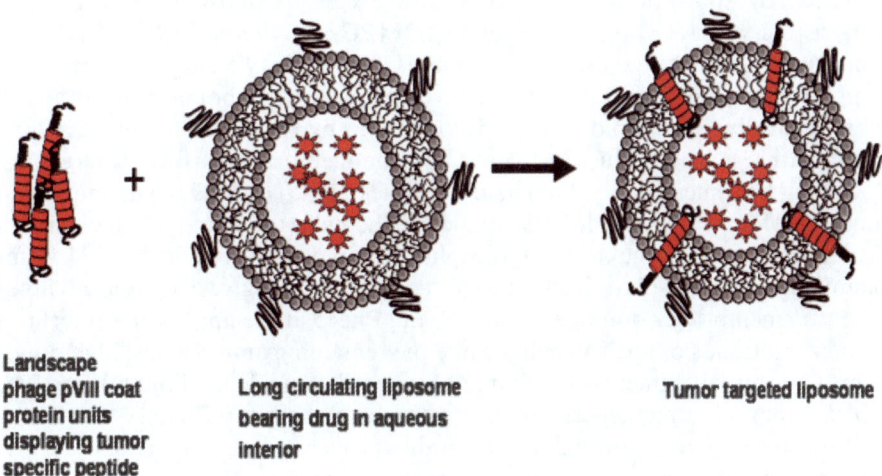

Landscape phage pVIII coat protein units displaying tumor specific peptide

Long circulating liposome bearing drug in aqueous interior

Tumor targeted liposome

Figure 4.9 Drug-loaded liposome targeted by the pVIII protein. The hydrophobic helix of the pVIII spans the lipid layer and binding peptide is displayed on the surface of the carrier particles. The drug molecules are shown as hexagons. Adapted from Jayanna *et al.*[94]

technology is thereby bypassed. No processing of the selected landscape phage is required: the phage themselves serve as the feedstock for the final product – coat protein genetically fused to the targeting peptide. The unique properties of phage major coat protein that enable this concept merit further attention and are discussed below.

The ability of the major coat protein pVIII to be incorporated into micelles and liposomes emerges from its natural function in phage assembly. During infection of the host *E. coli*, the 50-residue phage coat is dissolved in the bacterial cytoplasmic membrane, while viral DNA enters the cytoplasm.[76] New copies of the pVIII coat protein are synthesized as 73-residue water-soluble cytoplasmic precursors with 23-residue signal peptides at their N-termini. As the precursor protein is being inserted into the membrane, the signal peptide is cleaved off by a signal peptidase. As each progeny phage particle is extruded through the cell envelope, pVIII subunits are transferred from the membrane into the coat of the emerging phage. Thus, the major coat protein can change its conformation to accommodate distinctly different environments: membrane, intermediate particle and mature phage filament. This structural adaptability arises from the protein's unique architecture, which has been studied in detail in both mature particles, micelles and bilayer membranes.[77,78] The 50-residue mature protein is very hydrophobic and insoluble in water when separated from virus particles or membranes. In virus particles it forms a single, somewhat distorted α-helix with only the first four to five residues mobile and unstructured.[79] It is arranged in layers with a fivefold rotational symmetry and approximate twofold screw symmetry around the filament axis, as shown in Figure 4.10.

Liposomes displaying coat protein pVIII fixed in the lipid bilayers can be prepared by sonication of the virus with excess of phospholipids, such as dimyristoyl-*sn-glycero*-phosphocholine (DMPC).[80] Alternatively, pVIII protein can be reconstituted into phospholipids by dialysis, yielding liposomes with a molar lipid-to-protein ratio of ∼250.[81] The membrane-bound form of the fd pVIII coat protein in lipid bilayers is illustrated by Figure 4.11. The 16 Å long amphipathic helix (residues 8–18) rests on the membrane surface, whereas the 35 Å long transmembrane (TM) helix (residues 21–45) crosses the membrane at an angle of 26° up to residue Lys40, where the helix tilt changes. The helix tilt accommodates the thickness of the phospholipid bilayer, which is 31 Å for palmitoyloleoylphosphatidylcholine and -phosphatidylglycerol, typical lipids of *E. coli* membrane components. Tyr21 and Phe45 at the lipid–water interfaces delineate the ends of the TM helix, while the remaining amino acids, including a short C-terminal segment with charged lysines, lie outside the lipid bilayer. The TM and amphipathic helices are connected by a short turn (Thr19–Glu20).

Prior to being incorporated into emerging progeny phage particles, the pVIII coat protein integrates into inner membrane, a process that is believed to be mediated by interplay of electrophoretic influences (membrane potential), electrostatic forces (charges on membrane and protein) and hydrophobic interactions. The hydrophobicity of the TM domain is chiefly responsible for driving the insertion of the coat protein to allow for thermodynamic

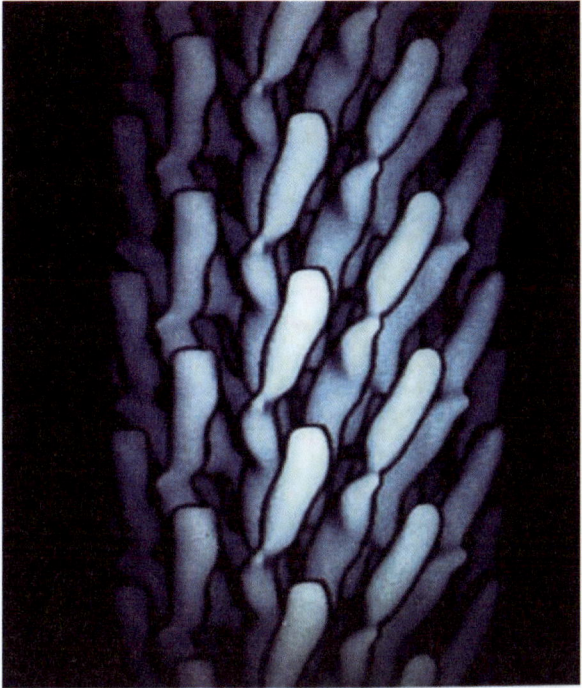

Figure 4.10 Filamentous phage. Segment of ~1% of phage virion with the array of pVIII proteins shown as electron densities. Micrograph and model courtesy of Irina Davidovich, Gregory Kishchenko and Lee Makowski.

equilibrium, whereas the membrane potential and charges on the protein in question are the major determinants of the topology of the membrane-associated protein.[82]

The spontaneous insertion of the mature pVIII protein into a lipid membrane can be envisioned as a three-step process. The first step involves binding of the protein to the membrane parallel to the plane of the membrane. In the second 'transition step', the hydrophobic region and one of the hydrophilic tails is inserted into the hydrocarbon core of the lipid bilayer. The third step would involve the release of the hydrophilic tail into the *trans* side. The structural characteristics of the membrane-bound coat protein after spontaneous insertion were shown to be predominantly α-helical.[83] Association of coat protein with lipid vesicles has been determined by fluorescence energy transfer (FET) between the coat protein's tyrosine and tryptophan residues (donors) and a diphenylhexatriene (DPH) in the lipids (acceptor).[84] Membrane penetration of the coat protein depends on the concentration of lipids available; almost complete penetration is seen at high lipid-to-protein ratios. Furthermore, lipid membranes with negative charges favor spontaneous insertion, whereas rigid lipid membranes preclude it. Coat protein insertion into liposomes follows an almost exclusive N-terminus$_{out}$ topological preference. A mutant of the major

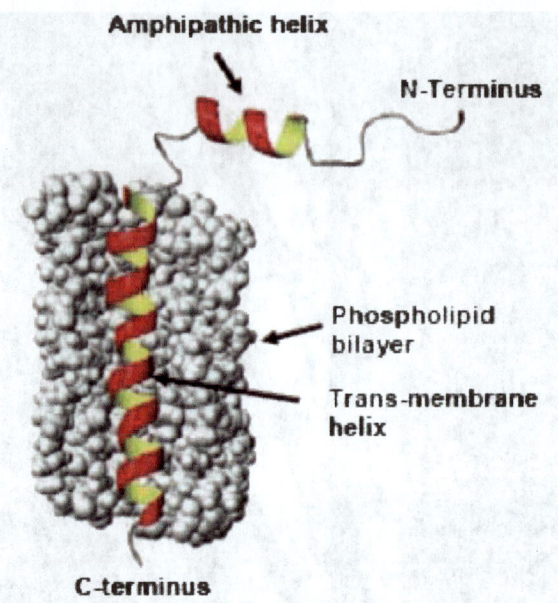

Figure 4.11   The model of pVIII in the lipid environment. Adapted from D. Stopar
et al., Biochim. Biophys. Acta Biomembr., 2003, **1611**, 5–15.

coat protein of bacteriophage Pf3 having three additional leucine residues in
the transmembrane region and complete replacement of charged amino acids
by aspargine (3L–4N) was shown to insert spontaneously into lipid vesicles.[85]
In an extension of this research, hybrid peptides comprising a fusion of a single
IgG binding unit of staphylococcal protein A attached to the N-terminus of the

---

Figure 4.12   Cancer cell-specific association of PC3-targeted rhodamine-labeled
liposomes. Flow cytometric analysis of a mixture of unlabeled target PC3
cells and green-labeled control HEK293 cells without liposomal treatment differentiated into two distinct populations (a), based on the
fluorescence intensity in the green channel (*y*-axis), consisting of labeled
control cells (R1) and unlabeled target cells (R3). Treatment with targeted or control rhodamine-labeled liposomes resulted in the appearance
of two new populations based on the fluorescence intensity in the red
channel (*x*-axis) including liposome-labeled PC3 cells (R4) and liposome-
labeled HEK293 cells (R2). A predominant shift of target PC3 cells along
the red channel was observed after binding PC3-specific liposomes (b)
that was not detected with other control liposomal formulations [lipo-
somes targeted with unrelated phage peptide termed 7b1 (c), liposomes
targeted with wild-type phage coat protein (d) and untargeted liposomes
(e)]. A representative experiment with liposomes modified with the 8–3
phage fusion protein is shown. Similar results were obtained with 9–8
PC3-specific phage fusion protein. Experiments were repeated in tripli-
cate. Adapted from Jayanna *et al.*[94]

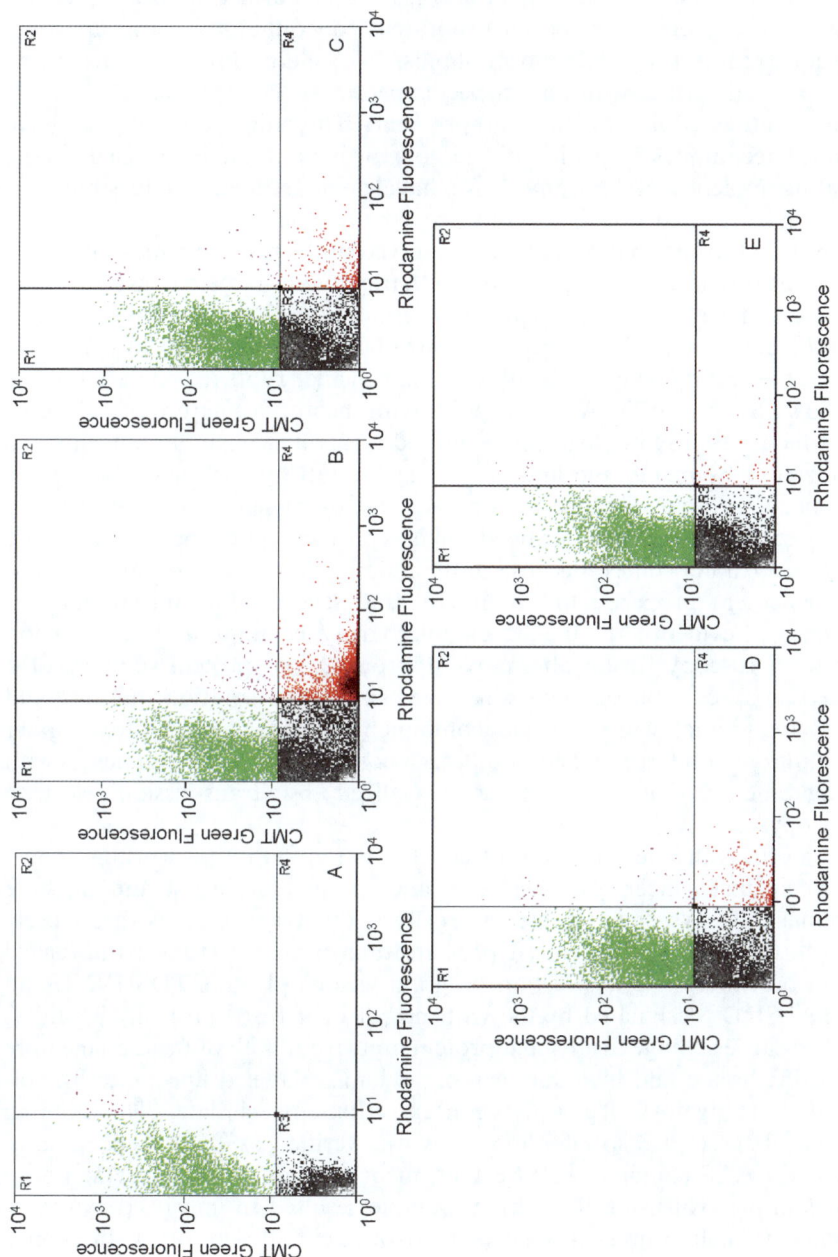

coat protein mutant was shown to insert spontaneously into liposomes, resulting in liposomes with a new emergent property: the ability to bind antibody.[86] Apart from spontaneous insertion into pre-formed liposome, recombinant coat proteins can be incorporated concomitantly with liposome formation from appropriate phospholipids,[81,87] yielding liposomes decorated on the external surface with target-specific peptides.[88]

Collaborative efforts of the authors' and Torchilin's laboratories have developed techniques by which phage coat protein could be applied as the navigating moieties of liposomes. We have demonstrated the feasibility of constructing targeted liposomes and proved that the targeting properties of peptides fused to the coat protein are preserved in the resulting liposomes.[84–93] For proof-of-concept experiments, a streptavidin-targeted system was employed. Phage bearing a streptavidin-binding peptide was converted to a new biorecognition affinity reagent: 'stripped phage'. Stripped phage is obtained by treating phage with chloroform, which disrupts the phage structure and allows the phage DNA to escape, leaving behind a mixture of coat proteins, which, because of their compromised structure, are more amenable to solubilization by membrane lipids.[94] Using the intrinsic self-assembly of the fusion phage proteins into the lipid membrane, discussed earlier, we incorporated streptavidin-targeted stripped phage proteins into the commercially available Doxil liposomes. Because of the modification, the liposome acquired a new emergent property: the ability to bind streptavidin and streptavidin conjugates, as evidenced by the protein microarray technique and transmission electron microscopy. In an alternative approach, coat protein subunits of a streptavidin-specific phage were solubilized in cholate solution, purified and allowed to interact with the lipid components of commercial Doxil under reconstitution conditions. The resulting liposomes formed complexes with streptavidin-coated gold beads as visualized by transmission electron microscopy.

In order to translate this proof-of-concept paradigm into a working model targeted against cancer, we selected phage specific for breast and prostate carcinoma cells. We chose to use PC3 cells as our target cells as they closely approach the profile of advanced prostate tumors that was the pathological niche of our interest. The selection program yielded phage DTDSHVNL and DVVYALSDD (designated by the sequence of their fused targeting peptides) described earlier. Phage fusion coat protein units from both of these phage were purified in cholate and introduced into rhodamine-labeled liposomes by gradient dialysis against decreasing concentrations of cholate. The homing potential of these PC3-specific liposomes was verified by fluorescence microscopy and FACS (Figure 4.12). Further, incorporation of PC3-specific phage peptides into Doxil using the same technique resulted in improved cytotoxic performance of liposome-encapsulated doxorubicin (Figure 4.13). In similar studies involving phage liposomes targeted to MCF-7 breast cancer cells, it was shown that the fusion phage protein allows the liposomes to escape the endosomal route of cell internalization and subsequent degradation, thereby resulting in higher cytotoxicity.[93] Fluorescence resonance energy transfer

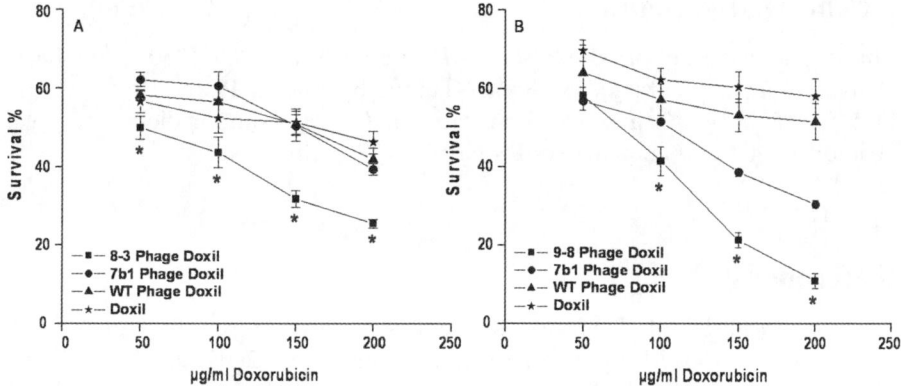

Figure 4.13 *In vitro* cytotoxicity results of PC3-specific doxorubicin-loaded liposomal formulations. Commercial Doxil was grafted with PC3-specific phage fusion proteins (8–3 and 9–8) or control phage fusion proteins (7b1, an unrelated peptide or coat protein from wild-type phage, WT) and incubated with PC3 cells for 48 h. The cytotoxicity of each preparation was expressed as percentage survival compared with untreated cells which were considered to be 100%. (a) 8–3 grafted Doxil; (b) 9–8 grafted Doxil. Star indicates $p < 0.05$, one-way ANOVA with Tukey's honestly significant difference *post hoc* test. Adapted from Jayanna *et al.*[93]

(FRET) was used to demonstrate the pH-dependent membrane fusion activity of the phage liposomes. Endosomal escape and cytosolic delivery of phage liposomes were visualized with fluorescence microscopy.

Current research is focused on the combinatorial production of targeted liposomes, drug loading into such targeted liposomes and assessing the therapeutic effects of such formulations in cell cultures and animal models.

4.8   Conclusion

Bacteriophage therapy, when first introduced in the early part of the twentieth century, was seen to be very promising. Later, poor clinical results due to a lack of adequate information about phage biology, inconsistent formulations, poorly planned studies and the discovery of new antibiotics pushed phage-based therapeutics into medical oblivion. Modern phage research, including phage display, has resulted in a resurgence of interest in these viruses *per se* and introduced many new potential applications of phage particles. The stability and amenability of phage to experimental manipulation have made them a workhorse of molecular biology and biological nanoengineering, with applications ranging from nanocircuitry to anticancer chemotherapy. Combined with the parallel developments in other fields of science, phage technology currently represents a cross-roads where techniques from diverse areas meet and fuse to provide innovative answers to the pressing problems of science.

Acknowledgements

This work was supported by the NIH grants 1 R01 CA125063-01 'Drug Delivery Carriers Targeted to Breast Tumor by Fusion Phage Proteins' and 1U54CA151881-0 'Center for Translational Cancer Nanomedicine' (Project 'Tumor Targeting Nanocarriers Using Phage Proteins').

References

1. G. L. Griffiths, M. J. Mattes, R. Stein, S. V. Govindan, I. D. Horak, H. J. Hansen and D. M. Goldenberg, *Clin. Cancer Res.*, 2003, **9**, 6567–6571.
2. H. M. Ellerby, D. E. Bredesen, S. Fujimora and V. John, *J. Med. Chem.*, 2008, **51**, 5887–5892.
3. H. M. Ellerby, W. Arap, L. M. Ellerby, R. Kain, R. Andrusiak, G. D. Rio, S. Krajewski, C. R. Lombardo, R. Rao, E. Ruoslahti, D. E. Bredesen and R. Pasqualini, *Nat. Med.*, 1999, **5**, 1032–1038.
4. W. Arap, W. Haedicke, M. Bernasconi, R. Kain, D. Rajotte, S. Krajewski, H. M. Ellerby, D. E. Bredesen, R. Pasqualini and E. Ruoslahti, *Proc. Natl. Acad. Sci. USA*, 2002, **99**, 1527–1531.
5. E. Henke, J. Perk, J. Vider, P. de Candia, Y. Chin, D. B. Solit, V. Ponomarev, L. Cartegni, K. Manova and N. Rosen, *Nat. Biotechnol.*, 2008, **26**, 91.
6. D. M. Gerlag, E. Borges, P. P. Tak, H. M. Ellerby, D. E. Bredesen, R. Pasqualini, E. Ruoslahti and G. S. Firestein, *Arthritis Res.*, 2001, **3**, 357–361.
7. M. G. Kolonin, P. K. Saha, L. Chan, R. Pasqualini and W. Arap, *Nat. Med.*, 2004, **10**, 625–632.
8. B. R. Liu, M. Yang, X. L. Li, X. P. Qian, Z. T. Shen, Y. T. Ding and L. X. Yu, *J. Pharm. Sci.*, 2008, **97**, 3170–3181.
9. C. Sun, J. S. H. Lee and M. Q. Zhang, *Adv. Drug Deliv. Rev.*, 2008, **60**, 1252–1265.
10. A. K. Iyer, G. Khaled, J. Fang and H. Maeda, *Drug Discov. Today*, 2006, **11**, 812–818.
11. H. Maeda, J. Wu, T. Sawa, Y. Matsumura and K. Hori, *J. Control. Release*, 2000, **65**, 271–284.
12. R. K. Jain, *Cancer Res.*, 1990, **50**, 814–819.
13. K. Fujimori, D. G. Covell, J. E. Fletcher and J. N. Weinstein, *Cancer Res.*, 1989, **49**, 5656–5663.
14. G. P. Adams, R. Schier, A. M. McCall, H. H. Simmons, E. M. Horak, R. K. Alpaugh, J. D. Marks and L. M. Weiner, *Cancer Res.*, 2001, **61**, 4750–4755.
15. G. P. Adams and R. Schier, *J. Immunol. Methods*, 1999, **231**, 249–260.
16. V. Cortez-Retamozo, N. Backmann, P. D. Senter, U. Wernery, P. De Baetselier, S. Muyldermans and H. Revets, *Cancer Res.*, 2004, **64**, 2853–2857.
17. R. M. Reilly, J. Sandhu, T. M. Alvarez-Diez, S. Gallinger, J. Kirsh and H. Stern, *Clin. Pharmacokinet.*, 1995, **28**, 126–142.

18. R. C. Ladner, A. K. Sato, J. Gorzelany and M. de Souza, *Drug Discov. Today*, 2004, **9**, 525–529.
19. O. H. Aina, T. C. Sroka, M.-L. Chen and K. S. Lam, *Pept. Sci.*, 2002, **66**, 184–199.
20. C. Borghouts, C. Kunz and B. Groner, *J. Pept. Sci.*, 2005, **11**, 713–726.
21. F. Nilsson, L. Tarli, F. Viti and D. Neri, *Adv. Drug Deliv. Rev.*, 2000, **43**, 165–196.
22. G. P. Smith, *Science*, 1985, **228**, 1315–1317.
23. J. R. Brigati and V. A. Petrenko, *Anal. Bioanal. Chem.*, 2005, **382**, 1346–1350.
24. R. Craig and S. Li, *Mini Rev. Med. Chem.*, 2006, **6**, 757–764.
25. A. Y. Ting, K. Witte, K. Shah, B. Kraybill, K. M. Shokat and P. G. Schultz, *Pept. Sci.*, 2001, **60**, 220–228.
26. S. Atwell and J. A. Wells, *Proc. Natl. Acad. Sci. USA*, 1999, **96**, 9497–9502.
27. C. F. Barbas, D. Hu, N. Dunlop, L. Sawyer, D. Cababa, R. M. Hendry, P. L. Nara and D. R. Burton, *Proc. Natl. Acad. Sci. USA*, 1994, **91**, 3809–3813.
28. B. Dreier, R. R. Beerli, D. J. Segal, J. D. Flippin and C. F. Barbas III, *J. Biol. Chem.*, 2001, **276**, 29466–29478.
29. J. L. Harris, E. P. Peterson, D. Hudig, N. A. Thornberry and C. S. Craik, *J. Biol. Chem.*, 1998, **273**, 27364–27373.
30. D. J. Matthews and J. A. Wells, *Science*, 1993, **260**, 1113–1117.
31. A. E. Nixon, *Curr. Pharm. Biotechnol.*, 2002, **3**, 1–12.
32. L. F. Wang and M. Yu, *Curr. Drug Targets*, 2004, **5**, 1–15.
33. L. Krumpe and T. Mori, *Int. J. Pept. Res. Ther.*, 2006, **12**, 79–91.
34. L. M. Mullen, S. P. Nair, J. M. Ward, A. N. Rycroft and B. Henderson, *Trends Microbiol.*, 2006, **14**, 141–147.
35. M. Paschke, *Appl. Microbiol. Biotechnol.*, 2006, **70**, 2–11.
36. M. Szardenings, *J. Receptors Signal Transduct.*, 2003, **23**, 307–349.
37. D. N. Krag, S. P. Fuller, L. Oligino, S. C. Pero, D. L. Weaver, A. L. Soden, C. Hebert, S. Mills, C. Liu and D. Peterson, *Cancer Chemother. Pharmacol.*, 2002, **50**, 325–332.
38. D. N. Krag, G. S. Shukla, G.-P. Shen, S. Pero, T. Ashikaga, S. Fuller, D. L. Weaver, S. Burdette-Radoux and C. Thomas, *Cancer Res.*, 2006, **66**, 7724–7733.
39. G. J. Hunter, D. H. Rowitch and R. N. Perham, *Nature*, 1987, **327**, 252–254.
40. V. A. Petrenko and G. P. Smith, *Protein Eng.*, 2000, **13**, 589–592.
41. V. Petrenko, *Expert Opin. Drug Deliv.*, 2008, **5**, 825–836.
42. P. K. Jayanna, D. Bedi, P. DeInnocentes, R. C. Bird and V. A. Petrenko, *Protein Eng. Des. Sel.*, 2010, **6**, 423–430.
43. D. Beckett and O. C. Uhlenbeck, *J. Mol. Biol.*, 1988, **204**, 927–938.
44. P. G. Stockley, N. J. Stonehouse and K. Valegard, *Int. J. Biochem.*, 1994, **26**, 1249–1260.
45. M. Wu, W. L. Brown and P. G. Stockley, *Bioconjug. Chem.*, 1995, **6**, 587–595.

46. R. L. Ochs, M. T. Reilly, J. W. Freeman and H. Busch, *Cancer Res.*, 1988, **48**, 6523–6529.
47. R. K. Busch, L. Perlaky, B. C. Valdez, D. Henning and H. Busch, *Cancer Lett.*, 1994, **86**, 151–157.
48. L. Perlaky, K. Smetana, R. K. Busch, Y. Saijo and H. Busch, *Cancer Lett.*, 1993, **74**, 125–135.
49. M. Wu, T. Sherwin, W. L. Brown and P. G. Stockley, *Nanomed. Nanotechnol. Biol. Med.*, 2005, **1**, 67–76.
50. R. A. Mastico, S. J. Talbot and P. G. Stockley, *J. Gen. Virol.*, 1993, **74**, 541–548.
51. K. Valegard, J. B. Murray, P. G. Stockley, N. J. Stonehouse and L. Liljas, *Nature*, 1994, **371**, 623–626.
52. I. Fogelman, V. Davey, H. D. Ochs, M. Elashoff, M. B. Feinberg, J. A. Mican, J. P. Siegel, M. Sneller and H. C. Lane, *J. Infect. Dis.*, 2000, **182**, 435–441.
53. Y. Wu, Y. Wan, J. Bian, J. Zhao, Z. Jia, L. Zhou, W. Zhou and Y. Tan, *Int. J. Cancer*, 2002, **98**, 748–753.
54. F. Eriksson, W. D. Culp, R. Massey, L. Egevad, D. Garland, M. A. A. Persson and P. Pisa, *Cancer Immunol. Immunother.*, 2007, **56**, 677–687.
55. E. W. Kovacs, J. M. Hooker, D. W. Romanini, P. G. Holder, K. E. Berry and M. B. Francis, *Bioconjug. Chem.*, 2007, **18**, 1140–1147.
56. I. Yacoby, M. Shamis, H. Bar, D. Shabat and I. Benhar, *Antimicrob. Agents Chemother.*, 2006, **50**, 2087–2097.
57. I. Yacoby, H. Bar and I. Benhar, *Antimicrob. Agents Chemother.*, 2007, **51**, 2156–2163.
58. H. Bar, I. Yacoby and I. Benhar, *BMC Biotechnol.*, 2008, **8**, 37.
59. H. Kobayashi, K. Shirakawa, S. Kawamoto, T. Saga, N. Sato, A. Hiraga, I. Watanabe, Y. Heike, K. Togashi, J. Konishi, M. W. Brechbiel and H. Wakasugi, *Cancer Res.*, 2002, **62**, 860–866.
60. Y. Mazor, I. Barnea, I. Keydar and I. Benhar, *J. Immunol. Methods.*, 2007, **321**, 41–59.
61. W. L. Yip, A. Weyergang, K. Berg, H. H. Tonnesen and P. K. Selbo, *Mol. Pharmacol.*, 2007, **4**, 241–251.
62. Y. Geng, P. Dalhaimer, S. Cai, R. Tsai, M. Tewari, T. Minko and D. E. Discher, *Nature Nanotechnology*, 2007, **2**, 249–255.
63. K. A. Kelly and D. A. Jones, *Neoplasia*, 2003, **5**, 437–444.
64. K. A. Kelly, S. R. Setlur, R. Ross, R. Anbazhagan, P. Waterman, M. A. Rubin and R. Weissleder, *Cancer Res.*, 2008, **68**, 2286–2291.
65. K. A. Kelly, P. Waterman and R. Weissleder, *Neoplasia*, 2006, **8**, 1011–1018.
66. L. A. Landon and S. L. Deutscher, *J. Cell. Biochem.*, 2003, **90**, 509–517.
67. T. Wang, G. G. D'Souza, D. Bedi, O. A. Fagbohun, L. P. Potturi, B. Papahadjopoulos-Sternberg, V. A. Petrenko and V. P. Torchilin, *Nanomedicine (Lond.)*, 2010, **5**, 563–574.
68. J. R. Newton, K. A. Kelly, U. Mahmood, R. Weissleder and S. L. Deutscher, *Neoplasia*, 2006, **8**, 772–780.

69. P. Holig, M. Bach, T. Volkel, T. Nahde, S. Hoffmann, R. Muller and R. E. Kontermann, *Protein Eng. Des. Sel.*, 2004, **17**, 433–441.
70. E. Koivunen, W. Arap, H. Valtanen, A. Rainisalo, O. P. Medina, P. Heikkila, C. Kantor, C. G. Gahmberg, T. Salo, Y. T. Konttinen, T. Sorsa, E. Ruoslahti and R. Pasqualini, *Nat. Biotechnol.*, 1999, **17**, 768–774.
71. T.-Y. Lee, H.-C. Wu, Y.-L. Tseng and C.-T. Lin, *Cancer Res.*, 2004, **64**, 8002–8008.
72. O. P. Medina, T. Soderlund, L. J. Laakkonen, E. K. J. Tuominen, E. Koivunen and P. K. J. Kinnunen, *Cancer Res.*, 2001, **61**, 3978–3985.
73. F. Pastorino, C. Brignole, D. Di Paolo, B. Nico, A. Pezzolo, D. Marimpietri, G. Pagnan, F. Piccardi, M. Cilli, R. Longhi, D. Ribatti, A. Corti, T. M. Allen and M. Ponzoni, *Cancer Res.*, 2006, **66**, 10073–10082.
74. H. Slimani, E. Guenin, D. Briane, R. Coudert, N. Charnaux, A. Starzec, R. Vassy, M. Lecouvey, Y. G. Perret and A. Cao, *J. Drug Target.*, 2006, **14**, 694–706.
75. L. Nobs, F. Buchegger, R. Gurny and E. Allemann, *Drugs Pharm. Sci.*, 2006, **158**, 123–148.
76. R. Webster, in *Phage Display: a Laboratory Manual*, ed. C. F. Barbas III, D. R. Barton, J. K. Scott and G. J. Silverman, Cold Spring Harbor Laboratory Press, Cold Spring Harbor, NY, 2001, pp. 1.1–1.37.
77. S. Lee, M. F. Mesleh and S. J. Opella, *J. Biomol. NMR*, 2003, **26**, 327–334.
78. A. C. Zeri, M. F. Mesleh, A. A. Nevzorov and S. J. Opella, *Proc. Natl. Acad. Sci. USA*, 2003, **100**, 6458–6463.
79. D. A. Marvin, R. D. Hale, C. Nave and M. Helmer-Citterich, *J. Mol. Biol.*, 1994, **235**, 260–286.
80. S. P. Fodor, A. K. Dunker, Y. C. Ng, D. Carsten and R. W. Williams, *Prog. Clin. Biol. Res.*, 1981, **64**, 441–455.
81. R. B. Spruijt, C. Wolfs and M. A. Hemminga, *Biochemistry*, 1989, **28**, 9158–9165.
82. D. Kiefer and A. Kuhn, *EMBO J.*, 1999, **18**, 6299–6306.
83. E. Thiaudiere, M. Soekarjo, E. Kuchinka, A. Kuhn and H. Vogel, *Biochemistry*, 1993, **32**, 12186–12196.
84. M. Soekarjo, M. Eisenhawer, A. Kuhn and H. Vogel, *Biochemistry*, 1996, **35**, 1232–1241.
85. A. N. J. A. Ridder, W. van de Hoef, J. Stam, A. Kuhn, B. de Kruijff and J. A. Killian, *Biochemistry*, 2002, **41**, 4946–4952.
86. T. Matsuo, T. Yamamoto, K. Niiyama, N. Yamazaki, T. Ishida, H. Kiwada, Y. Shinohara and M. Kataoka, in *Micro-Nano Mechatronics and Human Science*, 2007, pp. 73–78.
87. R. B. Spruijt and M. A. Hemminga, *Biochemistry*, 1991, **30**, 11147–11154.
88. P. K. Jayanna, V. P. Torchilin and V. A. Petrenko, *Nanomed. Nanotechnol. Biol. Med.*, 2009, **5**, 83–89.
89. V. A. Petrenko, in *Nanotechnology Conference and Trade Show, Santa Clara, CA*, Vol. 2, 2007, pp. 703–706.

90. V. A. Petrenko, S. N. Ustinov and I. Chen, in *European Nano Systems 2006, Paris*, Edited and published by TIMA Laboratory, Grenoble, France, pp. 89–94.
91. T. Wang, V. A. Petrenko and V. P. Torchilin, *Mol. Pharmacol.*, **7**, 1007–1014.
92. T. Wang, S. Yang, V. A. Petrenko and V. P. Torchilin, *Mol. Pharmacol.*, **7**, 1149–1158.
93. P. K. Jayanna, D. Bedi, J. W. Gillespie, P. DeInnocentes, T. Wang, V. P. Torchilin, R. C. Bird and V. A. Petrenko, *Nanomedicine (Lond.)*, 2010, **6**, 538–546.
94. M. Manning, S. Chrysogelos and J. Griffith, *J. Virol.*, 1981, **40**, 912–919.
95. V. A. Petrenko, G. P. Smith, X. Gong and T. Quinn, *Protein Eng.*, 1996, **9**, 797–801.

CHAPTER 5

Imaging with Bacteriophage-derived Probes

SUSAN L. DEUTSCHER[a] AND KIMBERLY A. KELLY[b]

[a] Biochemistry Department, University of Missouri, Columbia, MO 65211, USA and Harry S. Truman Medical Veterans Hospital, Columbia, MO 65212, USA; [b] Department of Biomedical Engineering and Robert M. Berne Cardiovascular Research Center, University of Virginia, Charlottesville, VA 22904, USA

5.1 Selection of Bacteriophage as Imaging Probes

5.1.1 Imaging Agents

It is becoming increasingly apparent that affinity ligands derived from screens have played an essential role in the development of molecularly targeted imaging agents. Molecular imaging could allow clinicians to visualize the expression and activity of disease-specific molecules, thereby providing relevant information in the diagnosis and treatment of patients. Over the last decade, the field of cellular and molecular imaging has witnessed extraordinary progress in the development of new technologies and imaging systems.[1]

Despite these advances, there still exists a paucity of molecularly targeted imaging agents. Antibodies, peptides and, more recently, nanoparticles are being developed for the targeted detection of diseases such as cancer where precise molecular information enables (1) precise positional information, (2) identification of metastatic spread, (3) prognostic information, (4) determination of therapeutic efficacy and (5) patient treatment stratification. Antibodies are the

RSC Nanoscience & Nanotechnology No. 17
Phage Nanobiotechnology
Edited by Valery A. Petrenko and George P. Smith
© Royal Society of Chemistry 2011
Published by the Royal Society of Chemistry, www.rsc.org

most common biological targeting vehicles for the specific delivery of an imaging tag, usually in the form of a radionuclide, to sites of disease.[2] However, when antibodies were used as the targeting molecule, the immunogenicity, cost and difficulty of producing an active conjugated form of these proteins was detrimental.[3]

Peptide constructs have been successfully employed as imaging probes because of their rapid blood clearance and tissue penetration, increased diffusion, non-immunogenic nature and ease of synthesis.[4] Highly specific peptide sequences with ideal pharmacokinetics can be obtained from phage display screens. Regulatory peptides that bind receptors overexpressed on tumors are being actively pursued in imaging studies. One of the best examples is octreotide, an eight amino acid cyclized peptide somatostatin derivative. [^{111}In]DTPA-octreotide (OctreoScan) was the first FDA-approved radiopharmaceutical for use in single photon emission computed tomography (SPECT) imaging of somatostatin receptor-positive tumors.[5] Furthermore, specific targeting of tumor cells using high-affinity and highly selective peptides conjugated to conventional chemotherapeutics such as doxorubicin permit the use of low doses, eliminating the toxic side-effects of such agents.[6]

Although peptides are excellent targeting moieties, their *in vivo* pharmacokinetics (*i.e.* short plasma half-life and rapid renal excretion) make them suboptimal as imaging agents. To improve *in vivo* pharmacokinetics and to produce targeted, multimodal, multivalent imaging agents, nanoparticles are also being explored as potential disease imaging probes. Nanoparticles are generally defined as dispersed, solid particles having at least one dimension of 100 nm or less and possessing a high specific surface-to-volume ratio. Fluorescent silver, radioactive gold, magnetic iron oxide and fluorocarbon nanoparticles have been developed for imaging purposes,[7] The most common inorganic nanoparticles in medicine and imaging are gold nanoparticles,[8] quantum dots (Qdots)[9] and magnetic iron oxide (MNP) nanoparticles.[10] Qdots are tunable semiconductor crystals with narrow emission bands, broad absorption spectra and high quantum yields.[11] MNPs have been used in magnetic resonance imaging (MRI) and are based on the their ability to modulate the uniformity of the magnetic field.[12] The majority of inorganic nanoparticles have not been used in live animals and humans due to cytotoxicity and cell delivery concerns.[13,14] One example of an FDA-approved nanoparticle is Abraxane, a biological-based nanoparticle with Taxol conjugated to bovine serum albumin.[15] Possible side-effects of nanoparticles differ dramatically due to the variety of materials and sizes of the particles being developed. Acute toxicity studies have shown pulmonary toxicity with carbon nanotubules and liver and kidney necrosis with ionic dendrimers and zinc, titanium and copper nanoparticles.[14] Phage represent ideal nanoparticles given that they are always the same size (monodisperse) and their size can be controlled due to their genetic patterning, they degrade into amino acids and DNA base pairs, they are not able to replicate in mammalian cells and they are cost effective.

5.1.2 Phage Nanoparticles

A new research direction in nanotechnology is the use of biological building blocks for the fabrication of nanoparticles. Bacteriophage (phage) are among the most promising new type of biological nanomaterials that have evolved from studies of their use in phage display.[16,17] Filamentous phage self-assemble into ~900×6 nm rod-like protein-encapsulated structures. Within the highly organized protein sheath is the genetic information to dictate its own production. Filamentous phage have previously been exploited for the assembly of semiconducting magnetic nanowires and lithium battery electrodes.[18–21] We assert that filamentous fd phage themselves can serve as valuable tumor-targeting and imaging agents.[22,23] Phage can function as self-replicating biological nanoparticle imaging agents in that they can be covalently attached to numerous tags or labels while simultaneously expressing multiple copies of foreign tumor or disease-targeting peptides. The resulting signal amplification is of enormous benefit to their use in disease detection and imaging. Further benefits to the use of phage as nanoparticles in imaging are that they have been physically well characterized, are resistant to harsh conditions and are organic and non-pathogenic. Our laboratories have shown that fd phage are well tolerated at high concentrations *in vivo* and generate only a weak immune response after many weeks of continued administration.[24] As such, they may be superior to that of other readily available inorganic nanoparticles for *in vivo* imaging.

5.1.3 Phage Display for Imaging Probe Discovery

In 1985, George P. Smith demonstrated that a foreign peptide sequence could be fused to coat protein (cp) III or VIII of filamentous phage and displayed on the surface of the phage.[16,25] Numerous phage-displayed libraries genetically encoding randomly displayed peptides of various lengths ensued. Each phage displays a single peptide of 6–45 amino acids, but a library as a whole may represent billions of peptides altogether. Phage, unlike higher organisms, have only one copy of each gene, so protein expression is not dependent on the interaction of multiple genes. Thus, each gene leads to one protein and each protein has one gene such that genotype equals phenotype. A clone isolated based on phenotype is easy to identify by sequencing the appropriate portion of the phage genome. The genetic encoding of a library allows the re-synthesis and re-screening of molecules with a desired binding activity. The resulting amplification of interacting molecules in subsequent rounds of 'affinity selection' can yield very rare, specific binders from a large expanse of molecules. There are many thousands of reports using *in vitro* and *in situ* phage display to select peptides that target numerous disease-associated antigens ranging through receptors, antibodies, vasculature components, carbohydrates and nucleic acids.[26–31] However, while many phage display selected peptides have been described that bind well to the desired antigen *in vitro*, there are surprisingly few examples of the peptides functioning to localize to the target and image *in vivo*.[32] For example, peptides that bind heat shock protein 90,[33,34] prostate specific antigen,[29] numerous growth factor receptors,[35–37] matrix metalloproteinases[38,39] and the Thomsen–Friedenreich (TF) carbohydrate antigen[40–42] exhibited excellent *in vitro* binding properties. Peptides used successfully to image tumors *in vivo* include the ErbB-2-targeting peptide 'KCCYSL',[43,44] vascular cell adhesion molecule-1 (VCAM-1)[45] and the galectin-3-targeting peptide 'ANTPCGPYTHDCPVKP'[46–48] developed in our laboratories. Reasons for the poor performance of many phage display selected peptides *in vivo* may include the method of selection, the hydrophobic nature of peptides displayed on phage and the failure of the peptides to function outside the phage framework.

Our laboratories have explored *in vivo* phage display as a means to identify phage and corresponding peptides, with optimal tumor-targeting properties in the context of a living animal.[1,49] We have shown that *in vivo* selected phage can serve as valuable first-line agents to determine if the phage and corresponding synthesized peptides would function as efficacious tumor-targeting and imaging agents. Moreover, phage decorated with tumor-homing peptides can function *in vivo* as biological nanoparticle imaging agents.[49] We developed *in vivo* phage display strategies to select phage that extravasate the vasculature and bind a variety of tumor cells *in vivo*. Intuition would suggest that the large size of phage precludes their ability to migrate through the vasculature and bind to antigens on tumor cells. However, the diversity of phage display libraries generated by George Smith, coupled with the power of *in vivo* phage display techniques, has allowed such phage to be obtained. Key components to *in vivo* selection of tumor-targeting phage include pre-clearing of libraries of vasculature-targeting phage, appropriate biodistribution times, detergent extraction of phage from target tissues and large-scale mass propagation and amplification of phage.[24] A schematic of *in vivo* phage selection of tumor-targeting phage is shown in Figure 5.1. Biodistribution studies of the fUSE5 fd phage library that displays 15 amino acid peptides on cpIII in three commonly used strains of mice (CF-1, SCID and nude) were performed to define better the rates and routes of clearance of the engineered phage. Approximately 70% of the injected phage were recovered in the blood in CF-1 mice at 15 min, whereas only ∼17% of the injected phage were detected in the blood in SCID and nude mice at 15 min. Analyses indicated that the immunocompetent mice were able to eliminate phage more efficiently than normal CF-1 mice. The longer half-life of phage in immunodeficient mice suggests that incubation times of greater than 15 min of phage in nude and SCID mice will allow for phage distribution to organs and tissues.

The biodistribution of the phage in all major organs and tissues was evaluated at 5, 15 and 30 min and 1, 6 and 24 h. As shown in Figure 5.2, the fUSE5 phage clear slowly through the reticuloendothelial system. At 30 min, the highest concentrations of phage were detected in the liver, spleen, blood and kidney. At longer time points, phage were only detectable in CF-1 mice in the spleen, kidney and lung.[24] In contrast, phage were identified in most major organs in nude and SCID mice for at least 24 h. Immunohistochemistry studies verified phage in the kidney and liver at 24 h post-injection (Figure 5.3).

We employed *in vivo* selections of fUSE5 libraries based on our knowledge of their pharmacokinetics in prostate tumor-bearing SCID mice to obtain

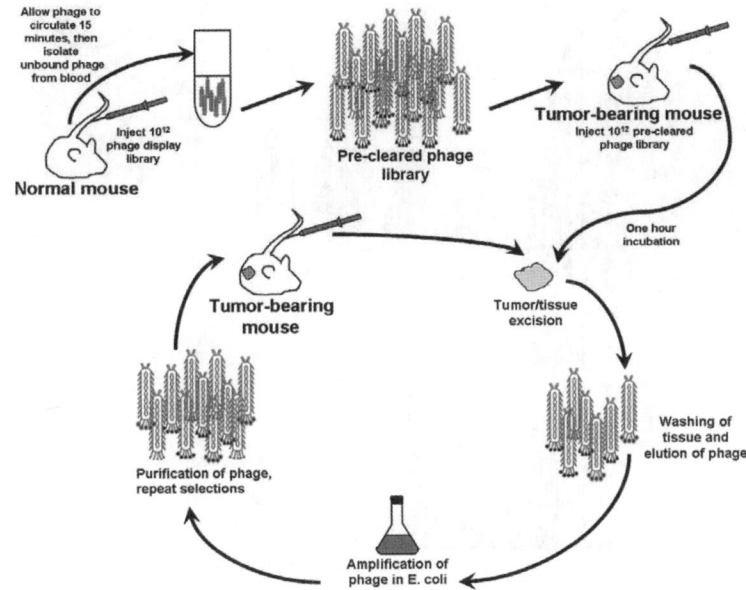

Figure 5.1   Schematic of *in vivo* phage display selection. A phage-displayed library is
first pre-cleared in a normal, non-tumor-bearing mouse. The resulting
phage population is then amplified and purified in preparation for the *in
vivo* selection of tumor-homing phage. A tumor-bearing mouse is injected
with 10^{12} virions of pre-cleared phage and allowed sufficient time for
target localization. The mouse is perfused and the tumor is removed.
Tumor tissue-bound phage are eluted with media containing CHAPs
detergent, amplified in host bacteria and purified. The resulting sub-
population of phage are then injected into a second tumor-bearing mouse
and the process repeated 2–4 more times.

prostate and other tumor imaging probes.[49] These phage and also many others
have been tagged with a variety of fluorophores and radionuclides for use in the
in vivo imaging of disease.

5.2   Radiolabled Phage as Imaging Agents

A major goal of our laboratories is to develop phage into radioimaging probes
for use in SPECT and positron emission tomography (PET) imaging mod-
alities. Initial studies with phage for *in vivo* imaging centered on their obvious
potential to image bacteria in mice. ^{99m}Tc-labeled and biotinylated phage
successfully imaged bacterial infections in mouse models of inflammation.[50,51]
The most straightforward approach would be to couple covalently a radiometal
bifunctional chelator on to the phage for direct labeling. Phage in these studies
were radiolabeled with ^{99m}Tc using mercaptoacetyltriglycine-modified phage.
Whereas infected regions accumulated $\sim 1\%$ injected dose (ID) g^{-1} of radioactivity, the liver accumulated between 10 and 40% ID g^{-1} of the radioactivity.
These results are consistent with the long clearance time of phage through the

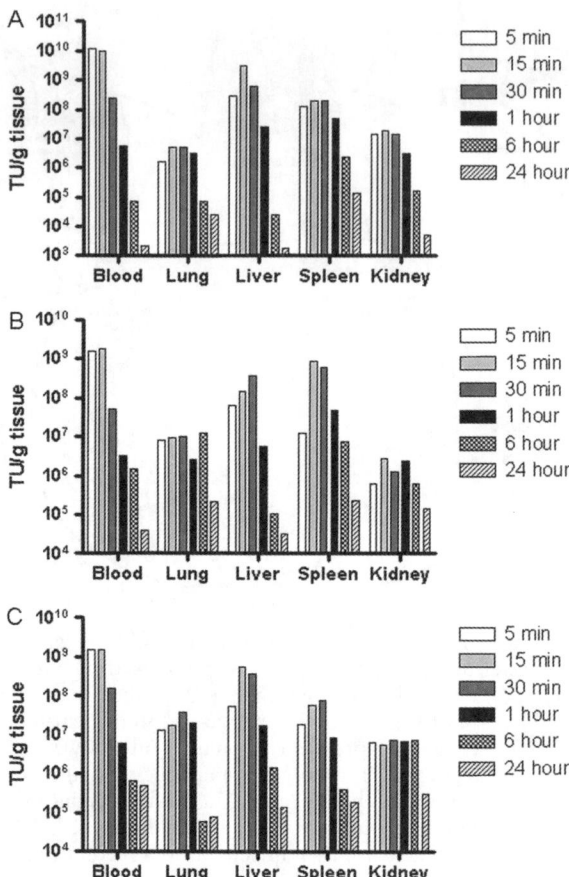

Figure 5.2   Biodistribution of a fUSE5 phage-displayed library in mice. Three different types of commonly used mice, CF-1 (A), nude (B) and SCID (C), were utilized for an investigation of the biodistribution of a fUSE5 phage-displayed library. Each mouse received a tail vein injection of 10^{12} virions of phage. The phage were allowed to circulate for 1 h before the mice were sacrificed and tissues harvested. The harvested tissues were then probed for infectious phage particles to determine the *in vivo* distribution.

RES.[24] Hence directly radiolabeled phage are likely to suffer from reduced imaging sensitivity of detection in the vicinity of the tumor. For therapeutic purposes, liver accumulation of radiolabeled molecules limits the maximum tolerated doses that can be administered without radiation-induced toxicity. Hence implementation of a pretargeting system with phage could allow for the clearance of the majority of the phage before injection of an imaging label. To this end, we developed both two- and three-step pretargeting strategies with biotinylated phage for the SPECT imaging of tumors. In the two-step system, biotinylated phage displaying tumor-homing peptides in combination with ^{111}In-radiolabeled diethylenetriaminepentaacetic acid (DTPA)–streptavidin were utilized to image melanoma and prostate cancer in mice.[52]

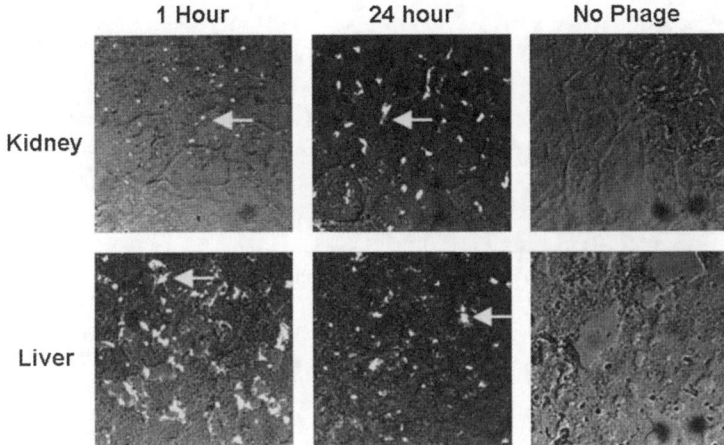

Figure 5.3   Immunofluorescent detection of phage virions within liver and kidney of SCID mice. SCID mice received tail vein injections of 10^{10} virions of fUSE5 phage-displayed library. Liver and kidney tissues were probed for retained phage virions at 1 and 24 h post-injection of the library. Resected tissue was frozen and sliced to facilitate histological investigation. Polyclonal anti-phage antibody and Alexa Fluor 488-conjugated secondary antibody were utilized for the visualization of phage (arrows point to phage shown in white).

A three-step pretargeting method was also developed utilizing biotinylated phage, avidin and [^{111}In]bisbiotin. These strategies are depicted in Figure 5.4.

The ability of prostate tumor-homing phage to act as SPECT imaging agents in a two-step versus three-step pretargeting approach were investigated. Biotinylated phage were generated *via* amine modification of the phage with NHS-PEO$_4$-biotin. SCID mice bearing PC-3 human prostate tumors received a tail vein injection of the prostate carcinoma tumor-homing, biotinylated phage known as G1.[4] The phage circulated for 4 h, during which time the biotinylated phage cleared the blood and localized to the tumor. The mouse utilized in the two-step pretargeting scheme received a second tail vein injection of [^{111}In]DTPA–streptavidin, followed by SPECT/CT imaging 24 h post injection of the radiolabel. The mouse utilized in the three-step pretargeting scheme received a second injection of avidin, which was allowed to circulate for 24 h in order to bind biotinylated phage and for excess avidin to clear the body. This step was followed by injection and circulation of [^{111}In]bisbiotin for ~30 min to allow time to circulate and bind avidin before SPECT/CT imaging. The biodistribution properties of [^{111}In]DTPA–streptavidin (step A.1, Figure 5.4) and [^{111}In]bisbiotin (step B.1, Figure 5.4), used in the two- and three-step pretargeting strategies, respectively, were investigated in SCID mice. As shown in Figure 5.5, the major organ of radioactive uptake was the kidneys for both compounds and there was additional uptake of 18% ID g^{-1} of [^{111}In]DTPA–streptavidin in the liver. The biodistribution properties of [^{111}In]DTPA–streptavidin used in the two-step pretargeting and [^{111}In]bisbiotin used in the three-step pretargeting with biotinylated G1 phage in PC-3 prostate tumor-bearing SCID mice were then evaluated.

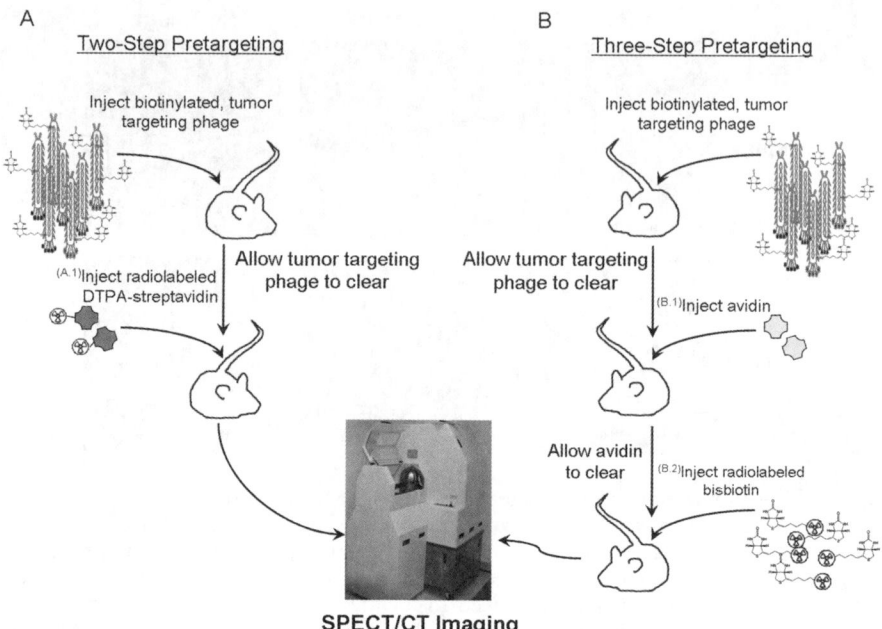

A
Two-Step Pretargeting

Inject biotinylated, tumor
targeting phage

(A.1)Inject radiolabeled
DTPA-streptavidin

Allow tumor targeting
phage to clear

B
Three-Step Pretargeting

Inject biotinylated, tumor
targeting phage

Allow tumor targeting
phage to clear

(B.1)Inject avidin

Allow avidin
to clear

(B.2)Inject radiolabeled
bisbiotin

SPECT/CT Imaging

Figure 5.4   Schematic of a two- and three-step pretargeting strategy for the *in vivo*
imaging of cancer in mice. SCID mice bearing PC-3 human prostate
tumors receive tail vein injections of the prostate carcinoma tumor-homing phage. The phage circulates within the mouse for 4 h, during which
time the biotinylated phage localizes to the tumor and clears the blood.
The mouse utilized in a two-step pretargeting scheme (A) then receives a
second tail vein injection of ^{111}In-radiolabeled DTPA–streptavidin, followed by SPECT/CT imaging 24 h post-injection of the radiolabel. The
mouse utilized in the three-step pretargeting scheme (B) receives a second
injection of avidin, which is allowed to circulate for 24 h to bind to the
biotinylated phage and for excess avidin to clear. The third injection of
^{111}In-radiolabeled bisbiotin is allowed to circulate for 30–40 min and clear
the body before SPECT/CT imaging.

   As shown in Figure 5.6, there was high radioactive uptake in the two-step
pretargeting approach in the liver, kidneys and intestines at 24 h, attributed to
the pharmacokinetics of [^{111}In]streptavidin. However, there was less than 1%
ID g^{-1} of radioactivity in the tumor. In contrast, there was good ($\sim 3\%$ ID g^{-1})
uptake of radiolabel in the tumor using the three-step approach (Figure 5.7).
Importantly, there was less kidney uptake (5% ID g^{-1} at 1 h) using the three-
step pretargeting tact and little radioactive accumulation in other organs
including the liver, compared with the two-step method (Figure 5.7). These
results indicate that a three-step pretargeting approach would produce the best
phage-based imaging agent.
   To test this, the ability of G1 phage to act as SPECT imaging agents of
human PC-3 prostate tumor heterotransplants in mice was investigated. As

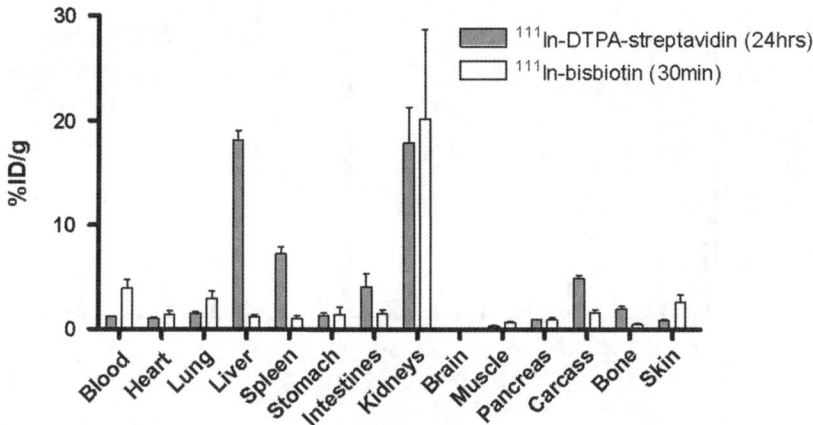

Figure 5.5 Biodistribution of radiolabeled streptavidin and biotin in non-tumor bearing SCID mice. SCID mice received tail vein injections of 3–185 kBq of [^{111}In]DTPA–streptavidin or [^{111}In]bisbiotin. The SCID mice receiving radiolabeled biotin were sacrificed 30 min post-injection followed by tissue harvesting. SCID mice receiving radiolabeled streptavidin were sacrificed 24 h post-injection and their tissues harvested. The harvested tissues were counted on a gamma counter to ascertain the amounts of radioactivity in each.

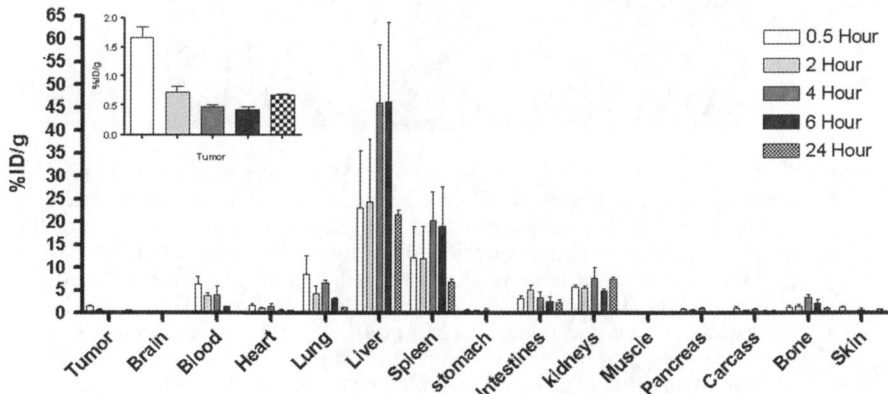

Figure 5.6 Biodistribution of pretargeted [^{111}In]DTPA–streptavidin utilized in a two-step pretargeting scheme. SCID mice bearing PC-3 human prostate carcinoma tumors received tail vein injections of 10^{11} virions of biotinylated prostate tumor-homing G1 phage; 4 h post-injection, the tumor-bearing mice received a second tail vein injection of 3–185 kBq of [^{111}In]DTPA–streptavidin. The mice were then sacrificed at 0.5, 2, 4, 6 or 24 h post-injection of the radiolabel followed by tissue harvesting and counting.

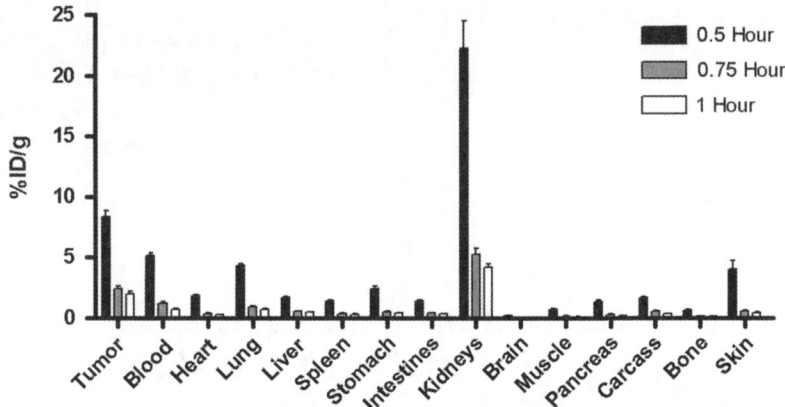

Figure 5.7   Biodistribution of pretargeted [^{111}In]bisbiotin utilized in a three-step
                 pretargeting scheme. SCID mice bearing PC-3 human prostate carcinoma
                 tumors received tail vein injections of 10^{11} virions of biotinylated G1
                 phage; 4 h post-injection, the tumor-bearing mice received a second tail
                 vein injection of 50 μg of avidin followed by an injection of 111–185 kBq of
                 [^{111}In]bisbiotin 24 h later. The mice were then sacrificed and tissues har-
                 vested for counting at 0.5, 0.75 or 1 h post-injection of the radiolabel.

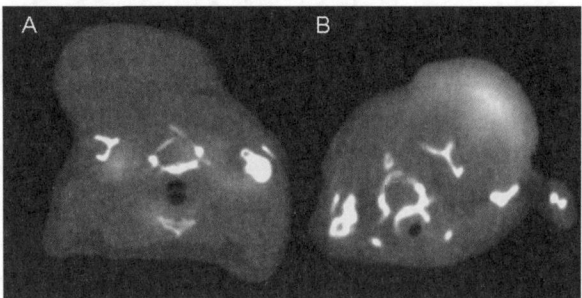

Figure 5.8   Imaging human PC-3 prostate tumors in SCID mice *via* two- and three-step
                 pretargeting strategies. SCID mice bearing PC-3 human prostate carcinoma
                 tumors were imaged using a two-step (A) or three-step (B) pretargeting
                 strategy. Mice were injected with 10^{11} virions of biotinylated G1 phage. The
                 biotinylated tumor-homing phage were allowed 4 h to localize to the tumor
                 and for the excess unbound phage to clear the body. The mouse in (A)
                 received a final injection of 7.4 MBq of [^{111}In]DTPA–streptavidin. SPECT/
                 CT imaging was performed 24 h post-injection. The mouse in (B) received a
                 second injection of 50 μg of avidin, followed 24 h later by a final injection of
                 7.4 MBq of [^{111}In]bisbiotin. SPECT/CT imaging was performed 30 min later.

shown in Figure 5.8, SPECT images of G1 phage using the two-step approach
failed to reveal tumoral accumulation. In contrast, the prostate tumor was
clearly imaged using the three-step scheme.

   These studies suggest that phage can act as bionanoparticles for the *in vivo*
SPECT radioimaging of cancer. Future studies will focus on the development

of phage pretargeting schemes utilizing biotin radiolabeled with PET imaging isotopes, including ^{18}F, ^{68}Ga and [64]Cu.[53] In addition, steps to reduce radiolabel kidney uptake would greatly improve the imaging and therapeutic efficacy of small-molecule radiolabeled molecules including Fab, peptides and biotin. High renal uptake and retention are a universal problem in that for imaging purposes, renal uptake greatly reduces the sensitivity of detection in the vicinity of the tumor. For therapeutic purposes, renal accumulation of radiolabeled molecules limits the maximum tolerated doses that can be administered without radiation-induced nephrotoxicity. Numerous approaches have been investigated to alter renal retention of radiopharmaceuticals, with mixed success, including changing peptide sequence and charge, radionuclide, chelate, administration of basic compounds and plasma substitutes.[54,55] Steps to reduce kidney uptake of [^{111}In]bisbiotin are currently being investigated.

5.3 Optical Molecular Imaging with Phage

Nowhere has the versatility of phage been taken advantage of more than in utilizing them as targeted optical imaging probes. Whereas fluorescently labeled phage can be used for long-term animal imaging studies, radiolabeled phage may cause significant organ damage. In addition, the optically visible targeted phage may be of particular assistance to surgeons to allow them to remove all of the disease and have disease-free margins of resection. However, the sensitivity of *in vivo* fluorescence imaging is not nearly as robust as that achieved using SPECT or PET *in vivo* imaging.

Phage affinity selection procedures have been successfully utilized to select peptides that bind to various fluorophores and even quantum dots. Marks *et al.* were able to identify the peptide sequences RTIWEPKEASNHT and TWTWPEISE that bind to Texas Red and X-Rhod calcium sensors with high affinity.[56] In an extension of this work, we were able to find peptides that bind tightly to near-infrared fluorophores.[57] Far-red and near-infrared fluorophores are useful for *in vivo* imaging applications, as light penetrates tissue more efficiently and because tissue autofluorescence is much lower in this range. These peptide have utility in addition to allowing the phage to be 'decorated' with fluorophores. The peptides themselves represent small biological ligands that, when fused to proteins, provide a minimally invasive way of tagging proteins and permit precise site-specific identification of intracellular proteins carrying the peptide. In this way, these proteins may provide the benefits of fluorescent proteins in studying diverse cellular processes without the disadvantages, namely those of instability of fluorescence and potential steric hindrance due to the size of the fluorescent protein.

Another intriguing use of the ability to select for peptides of specific binding activity is that of using phage-derived peptides to bind to ZnS Qdots. Although not used yet for imaging, the authors were able to array quantum dots at the end of the M13 phage by expressing the peptide CNNPMHQNC as a pIII fusion protein.[58] In one preparation, the phage formed a phage–ZnS self-supporting film that can be stored at room temperature without loss of titer. This novel nanoscale material and its fabrication have important implications for the construction of next-generation optical, electronic and magnetic materials and devices.

Although the techniques described above allow the identification of peptides specific for inorganic molecules, the non-covalent interaction between the peptide and the fluorescent moiety make it difficult to ensure that the two always remain coupled. Therefore, techniques have been employed to modify directly the pVIII phage coat protein through standard and robust chemistries. Jaye *et al.* were able to produce FITC and Alexa Fluor 647 labeled phage that retained their ability to bind to their selected targets of granulocytes and monocytes.[59] The authors were able to use the labeled phage clones for flow cytometry and also in immunofluorescence studies. Since a typical phage screen can produce tens to hundreds of hits, this approach allows the higher throughput screening of clones without utilizing a secondary or tertiary binding agent. A further hurdle for the development of *in vivo* imaging agents is to ensure that the peptide identified can recognize its target *in vivo*. For this purpose, phage have been directly modified with fluorophores throughout the spectrum and have been utilized for multi-wavelength imaging, allowing the direct visualization of several targets in one experiment.[1] After *in vivo* validation, peptides have been synthesized then conjugated to nanoparticles and used to image disease, including atherosclerosis and cancer. In one example, Kelly *et al.*[1] utilized phage display techniques to identify a series of peptides specific for pancreatic ductal adenocarcinoma (PDAC), which is a devastating disease with a mean survival of 3–6 months after diagnosis. In order to develop novel biomarkers to early PDAC, the authors generated cell lines from mice with localized PDAC and used these cells in a phage display selection to identify a pool of internalizing phage peptides specific for PDAC cells. The subtraction was performed with primary duct cells from wild-type mice, cells that are the normal counterparts to the cancer cells. From the screen, two clones, 15 (TMAPSIK) and 27 (LLPSGKP) demonstrated ideal affinity and specificity for the target PDAC cells. The binding affinity and specificity were further demonstrated *via* flow cytometry of mouse PDAC and normal ductal cells labeled with FITC-phage 15 and 27; these clones showed, respectively, 106- and 112-fold specificity for PDAC cells (Figure 5.9). To test these two peptides as a potential diagnostic agent for PDAC, phage clone 27 and clone 15 were labeled with the fluorochromes VT680 and Texas Red, then injected into mice from the Kras p53 PDAC model or into control littermates. This technique permitted the imaging of PDAC in mice (Figure 5.10) and subsequent immunohistochemistry demonstrated phage uptake by the tumor cells. In contrast, control phage with an unrelated peptide sequence injected into the same animal model showed a virtual absence of staining.

Since these peptides were able to direct phage to the tumors, the peptides were synthesized and used to develop a non-biologic, synthetic imaging agent. Peptide 27 was synthesized and attached to a magnetofluorescent nanoparticle (MFNP). The resultant MRI/optically detectable agent was tested in mice harboring PDAC tumors. By 24 h after intravenous administration of the

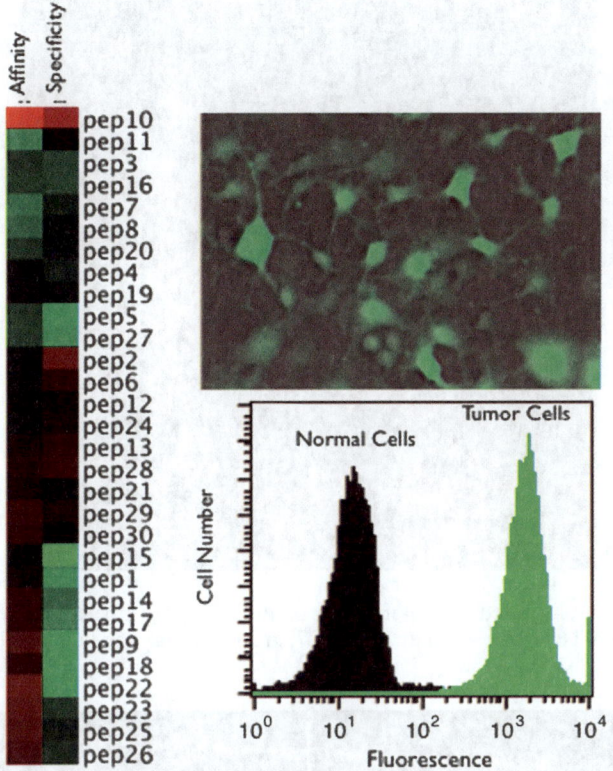

Figure 5.9 Isolation and validation of PDAC-specific peptides. the heat map depicts the affinity and specificity of individual phage clones. Data are displayed in terms of higher rankings (green) to lower rankings (red). FITC-labeled clone 27 is specific (FACs) and allows imaging of PDAC cells. Figure adapted from reference 60.

targeted nanoparticle, intravital confocal microscopy detected discrete areas of fluorescence in the abdominal region of these mice, suggestive of agent uptake (Figure 5.11). The signal was absent from the normal pancreas and also from regions of ductal metaplasia. The agent was specifically present in the tumor tissue rather than being retained in the vasculature, since a vasculature agent administered 10 min before injection failed to colocalize.

Phage themselves are becoming highly versatile *in vivo* imaging agents as they are economical, replenishable and have no intrinsic tropism for mammalian cells.[1] Kelly *et al.* demonstrated that phage displaying the VHSPNKK peptide imaged VCAM-1-expressing endothelial cells in a murine tumor necrosis factor-α (TNFα)-induced inflammatory ear model *via* intravital confocal microscopy (Figure 5.3). In the TNFα ear model used, 24 h prior to phage injection, the left ear of a mouse received a subcutaneous injection of TNFα while the right ear received no injection. VCAM-1 phage directly labeled with VT680 were then injected into the mouse and imaged 6 h post-injection by intravital

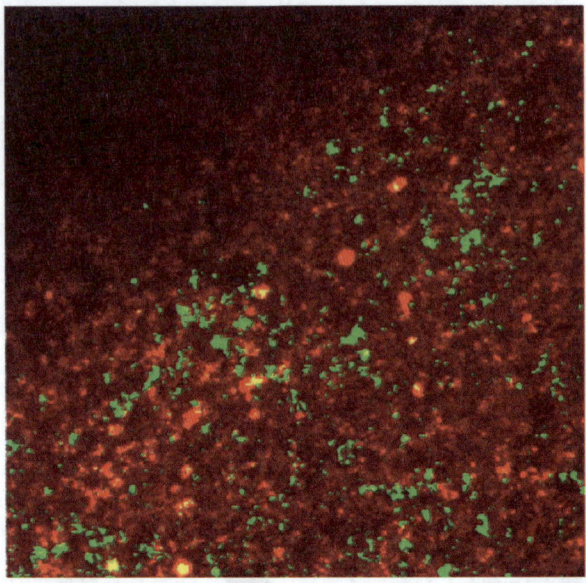

Figure 5.10   Multiplexed *in vivo* imaging of PDAC using labeled phage clones. Red, VT680-labeled phage clone 27; green, rhodamine-labeled phage clone 15.

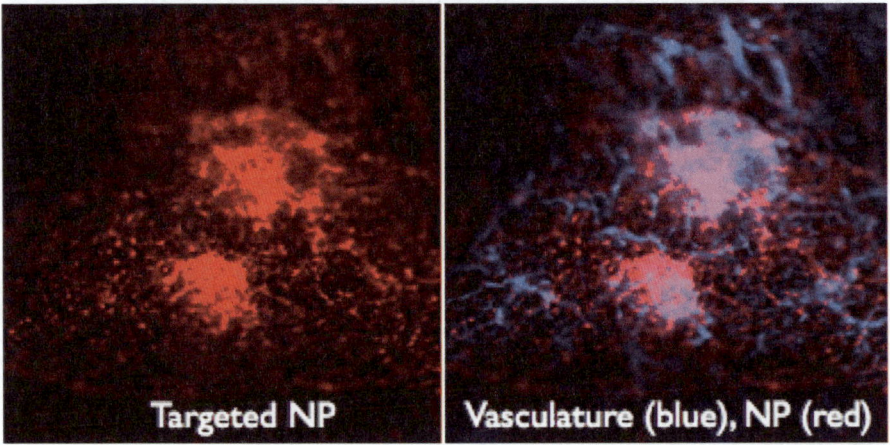

Figure 5.11   *In vivo* imaging of PDAC and tumor vasculature. PDAC was detected using a fluorescent nanoparticle that is targeted to PDAC via tumor-targeted peptide identified from phage display selection. Figure adapted from reference 60.

confocal microscopy. The result was the highly specific *in vivo* molecular imaging of VCAM-1 in mouse endothelium.[1] We have employed our *in vivo* selected prostate tumor targeting phage as optical imaging agents. The vasculature-extravasating phage were fluorescently labeled with the NIRF AF680 and allowed the successful non-invasive optical imaging of prostate tumors in immunocompromised mice.[47] This approach holds promise for future clinical applications as phage have been used clinically as anti-microbes and also *in vivo* selection has been done in humans for the identification of cancer-specific peptides.

One of newest applications of phage is to integrate tumor targeting and genetic (viral) imaging in order to deliver and image specific transgenes. Pasqualini's group developed chimeric fd-tet phage (displaying RGD integrin-targeting sequences)–adeno-associated virus contructs which were evaluated for not only tumor targeting and imaging but also herpes simplex virus thymidine kinase gene expression in mouse models of Karposi's sarcoma and bladder and prostate carcinoma, as monitored by PET.[60]

5.4 Conclusion

In summary, the M13 bacteriophage has proven to be a versatile platform for the generation of molecularly targeted *in vivo* imaging agents. The generation of diverse phage-displayed libraries and the creativity of investigators using them for selection procedures in combination with the realization that the phage capsid proteins can be modified genetically or chemically have all aligned to provide a new source of selectable, targeted imaging agents. Bacteriophage represent ideal monodisperse nanoparticles in that they self-assemble, are composed of biodegradable material and are able to be genetically modified to target almost all biological and even non-biological materials. Unlike synthetic nanoparticles, phage particles are inexpensive to produce and their components are non-toxic, having already been tested for safety in preclinical and clinical trials. The merging of phage as not only vehicles for selection of antibodies, peptides and proteins but also their emergence as a nanomaterial that can be easily genetically manipulated has implications not only in the field of molecular imaging but also in nanomedicine, nanodevices, biosensors and gene transfer.

References

1. R. Weissleder and M. J. Pittet, *Nature*, 2008, **452**, 580–589.
2. M. M. Goldenberg, *Clin. Ther.*, 1999, **21**, 309–318.
3. J. Klastersky, *Curr. Opin. Oncol.*, 2006, **18**, 316–320.
4. J. C. Reubi, *Endocr. Rev.*, 2003, **24**, 389–427.
5. W. H. Bakker, E. P. Krenning, J. C. Reubi, W. A. Breeman, B. Setyono-Han, M. de Jong, P. P. Kooij, C. Bruns, P. M. van Hagen and P. Marbach *et al.*, *Life Sci.*, 1991, **49**, 1593–1601.
6. W. Arap, R. Pasqualini and E. Ruoslahti, *Science*, 1998, **279**, 377–380.
7. K. K. Jain, *Clin. Chem.*, 2007, **53**, 2002–2009.

8. R. Kannan, V. Rahing, C. Cutler, R. Pandrapragada, K. K. Katti, V. Kattumuri, J. D. Robertson, S. J. Casteel, S. Jurisson, C. Smith, E. Boote and K. V. Katti, *J. Am. Chem. Soc.*, 2006, **128**, 11342–11343.

9. A. Shiohara, A. Hoshino, K. Hanaki, K. Suzuki and K. Yamamoto, *Microbiol. Immunol.*, 2004, **48**, 669–675.

10. D. E. Sosnovik, M. Nahrendorf and R. Weissleder, *Circulation*, 2007, **115**, 2076–2086.

11. B. Ballou, B. C. Lagerholm, L. A. Ernst, M. P. Bruchez and A. S. Waggoner, *Bioconjug. Chem.*, 2004, **15**, 79–86.

12. L. Josephson, in *Biological and Biomedical Nanotechnology*, eds. A. Lee and L. Lee, Springer, Berlin, 2006, pp. 227–237.

13. F. X. Gu, R. Karnik, A. Z. Wang, F. Alexis, E. Levy-Nissenbaum, S. Hong, R. S. Langer and O. C. Farokhzad, *Nanotoday*, 2007, **2**, 14–21.

14. S. T. Stern and S. E. McNeil, *Toxicol. Sci.*, 2008, **101**, 4–21.

15. V. Roy, B. R. Laplant, G. G. Gross, C. L. Bane and F. M. Palmieri, *Ann. Oncol.*, 2009, **20**, 449–453.

16. G. P. Smith, *Science*, 1985, **228**, 1315–1317.

17. R. Y. Sweeney, E. Y. Park, B. L. Iverson and G. Georgiou, *Biotechnol. Bioeng.*, 2006, **95**, 539–545.

18. Y. Huang, C. Y. Chiang, S. K. Lee, Y. Gao, E. L. Hu, J. De Yoreo and A. M. Belcher, *Nano Lett.*, 2005, **5**, 1429–1434.

19. A. S. Khalil, J. M. Ferrer, R. R. Brau, S. T. Kottmann, C. J. Noren, M. J. Lang and A. M. Belcher, *Proc. Natl. Acad. Sci. USA*, 2007, **104**, 4892–4897.

20. C. Mao, D. J. Solis, B. D. Reiss, S. T. Kottmann, R. Y. Sweeney, A. Hayhurst, G. Georgiou, B. Iverson and A. M. Belcher, *Science*, 2004, **303**, 213–217.

21. P. J. Yoo, K. T. Nam, J. Qi, S. K. Lee, J. Park, A. M. Belcher and P. T. Hammond, *Nat. Mater.*, 2006, **5**, 234–240.

22. L. A. Landon, J. Zou and S. L. Deutscher, *Curr. Drug Discov. Technol.*, 2004, **1**, 113–132.

23. J. Newton and S. L. Deutscher, *Handb. Exp. Pharmacol.*, 2008, 145–163.

24. J. Zou, M. T. Dickerson, N. K. Owen, L. A. Landon and S. L. Deutscher, *Mol. Biol. Rep.*, 2004, **31**, 121–129.

25. J. Yu and G. P. Smith, *Methods Enzymol.*, 1996, **267**, 3–27.

26. S. Fong, M. V. Doyle, R. J. Goodson, R. J. Drummond, J. R. Stratton, L. McGuire, L. V. Doyle, H. A. Chapman and S. Rosenberg, *Biol. Chem.*, 2002, **383**, 149–158.

27. M. Houimel, P. Schneider, A. Terskikh and J. P. Mach, *Int. J. Cancer*, 2001, **92**, 748–755.

28. K. Kelly, H. Alencar, M. Funovics, U. Mahmood and R. Weissleder, *Cancer Res.*, 2004, **64**, 6247–6251.

29. M. Pakkala, A. Jylhasalmi, P. Wu, J. Leinonen, U. H. Stenman, H. Santa, J. Vepsalainen, M. Perakyla and A. Narvanen, *J. Pept. Sci.*, 2004, **10**, 439–447.

30. V. I. Romanov, D. B. Durand and V. A. Petrenko, *Prostate*, 2001, **47**, 239–251.
31. A. J. Zurita, P. Troncoso, M. Cardo-Vila, C. J. Logothetis, R. Pasqualini and W. Arap, *Cancer Res.*, 2004, **64**, 435–439.
32. M. J. Blend, J. J. Stastny, S. M. Swanson and M. W. Brechbiel, *Cancer Biother. Radiopharm.*, 2003, **18**, 355–363.
33. Y. Kim, A. M. Lillo, S. C. Steiniger, Y. Liu, C. Ballatore, A. Anichini, R. Mortarini, G. F. Kaufmann, B. Zhou, B. Felding-Habermann and K. D. Janda, *Biochemistry*, 2006, **45**, 9434–9444.
34. C. I. Vidal, P. J. Mintz, K. Lu, L. M. Ellis, L. Manenti, R. Giavazzi, D. M. Gershenson, R. Broaddus, J. Liu, W. Arap and R. Pasqualini, *Oncogene*, 2004, **23**, 8859–8867.
35. R. Binetruy-Tournaire, C. Demangel, B. Malavaud, R. Vassy, S. Rouyre, M. Kraemer, J. Plouet, C. Derbin, G. Perret and J. C. Mazie, *EMBO J.*, 2000, **19**, 1525–1533.
36. H. Fan, Y. Duan, H. Zhou, W. Li, F. Li, L. Guo and R. W. Roeske, *IUBMB Life*, 2002, **54**, 67–72.
37. N. J. Skelton, Y. M. Chen, N. Dubree, C. Quan, D. Y. Jackson, A. Cochran, K. Zobel, K. Deshayes, M. Baca, M. T. Pisabarro and H. B. Lowman, *Biochemistry*, 2001, **40**, 8487–8498.
38. J. Chen, C. H. Tung, J. R. Allport, S. Chen, R. Weissleder and P. L. Huang, *Circulation*, 2005, **111**, 1800–1805.
39. W. Pan, M. Arnone, M. Kendall, R. H. Grafstrom, S. P. Seitz, Z. R. Wasserman and C. F. Albright, *J. Biol. Chem.*, 2003, **278**, 27820–27827.
40. L. A. Landon, E. N. Peletskaya, V. V. Glinsky, N. Karasseva, T. P. Quinn and S. L. Deutscher, *J. Protein Chem.*, 2003, **22**, 193–204.
41. L. A. Landon, J. Zou and S. L. Deutscher, *Mol. Divers.*, 2004, **8**, 35–50.
42. E. N. Peletskaya, V. V. Glinsky, G. V. Glinsky, S. L. Deutscher and T. P. Quinn, *J. Mol. Biol.*, 1997, **270**, 374–384.
43. N. G. Karasseva, V. V. Glinsky, N. X. Chen, R. Komatireddy and T. P. Quinn, *J. Protein Chem.*, 2002, **21**, 287–296.
44. S. R. Kumar, T. P. Quinn and S. L. Deutscher, *Clin. Cancer Res.*, 2007, **13**, 6070–6079.
45. K. A. Kelly, M. Nahrendorf, A. M. Yu, F. Reynolds and R. Weissleder, *Mol. Imaging Biol.*, 2006, **8**, 201–207.
46. S. L. Deutscher, S. D. Figueroa and S. R. Kumar, *Nucl. Med. Biol.*, 2009, **36**, 137–146.
47. S. R. Kumar and S. L. Deutscher, *J. Nucl. Med.*, 2008, **49**, 796–803.
48. J. Zou, V. V. Glinsky, L. A. Landon, L. Matthews and S. L. Deutscher, *Carcinogenesis*, 2005, **26**, 309–318.
49. J. R. Newton, K. A. Kelly, U. Mahmood, R. Weissleder and S. L. Deutscher, *Neoplasia*, 2006, **8**, 772–780.
50. M. Rusckowski, S. Gupta, G. Liu, S. Dou and D. J. Hnatowich, *J. Nucl. Med.*, 2004, **45**, 1201–1208.

51. M. Rusckowski, S. Gupta, G. Liu, S. Dou and D. J. Hnatowich, *Nucl. Med. Biol.*, 2008, **35**, 433–440.
52. J. R. Newton, Y. Miao, S. L. Deutscher and T. P. Quinn, *J. Nucl. Med.*, 2007, **48**, 429–436.
53. C. J. Anderson and M. J. Welch, *Chem. Rev.*, 1999, **99**, 2219–2234.
54. E. J. Rolleman, M. de Jong, R. Valkema, D. Kwekkeboom, B. Kam and E. P. Krenning, *J. Nucl. Med.*, 2006, 47, 1730–1; author reply 1731.
55. J. E. van Eerd, E. Vegt, J. F. Wetzels, F. G. Russel, R. Masereeuw, F. H. Corstens, W. J. Oyen and O. C. Boerman, *J. Nucl. Med.*, 2006, **47**, 528–533.
56. K. M. Marks, M. Rosinov and G. P. Nolan, *Chem. Biol.*, 2004, **11**, 347–356.
57. K. A. Kelly, J. Carson, J. R. McCarthy and R. Weissleder, *PLoS ONE*, 2007, **2**, e665.
58. S. W. Lee, C. Mao, C. E. Flynn and A. M. Belcher, *Science*, 2002, **296**, 892–895.
59. D. L. Jaye, C. M. Geigerman, R. E. Fuller, A. Akyildiz and C. A. Parkos, *J. Immunol. Methods*, 2004, **295**, 119–127.
60. K. A. Kelly, N. Bardeesy, P. Anbazhagan, S. Gurumurthy, J. Berger, H. Alencar, R. DePinho, U. Mahmood and R. Weissleder, *PLoS ONE*, 2008, **5**, e85.
61. A. Hajitou, M. Trepel, C. E. Lilley, S. Soghomonyan, M. M. Alauddin, F. C. R. Marini, B. H. Restel, M. G. Ozawa, C. A. Moya, R. Rangel, Y. Sun, K. Zaoui, M. Schmidt, C. von Kalle, M. D. Weitzman, J. G. Gelovani, R. Pasqualini and W. Arap, *Cell*, 2006, **125**, 385–398.

CHAPTER 6

Phage-based Pathogen Biosensors

SUIQIONG LI,[a] RAMJI S. LAKSHMANAN,[b] VALERY A. PETRENKO[c] AND BRYAN A. CHIN[a]

[a] Materials Research and Education Center, Auburn University, Auburn, AL 36849, USA; [b] Department of Chemical and Biological Engineering, Drexel University, Philadelphia, PA 19104, USA; [c] Department of Pathobiology, College of Veterinary Medicine, Auburn University, Auburn, AL 36849, USA

6.1 Introduction

6.1.1 Threat of Pathogenic Microorganisms

Every year over 76 million Americans become ill due to food-borne illness, or 'food poisoning' as it is more commonly known. As a result of these food-borne illnesses, it is estimated that 325 000 hospitalizations and 5000 deaths will occur.[1] Food-borne illnesses are primarily caused by food contaminated with bacteria, viruses and parasites. Illnesses experienced may be as mild as diarrhea and a slightly elevated temperature to severe, life-threatening neurological ailments.

The large number of food-borne illnesses that occur each year can be attributed to changing human demographics, human behavior, mass transportation of foods and microbial adaptation.[2,3] The United States consists of an overall aging population. Improved medical care has increased the median age of humans;[2,3] however, people of older age are at a higher risk to food-borne illness due to weaker immune systems. Changes in human lifestyle and

RSC Nanoscience & Nanotechnology No. 17
Phage Nanobiotechnology
Edited by Valery A. Petrenko and George P. Smith
Published by the Royal Society of Chemistry, www.rsc.org

food consumption behavior are also to be blamed. International cuisine and frozen food products directly lead to longer transportation and storage times, which allows for greater chances of spoilage. The production of foods in foreign countries where health standards vary significantly also has led to a plethora of additional contamination sources. Contamination of food may occur at any stage during processing, production, packaging, transportation or storage. The risks of food contamination can be decreased by the use of existing technologies. The US Food and Drug Administration (FDA) and Centers for Disease Control (CDC) have identified the problem of food-borne illnesses as one of the most serious yet avoidable problems. With better awareness and proper hygienic practices, the losses due to food-borne illnesses can be reduced drastically.

In spite of the significant impact of food-borne illnesses on society, it has never been a major issue of concern to the general public. Food supply bioterrorism has invoked greater interest among researchers and authorities as a result of the looming terrorism threats following the September 11, 2001, US attacks and the anthrax mailings thereafter.[4] The food supply system is most vulnerable to a bioterrorism attack. The first deliberate and largest act of bioterrorism in the modern history of the United States was in 1984,[5] when two members of the Rajneesh cult targeted voters in Oregon to influence the outcome of the elections in their favor. These members contaminated the local restaurant salad bars with *Salmonella typhimurium* (serotype Typhimurium),[5] causing sickness in about 800 people. Reviews of the more important biological warfare agents, the health risks posed by them and protection measures can be found in the literature.[6–9]

6.1.2   Pathogen Detection Techniques

Effective and rapid monitoring of food, food production/processing facilities and the environment for various pathogenic microorganisms is the crucial key for the prevention of infectious diseases. Effective biological detection methods should meet a number of criteria, such as rapid response time and high sensitivity and specificity.[10–12] Rapid detection is critical for environmental biocontamination monitoring, effective disease control and rapid response to bioterrorist attacks. In addition, extremely sensitive detection methods are required since some pathogenic organisms, such as *Escherichia coli* O157:H7, are infectious at doses as low as a few cells.[10,12,13] At the same time, from a commercial point of view, microbiological detection methods should be inexpensive, robust and portable. For a pathogen detection method to be industrially successful, the detection test equipment must be taken outside laboratory confines and be used with minimal need for skilled personnel.[12,14]

Various biosensing techniques have been developed and applied in daily microbiological analysis. However, real-time biological monitoring remains a challenge. The ever-growing need for rapid detection of pathogenic microorganisms has resulted in increased interest in research and development of biosensor systems. In this chapter, we review the current development of phage-based biosensors and their applications for the detection of biological threat agents.

6.2 Current Trends and Existing Methodologies for Pathogen Detection

6.2.1 Conventional Pathogen Detection Techniques

The presence of microorganisms was reported for the first time in the 17th century by Antonie van Leeuwenhoek.[15] In 1676, van Leeuwenhoek used self-crafted lenses to view microorganisms that he named animalcules (now known as bacteria).[15] He used beet stains to view the microorganisms under the microscope. The next significant development in the identification of bacteria was during the period 1881–1885. During these four years, Robert Koch and his associates, Angelina Fannie Hesse, Walter Hesse and Julius Petri, isolated bacteria responsible for anthrax, tuberculosis and cholera.[16–19] They used Methylene Blue staining to view the microorganisms with the help of a compound microscope. Koch's group was also responsible for the development of culturing methods for microorganisms, including the discovery of agar gel as a culture medium and the invention of the Petri dish.[15]

The Gram staining method is one of the oldest and most widely used techniques for classifying bacteria even today. This method was originally devised by a Danish doctor, Hans Christian Joachim Gram, in 1884 and is used to classify bacteria by its ability to retain a Crystal Violet–iodine complex.[15] The bacteria are classified as Gram-negative or Gram-positive based on the chemical and physical properties of the bacterial cell wall. In Gram-positive bacteria, the stain does not penetrate the thick cell wall (peptidoglycan is present) and the cells appear violet. Gram-negative bacteria have a thinner cell wall (peptidoglycan is absent) and also have an outer layer with lipids that allow the stain to penetrate the cell wall and the cells appear pink. The problem with this method of classifying bacteria is that there are exceptions presented by some species. These bacterial species, even in the presence of peptidoglycan in the cell wall, are penetrated by the stain and thus result in an inaccurate classification of bacteria. The Gram staining method is also messy due to the use of staining dyes and is often prone to operator error.

The traditional methods described above for the identification of bacteria typically depend on amplification and enrichment of the microbe in favorable media. Enrichment and sampling can take from 4 to 7 days, depending on the growth rate of the microorganism.[16–19] This results in an increased turnaround time for these methods.

Table 6.1 shows a timeline of important contributions in microbiology leading to the development of the now modern techniques of bacterial identification. The widely used modern microbiological methods for

Table 6.1 Timeline showing some of the important contributions in bacterio-
logy.

Year	Researchers	Contribution
1676	Antonie van Leeuwenhoek	Used lenses to view single-celled organisms called 'animalcules'
1881	Robert Koch	Isolation of anthrax, tuberculosis and cholera and viewed using beet stains
1883	Angelina Fannie and Walter Hesse	Development of agar gels as a medium for culturing microorganisms
1883	Julius Petri	Invention of Petri dish for culturing microorganisms
1884	Hans Christian Gram	Developed a method for distinguishing between two major classes of bacteria
1891	Paul Ehrlich	Proposed that antibodies are responsible for immunity
1953	James Watson and Francis Crick	Described the structure of DNA
1971	Peter Perlmann and Eva Engvall	Development of ELISA
1983	Kary Mullis	Development of PCR

identification of bacteria known as the polymerase chain reaction (PCR) and enzyme-linked immunosorbent assay (ELISA) tests are discussed in sections that follow.

6.2.2   Polymerase Chain Reaction (PCR)

The PCR, developed in the mid-1980s, presented several advantages in terms of speed, specificity and sensitivity for the identification of bacteria. In the PCR test, a specific target DNA segment of interest in a bacterial sample is amplified to form millions of copies of the target DNA segment. Since PCR is a nucleic acid-based diagnostic method, the identification of bacteria (even when present in small quantities) can be carried out with relative ease.[20] The main ingredients used for this process are an excess of free nucleotides, primers and an enzyme, Taq polymerase. The primers are usually around 20 nucleotides long and complement the end of the target DNA segment of interest. The Taq polymerase is an enzyme that is capable of withstanding the high temperatures used in the PCR process.

A typical PCR procedure consists of three main steps: denaturing, annealing and extension/polymerization. The denaturing step is carried out by heating to a temperature of 94–98 °C to break the hydrogen bonds between the two strands of DNA. This step yields single strands of the DNA. In the annealing step, the temperature is lowered to 50–65 °C. At this temperature, the primers are attached to the single strands of DNA obtained from the denaturing step. The primer sequence is selected so that nucleotides present on the primer are complementary to the target DNA. In the final step, extension of the primer with the help of added free nucleotides is carried out at a temperature of 75–80 °C. At the end of this step, two copies of the target DNA are created. The two copies produced by this first cycle are then returned to the denaturing step for the next cycle. After 25–30 cycles of this procedure, millions of copies of the target DNA have been produced, thus amplifying the initial small quantity of target DNA.

There are numerous reports on the detection of bacterial pathogens using PCR in water,[21] milk,[22] chicken fecal samples[23] and ground beef.[24] PCR has also been used to detect *Bacillus anthracis*[25–27] in a variety of media. Due to the high sensitivity of this method, even minor contamination of sample DNA can yield erratic results. The selection of a suitable primer is also a challenge due to the need for prior knowledge of the target DNA sequence. The requirement for extremely clean and controlled environments and the need for trained personnel limit this methodology to laboratory use.

6.2.3 Enzyme-linked Immunosorbent Assay (ELISA)

The first immunoassay used for identification of bacterial pathogens was radioimmunoassay (RIA). This method involved tagging an antibody with a radioactive element. The amount of radiation of the radioactive species was then used to determine the concentration of analyte present. In the search for a safer alternative to this method (elimination of radioactive elements), ELISA was developed.[28] In the ELISA method, the radioactive elements are replaced by enzymes tagged to an antibody. The change in color produced by an enzyme-mediated reaction is measured, and the amount of target analyte present in a sample can be related to the amount of color change produced. There are three different formats for performing ELISA: 'indirect' ELISA, sandwich ELISA and competitive ELISA.[28] A schematic of the three types of ELISA procedures is shown in Figure 6.1.

In an indirect ELISA, the capture antigen is coated on a microtiter plate. The serum containing the antibody is added and is washed with buffer solutions to remove any unbound antibody. A secondary antibody specific to the serum antibody is then added and allowed to attach. This step is followed by the attachment of an enzyme-conjugated antibody to the secondary antibody. After the excess enzyme-conjugated antibodies have been washed away, a substrate is added that will produce a color change that can be detected and quantified using optical methods.

In a sandwich ELISA, the microtiter plate is coated with an antibody with high specificity to the target antigen of interest. The non-specific sites on the microtiter plate are blocked. The serum containing the antigen is then added and allowed to attach to the coated antibody. This is followed by a washing step to remove any unbound or loosely bound antigens. An enzyme-conjugated antibody (specific to the target antigen) is added. Another wash is used to remove any unbound or loosely bound enzyme-conjugated antibodies. Finally,

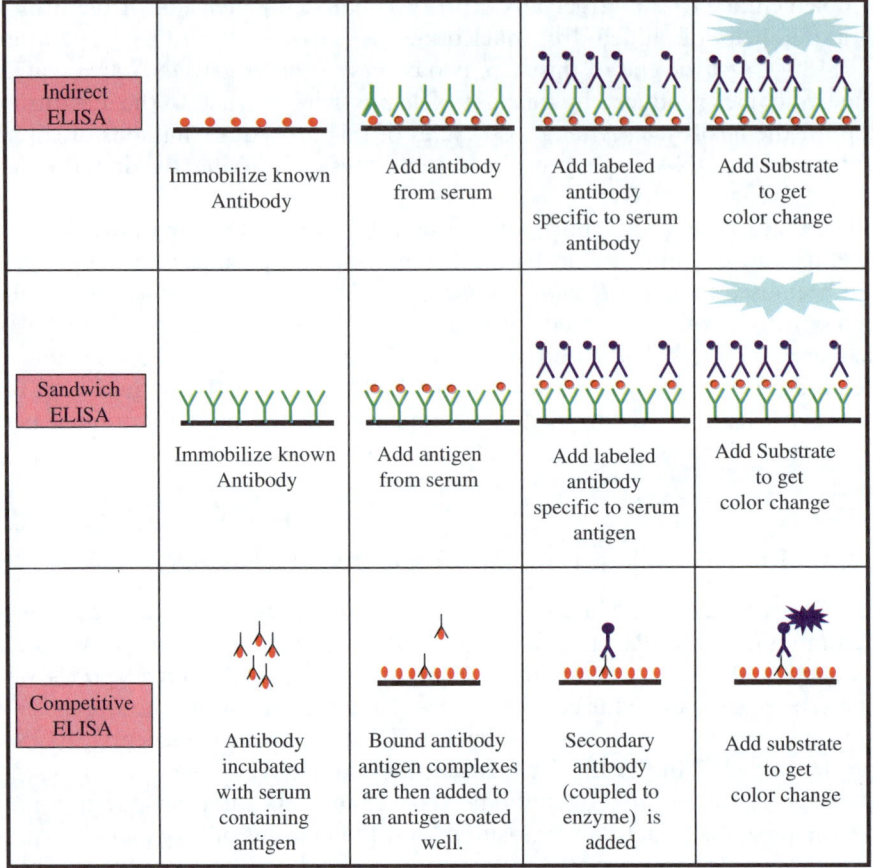

Figure 6.1 Schematic showing basic steps involved in a typical indirect ELISA, sandwich ELISA assay and competitive ELISA.

a substrate is added that will react with the enzyme to produce a color change depending on the concentration of the antigen present in the serum.

In a competitive ELISA, the steps are somewhat different. An unlabeled antibody is incubated in the presence of antigen. The antibody–antigen complexes obtained are added to an antigen-coated microtiter plate. Thus, the greater the amount of antigen in the sample, the smaller is the amount of antibody bound to the well. The secondary antibody (coupled to the enzyme) that is specific to the primary antibody is then added, and the substrate is added to produce a color change. In competitive ELISA, unlike indirect or sandwich ELISA, a sample with a higher antigen concentration will produce a weaker signal. Competitive and sandwich ELISA are usually used to determine the amount of antigen present in a serum/sample, whereas indirect ELISA is usually used to estimate the amount of antibody present in a serum sample.

6.2.4 Biosensor Techniques

The ever-growing need for early, rapid, sensitive and specific identification of microorganisms present in food, environmental and clinical samples has attracted considerable interest in research towards the development of novel diagnostic methodologies. The primary focus of research in the field of biosensors is to enable rapid detection of target pathogens outside laboratory confines with a minimal need for skilled personnel. Biosensors are now being used in a wide variety of fields such as food safety, detection of liquid contamination, clinical diagnostics, agriculture and the fight against bioterrorism.[12,29–32]

A diagnostic biosensor is composed of two major elements: a transducer and a biomolecular recognition element. A biosensor is defined as a device that incorporates a biological sensing element and a transducer for the detection of an analyte/target of interest. A schematic diagram of a typical biosensor is shown in Figure 6.2. The detection of an analyte by the biorecognition element

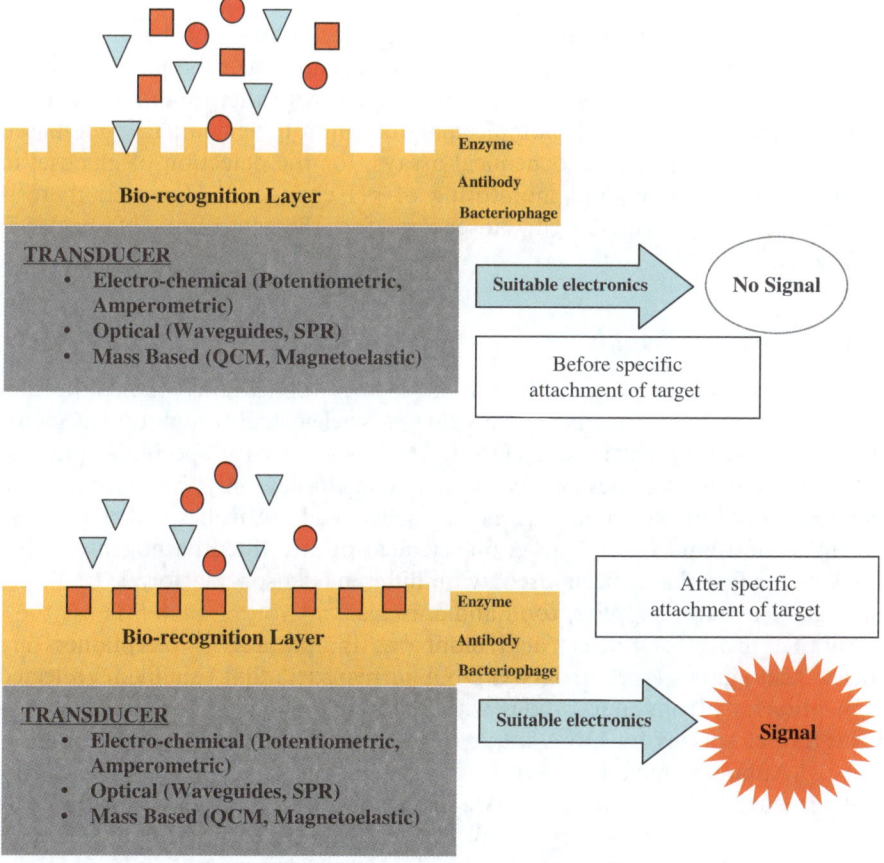

Figure 6.2 Schematic showing the basic components of a diagnostic biosensor.

is due to a specific interaction that generates a chemical or physical change. The transducer can directly or indirectly convert this interaction reaction into a measurable effect. Historically, biosensor research evolved from the discovery of the oxygen electrode by Leland C. Clark Jr.[33] He was the first to develop a biosensor for glucose detection. Since then, researchers have developed a plethora of recognition elements (enzymes, antibodies, bacteriophage) and transduction elements (optical, magnetic, electrochemical, mass-based) to build novel biosensor systems for a wide range of applications.

6.2.5   Biomolecular Recognition Element

The purpose of the biomolecular recognition element is to recognize and interact with a specific target species in the analyte. Typically used biorecognition elements are enzymes, nucleic acids, antibodies and bacteriophage.

6.2.5.1   Enzymes

Enzymes have been used as a biomolecular recognition probe for various biosensors. Enzymes usually aid in catalytic reactions to produce measurable signals. The catalytic power of enzymes, coupled with their high specificity of action, make them an ideal biomolecular recognition element. Enzyme-based biosensors are used in electrochemical assays, for the detection of glucose, in aromatic hydrocarbons and monitoring of pH changes.[34] However, there is limited use of enzyme-based biosensors for the detection of bacterial pathogens.[34]

6.2.5.2   Nucleic Acids

Nucleic acid bioprobes have found increasing applications in health care, agriculture and environmental monitoring.[12] Nucleic acid recognition elements are segments of a nucleic acid (DNA/RNA) strand with specific sequences. Since different nucleotides on the nucleic acid duplex only form bonds with counterparts of appropriate sequences, nucleic acid bioprobes can specifically recognize, and bind to, the target nucleic acid of interested pathogens. DNA/ RNA bioprobes have been used with different sensor platforms to detect pathogens in water supplies, food and animals.[35–37]

Another group of nucleic acid bioprobes is aptamers.[34,38] Aptamers are nucleic acid ligands (such as RNA, DNA or peptides) that usually are selected from libraries of oligonucleotides with random sequences. Instead of recognizing target species by DNA sequences, aptamers bind target molecules by shape.[34] Aptamers have been synthesized against a broad range of molecules, proteins and whole cells and have exhibited high affinity and specificity towards the targets of interest. Acoustic wave biosensors using aptamers as biomolecular recognition elements have been used to detect IgE antibodies[39] and HIV-1 Tat protein,[40] and to monitor the blood-coagulation cascade.[41] Enzyme and nucleic acid bioprobes are not the focus of this chapter and more details of these bioprobes have been reviewed by several authors.

6.2.5.3 Antibodies

Antibodies have been the most popular recognition element for applications in biosensors, particularly for the detection of bacterial microorganisms.[12] Antibodies specific to a target bacterium of interest are produced by injecting the target bacteria into an efficient antibody-producing animal. The animal's immune system then builds a family of polyclonal antibodies with mildly varying specificities and affinities to the injected bacteria. From this family of antibodies, the antibodies that yield high specificity and affinity to the target bacteria are extracted. This process of immunization and extraction is time consuming and requires specialized facilities. In addition to the complexity of extraction and purification, antibodies require controlled laboratory environments and are extremely sensitive to temperature and pH changes. Hence there is need for a more robust biorecognition element that can withstand conditions that are found in the field.

6.2.5.4 Bacteriophage

Bacteriophage are naturally evolved viruses that attach and infect their host bacteria at highly specific receptors. This principle has been the basis of a method known as phage typing. The phage-typing method involves the infection and lysis of the bacterial cell. This method has been widely used for epidemiological applications and bacterial detection.[42,43] However, the lytic nature of the phage used in phage typing limits their applications as biorecognition probes. A study of the genetics of filamentous phage and recent advances in phage display techniques have resulted in the development of recombinant phage, which are genetically modified to display peptides fused to some or all copies of the coat proteins. The phage clones displaying peptides highly specific to a target analyte of interest are separated using affinity selection procedures and identified by DNA sequencing.[44-51]

Filamentous phage, commonly used for display, are flexible, thread-shaped bacterial viruses, typically 6 nm wide and 800–900 nm long. The genetic material (single-stranded circular DNA) is enclosed within a tube containing the outer coat proteins. The outer coat is formed by thousands of equal copies of the major coat protein pVIII and the ends are capped by minor coat proteins (pIII, pVI, pVII and pIX). Landscape phage are created by modifying the genetic material in such a manner that they display numerous copies of peptides in a repeating fashion on the viral surface. It is called a 'landscape' phage because the viral surface constricts the peptides to a defined formation.[47,52] The surface area density of phage is 300–400 $m^2 g^{-1}$, of which more than 50% of the surface is composed of peptides that form the active binding sites.[50] Studies have shown that, unlike antibodies, phage are stable in harsh environments and

Table 6.2   ICTV classification of phages based on structural morphology and
genetic material enclosed.[53]

Family	Morphology	Nucleic acid
Myoviridae	Non-enveloped, contractile tail	Linear dsDNA
Siphoviridae	Non-enveloped, long non-contractile tail	Linear dsDNA
Podoviridae	Non-enveloped, short non-contractile tail	Linear dsDNA
Tectiviridae	Non-enveloped, isometric	Linear dsDNA
Corticoviridae	Non-enveloped, isometric	Circular dsDNA
Lipothrixviridae	Enveloped, rod-shaped	Linear dsDNA
Plasmaviridae	Enveloped, pleomorphic	Circular dsDNA
Rudiviridae	Non-enveloped, rod-shaped	Linear dsDNA
Fuselloviridae	Non-enveloped, lemon-shaped	Circular dsDNA
Inoviridae	Non-enveloped, filamentous	Circular ssDNA
Microviridae	Non-enveloped, isometric	Circular ssDNA
Leviviridae	Non-enveloped, isometric	Linear ssRNA
Cystoviridae	Enveloped, spherical	Segmented dsRNA

can be stored indefinitely at moderate temperatures with minimal loss of
activity.[44,48] Table 6.2 shows the International Committee on Taxonomy of
Viruses (ICTV) classification of phage[53] based on their morphology and the
kind of genetic material that they enclose. Filamentous phage (fd) used for the
preparation of landscape phage[48,51,52] belong to the class Inoviridae.

Procedures used in the affinity selection of bacteria-binding landscape phage
have been described in detail by Sorokulova *et al.*[51] In this method, the target
antigen is added and spread on a Petri dish. The f8/8 landscape phage library is
then added to the target antigen-coated Petri dish and is incubated for 1 h at
room temperature. The unbound phage from the Petri dish are washed with a
mixture of TBS and Tween. An elution buffer is then used to elute the phage
bound to the target antigen attached to the Petri dish. The elution process is
followed by neutralization of the eluate. The phage recovered from the eluate
are multiplied and used in the subsequent cycles of the selection process. The
whole procedure of selection is repeated several times to provide the clones
highly specific to the target antigen. The phage filaments produced at the end of
each cycle are characterized using DNA sequencing techniques to identify the
amino acid sequences of the displayed peptides. Common motifs in the amino
acid sequences are identified for the clones that have a larger affinity towards
the target antigen. A schematic of the affinity selection procedure is shown in
Figure 6.3.

6.3   Whole Filamentous Bacteriophage Particles as a Biorecognition Probe

After binding peptides or antibodies have been selected using the phage display
technique, they can be chemically synthesized or their genes can be transferred
to a high-expression system in order to produce recombinant proteins as

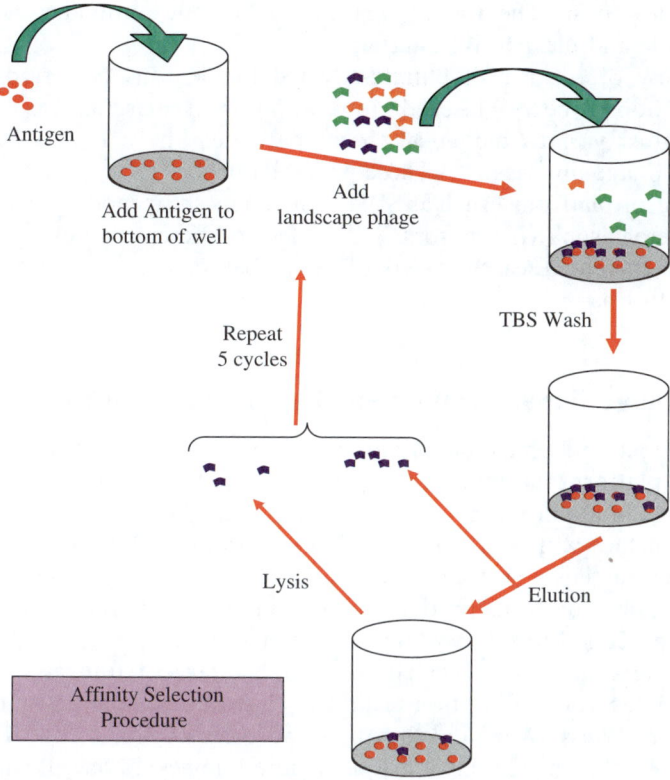

Figure 6.3 Schematic depicting selection procedure for phage.

biorecognition elements. There are numerous reports of phage display-selected recombinant antibodies against various antigens.[46,54] However, peptides and antibodies selected as fusions with phage can lose their binding properties when prepared as individual molecules. Additionally, antibodies can easily degrade in harsh environments.

Recently, researchers have begun to use directly the selected whole landscape phage particles as the bioprobes in biosensors.[48,55–61] As a biorecognition element, phage have several key advantages over antibodies. The phage serve as a three-dimensional support for bioreceptors against the target pathogen. Most of the surface area of phage displays selected peptides that provide the active binding sites.[50] The three-dimensional recognition surface with multiple binding sites provides strong multivalent interactions with the target pathogens. The structures of filamentous phage are very robust and resistant to degradation. Phage have been found to retain their infectivity after exposure to organic solvents, including 20% 2-propanol and 30–55% ethanol.[62,63] Thermal stability studies of phage have shown that recombinant phage are resistant to heat up to 80 °C,[44] whereas antibodies are known to lose their activity at 25 °C in less than 20 days.[64–67] Furthermore, phage can be produced in large quantities at a relatively low cost. The use of phage as substitute antibodies is a stable, reproducible and inexpensive alternative.[45]

The study of the use of affinity-selected filamentous bacteriophage as a biorecognition probe for biosensor applications is an emerging area of research. Phage against various biological threat pathogens have been selected from landscape phage libraries.[52,57] These phage bioprobes have been coupled with various sensor platforms, such as SPR,[68] thickness shear mode quartz sensors[56] and magnetoelastic (ME) sensors,[58–61] to detect different pathogens. Below we review the current research results of phage-based biosensors based on their sensor platforms.

6.3.1   Phage Immobilization on Biosensor Platforms

To form functional phage-based biosensors, phage need to be immobilized on the sensor platform surface and be used to capture/react with the target species. Ideally, the immobilized phage should be strongly attached to the sensor platform surface with high coverage and uniform distribution. Desorption of phage from the sensor surface should not occur in the analyte solution (such as aqueous buffers or food products containing target pathogens) during the detection process. More importantly, the immobilized phage should maintain their binding affinity and specificity towards the target pathogen.

Several phage immobilization techniques have been explored by the groups of Petrenko, Weiss, Wan, Lakshmanan and others.[48,56,58–60,69,70] These techniques can be classified into two distinct methodologies. In one, an anchor layer was coated on the sensor platform surface and then the phage were attached to the anchor layer. In the other, the phage were directly immobilized on the gold surface by physisorption.

6.3.1.1   *Phage Immobilization Through Anchor Layer Attachment*

Langmuir–Blodgett (LB)-coated monolayers have been used first to immobilize phage on sensor platforms. To attach phage, Petrenko and Vodyanov deposited biotin-modified phospholipid monolayers on the gold surface of the sensor using the LB method and then treated the monolayer with streptavidin.[48] The biotinylated phage were then immobilized on the sensor through biotin–streptavidin coupling. By using this LB-coated monolayer, phage selected for affinity to β-galactosidase were immobilized on quartz crystal microbalance (QCM) sensors. Experiments showed that phage immobilized on QCM sensors using this method exhibits a better affinity to β-galactosidase than the same phage used in ELISA.[48]

Functionalized self-assembled monolayers (SAMs) have been widely used to immobilize biological molecules. Several groups have reported the immobilization of phage through SAMs.[70–73] Klenerman and co-workers first immobilized phage on a QCM surface through a SAM.[71] They coated the QCM

surface with a SAM of soluble starch modified by mercaptodecanoic acid. The starch contains branch polymers of maltose and thus immobilized the phage displaying maltose-binding protein on the QCM surface. However, in their experiments, the phage used was not the bioprobe that is used in other bio-sensors discussed in this review. Increasing the oscillation amplitude of the QCM surface results in the detachment of phage from the QCM surface and generation of an acoustic emission. Klenerman and co-workers utilized this principle to measure the number of adsorbed phage particles and the corresponding phage concentration in solution.

M13 phage was covalently immobilized on a gold surface through a mono-layer of *N*-hydroxysuccinimide (NHS) ester-functionalized dithiol by Weiss, Penner and co-workers.[69,72] In this method, the monolayer was formed on the gold surface through self-assembly. The M13 phage was attached to the monolayer via an amide bond. A phage coverage of 10^{11} particles cm^{-2} was achieved using this method. The covalently immobilized phage retains the same binding affinity of the free M13 phage towards the target analyte (an anti-wild-type P8 antibody). An impressive advantage of covalent immobilization is the stability of the phage layer as the bioprobe. Experiments were conducted to detect P8 antibody using the QCM covalently coated with M13 phage. In the experiments, the P8 antibody was injected into the flow cell and measurements of the binding of P8 antibody to the QCM were made for 2000 s. Then the bound antibody was removed by injecting 0.5 M HCl into the flow cell followed by injection of the next higher concentration of P8 antibody. Measurement of the binding of the higher concentration of P8 antibody was then made using the QCM for 2000 s. This procedure of injection, binding of antibody and then stripping of bound antibodies was carried out repeatedly. The QCM produced a linear response with P8 concentration for more than 14 h. Recently, Zhu *et al.*[70] and Liu *et al.*[73] reported the immobilization of phage on bio-sensor platforms through 3-aminopropyltrimethoxysilane (3-APS) SAMs and 11-mercaptoundecanoic acid (11-MUA) SAMs, respectively.

6.3.1.2 Direct Phage Immobilization Using Physisorption

Vodyanoy, Petrenko and co-workers[56,74] developed a much simpler approach – physisorption – to immobilize the phage directly on the gold surface of the sensor platform. In this method, phage were immobilized on the clean gold surface of the sensor by incubating the sensor in the phage suspensions in buffer solution for about 1 h at room temperature. Based on physical adsorption, this immobilization technique avoids complex surface modification procedures, which considerably simplifies the immobilization process and reduces the immobilization time. Using physisorption, phage were successfully immobilized on different sensor platforms, such as SPR,[68] QCM[56,74] and magnetoelastic sensor platforms,[58–60] to detect different pathogens. For example, phage were immobilized on an SPR biosensor to detect β-galactosidase by Nanduri *et al.*[68] In this experiment, the optical thickness of the phage layer was measured to be 3 nm or approximately the thickness of one compact phage monolayer. Olsen *et al.* used QCM coated with phage by physisorption to detect *S. typhimurium* and a detection limit of 100 cells ml^{-1} was achieved.[56] Using this method, Chin's group immobilized landscape phages on magnetoelastic biosensors and successfully used them to detect *B. anthracis* spores[59,60] and *S. typhimurium* bacteria,[58,61] as described in detail below.

6.3.1.3 Effect of Salt and Phage Concentrations on Physisorption Immobilization of Phage

The morphology of the immobilized phage on the biosensor surface greatly affects the binding affinity of the phage towards the target pathogen and, thus, the performance of the biosensor. Filamentous phage is a rod-like polyelectrolyte with four net negative charges for each protein coat of a single phage.[75] Various experiments have shown that like-charged phage will aggregate to form bundles due to electrostatic interactions.[76–80] Olsen *et al.*[56] discussed two possible morphologies of immobilized phage (assuming no bundle formation) and calculated an optimum concentration of phage needed for immobilization. It was hypothesized that, instead of individual phage being immobilized on the sensor surface, phage bundles could form and be immobilized on the sensor surface. This leads to the question of whether a uniform distribution of phage filaments or a uniform distribution of phage bundles would yield a higher sensor binding affinity. Also, there is limited understanding[81] of the binding interaction between the gold surface and phage filaments.

The concentration of phage and the concentration of salt in the suspension used to deposit the phage are two factors that may control the final immobilized phage morphology. To understand the effect of these two factors on the physisorption immobilization of phage, Huang *et al.*[82] performed a systematic set of experiments immobilizing JRB7 phage (binding target *B. anthracis* spores) on the gold-coated surfaces of magnetoelastic sensors. After preparation of the biosensors, the binding capacity of the immobilized phage layers was determined. The set of conditions investigated included three phage concentrations (1.05×10^{12}, 1.05×10^{11} and 1.05×10^{10} vir ml^{-1}) and five different NaCl (Na$^+$ is the counter ion) concentrations (140, 280, 420, 560 and 840 mM) for each phage concentration. After phage immobilization, the sensors were rinsed with distilled water three times to remove any loosely bound phage and any salt remaining from the buffer solution. In order to prevent non-specific binding of spores on the gold surface, the phage-immobilized sensors were immersed in 1 mg ml^{-1} bovine serum albumin (BSA) for 1 h, followed by rinsing with distilled water. The biosensors prepared with different salt/phage concentrations were then exposed to a *B. anthracis* spore solution of a single concentration (5×10^8 cfu ml^{-1} in water). The biosensor response (changes in resonance frequency) and a count of the number of spores bound to the sensor surface as measured by scanning electron microscopy (SEM) were used to measure the sensors' binding affinity.

Figure 6.4a–e show the binding response of the biosensors prepared with phage suspension at a phage concentration of 10^{11} vir ml^{-1} in five salt

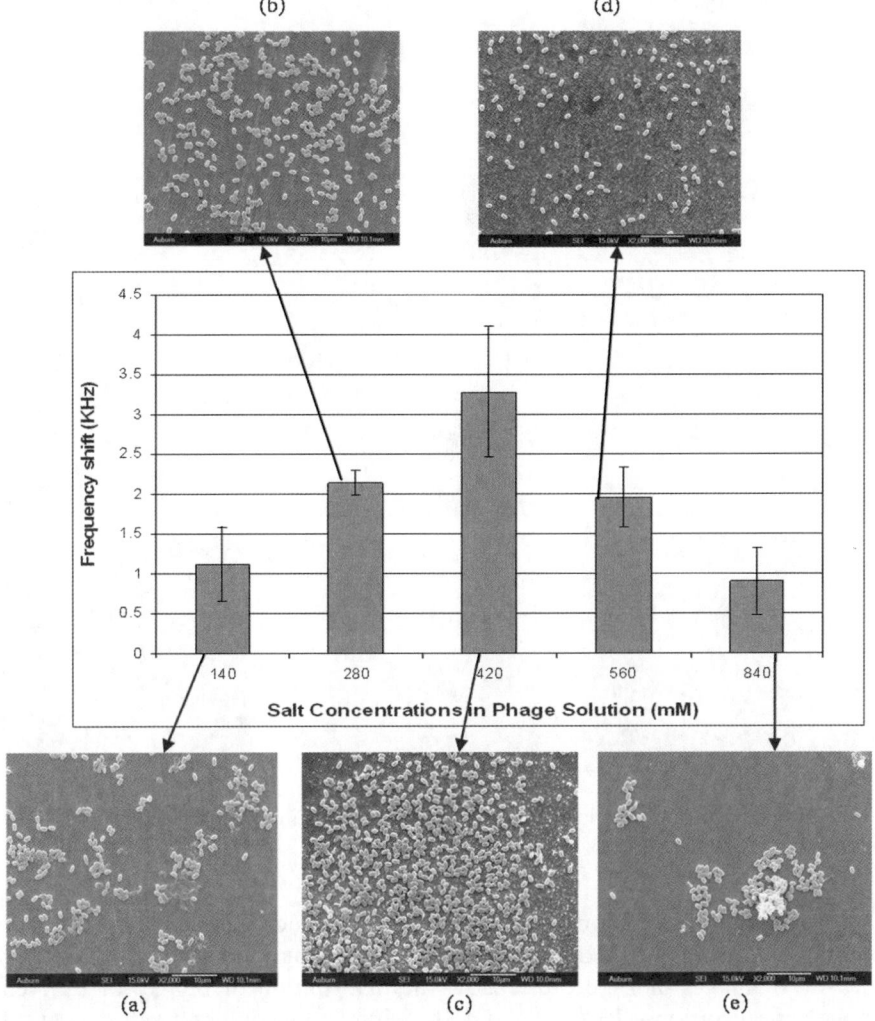

Figure 6.4 Frequency shift of sensors *versus* salt concentrations in 10^{11} vir ml^{-1} phage solution and SEM images of sensors covered with 10^{11} vir ml^{-1} phage in solutions with NaCl concentrations of (a) 140, (b) 280, (c) 420, (d) 560 and (e) 840 mM.[82]

concentrations (140–840 mM). With an increase in the salt concentration from 140 to 420 mM, an increase in the number of *B. anthracis* spores bound to the sensor was observed, with a maximum at a salt concentration of 420 mM. However, a further increase in salt concentration resulted in a decrease in the number of *B. anthracis* spores bound. As a result, phage at a concentration of 10^{11} vir ml^{-1} with 420 mM NaCl in 1×TBS was found to yield the most favorable conditions for binding of spores.

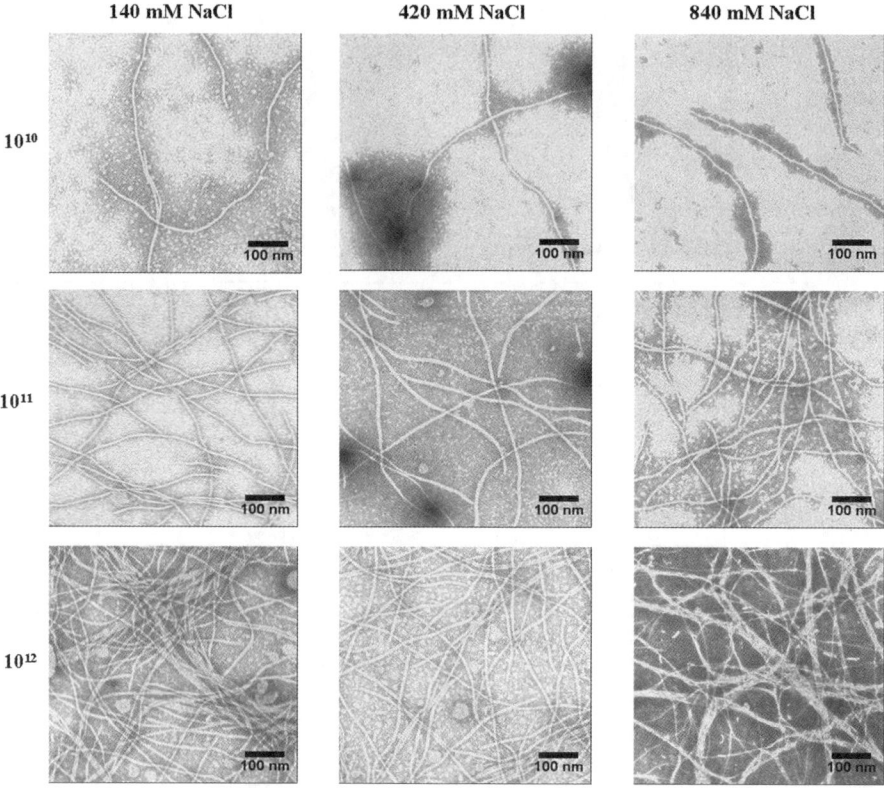

Figure 6.5 Transmision electron microscopy (TEM) images of phage in different NaCl solutions.[82]

A transmission electron microscopy (TEM) study showed that the low binding affinity might be due to the formation of bundles of phage filaments. The TEM images in Figure 6.5 show the distribution of phage for different phage concentrations (10^{10}, 10^{11} and 10^{12} vir ml^{-1}) at different salt concentrations (140, 420 and 840 mM). For a phage concentration of 10^{10} vir ml^{-1}, the distribution of the phage filaments is very low and there is no bundle formation regardless of salt concentration. However, the low phage concentration results in a low phage coverage on the sensor surface. On the other hand, when the phage concentration is high (10^{12} vir ml^{-1}), the phage filaments in the solution cluster together, forming bundles. Similarly, varying the salt concentration affects the formation of the phage bundles. On increasing the concentration of salt to 420 mM, the phage filaments tend to disaggregate, but a further increase in the salt content again results in the formation of bundles. Salt-induced aggregation of filamentous phage has been studied and documented by other researchers.[75,78,83,84] In most instances, the effect of the presence of divalent counter ions in wild-type fd phage suspensions was studied. Evidence from previous studies[77,85,86] shows that decreasing the salt concentration in the phage suspension will decrease the axial charge density of the filaments, promoting aggregation or 'clustering'. Likewise, as the salt concentration is increased, the axial charge density of the phage filaments increases, leading to less aggregation up to some optimum level and, hence, a more even distribution of filaments in solution.

Lakshmanan *et al.*[87] also immobilized E2 phage on an ME sensor surface using physical adsorption and performed the same study under conditions used in the JRB7 phage immobilization described above. The experiments showed results similar to those for the JRB7 phage-based ME biosensors. Figure 6.6a–f are SEM images that show the distribution of E2 phage on the sensor surface for the phage immobilized using the suspension in different NaCl concentrations. At lower Na$^+$ concentrations (280 and 420 mM NaCl), a uniform distribution of individual phage filaments was observed (Figure 6.6a and b). However, an increase in the counter ion concentration resulted in the formation of phage bundles, as can be seen in Figure 6.6c and d. The E2 phage biosensor with phage immobilized in solutions of different NaCl concentrations were exposed to 5×10^8 cfu ml^{-1} *S. typhimurium* solution. The binding distribution of *S. typhimurium* on the biosensor surface for the different concentrations of Na$^+$ revealed that the best binding of bacteria was obtained for the phage suspension with ≤ 420 mM Na$^+$ concentration. For low Na$^+$ concentrations (≤ 420 mM), the uniform distribution of individual phage filaments led to uniform binding of *S. typhimurium*. When phage bundles are formed at higher counter ion concentrations (> 420 mM), the distribution of these bundles on the biosensor surface is not uniform. The localized distribution of phage bundles led to the localized binding of *S. typhimurium*, which resulted in an overall reduction in the number of *S. typhimurium* attached to the sensor surface.

The results obtained by Huang and Lakshmanan using JRB7 phage and E2 phage-based ME biosensors show that the formation of phage bundles reduces the binding affinity of the phage bioprobe layer towards the target pathogens. However, there are several factors that need to be taken into consideration, mainly because the observed nature of the distribution of phage was considerably different for individual phage filaments and phage bundles for the same immobilization method. From the observations made, when the phage bundles formed, they seemed to cluster at random regions on the sensor surface rather than distributing uniformly. This resulted in the target pathogen being bound only to parts of the sensor surface. Hence it would be inaccurate, at this juncture, to negate the advantage that a phage bundle could provide by having more binding receptors than an individual phage. A phage bundle typically consists of more than 1000 individual phage filaments. This would result in a large number of binding receptors being concentrated on each phage bundle. It is hypothesized that if a uniform, dense distribution of phage bundles could be deposited on the biosensor surface, then a much higher binding affinity might be achieved. Additional work is required to achieve a uniform distribution of the bundles and to ascertain the nature and conditions that would optimize the performance of phage as a biorecognition probe for sensor applications.

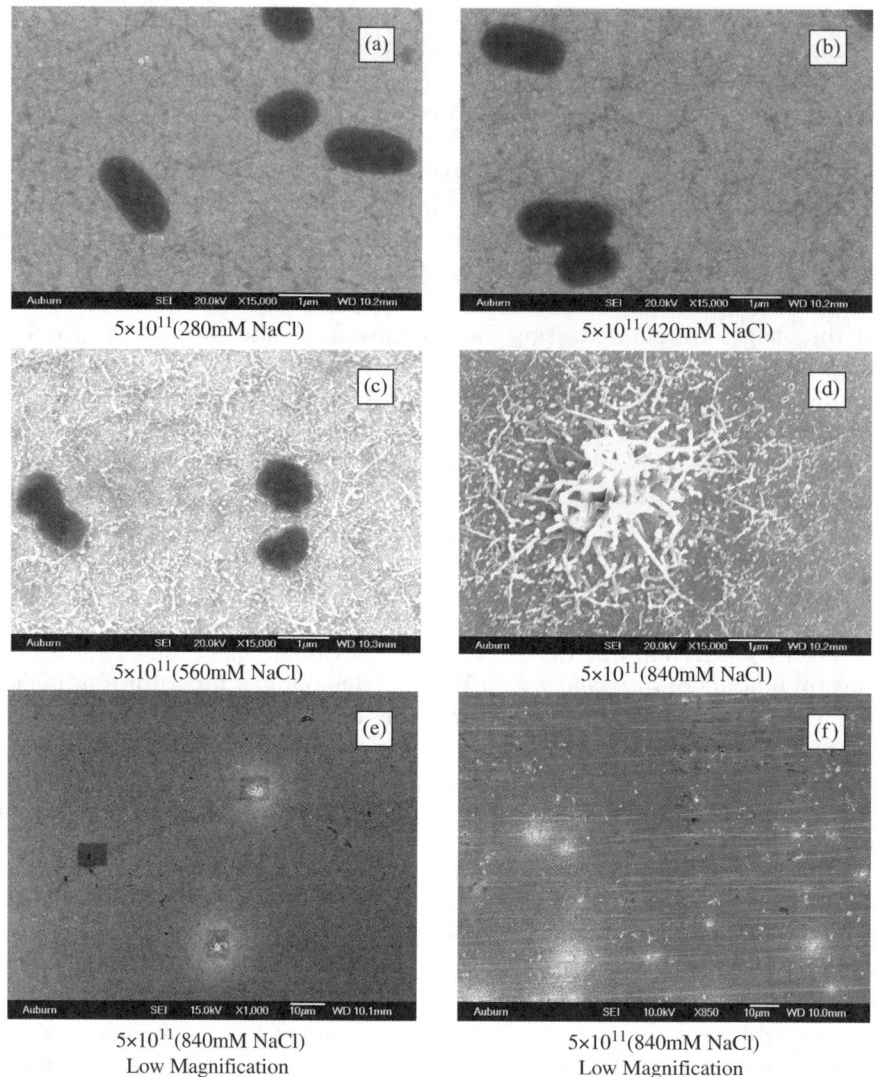

5×10^{11}(280mM NaCl)

5×10^{11}(420mM NaCl)

5×10^{11}(560mM NaCl)

5×10^{11}(840mM NaCl)

5×10^{11}(840mM NaCl)
Low Magnification

5×10^{11}(840mM NaCl)
Low Magnification

Figure 6.6   Scanning electron microscopy (SEM) images showing the nature of phage
distribution in the presence of different Na^{+} ion concentrations of (a) 280,
(b) 420, (c) 560 and (d) 840 mM and (e and f) low-magnification images of
sensors exposed to phage with 840 mM NaCl.[87]

6.3.2   Current Trends in Development of Phage-based Biosensors

6.3.2.1   *Phage-based Electrochemical Biosensors*

Electrochemical transducers are capable of converting the interaction
between the target species and the biorecognition probe into measurable
electrical signals (usually due to the movement of electrons and ions).

The magnitude of the change in the electrical signal is related to the amount of analyte present. Electrochemical transducers are categorized as amperometric and potentiometric.

6.3.2.1.1 Phage-based Amperometric Biosensors Utilizing Phage Typing.

As the term suggests, amperometric transducers measure the current produced by the biological interaction of interest. This is done by applying a constant potential across the working and reference electrodes and measuring the changes in the current produced. The applied potential ensures the electron transfer reaction of the electroactive biological species by gaining or losing an electron. The current thus produced can be related to the recognition interaction and, hence, the amount of target analyte present. Amperometric transducers have found a wide range of applications in gas analysis and che-mical/biological detection, primarily due to ease of operation and high sensitivity. The use of amperometric methods to detect bacteria by the release of intrinsic enzyme markers from the interior of the cells has been widely studied.[88–91] The released enzyme markers are complemented with a suitable substrate to produce electroactive reactions at the electrode. Early amperometric sensors relied on permeabilization (using chemicals) of the antigens for the release of intrinsic enzymes. In their earlier work, Mittleman *et al.*[88] established the release of β-galactosidase from *E. coli* by permeabilization. The interaction of β-galactosidase with substrate *p*-aminophenyl α-D-galacto-pyranoside releases *p*-aminophenol, which oxidizes at the electrode to produce current. The work by Neufeld *et al.*[89,92] and Yemini *et al.*[93] used these principles to detect various pathogens (*Bacillus cereus*, *Mycobacterium smegmatis* and *Escherichia coli*).

Amperometric sensors in combination with phage typing[90] have been used for the specific detection of various bacteria. In this method, the enzyme markers from the interior of bacterial cells are released by a phage-mediated lysis of the analyte. The released intrinsic enzyme then reacts with the substrate. The product of this reaction (usually *p*-aminophenol) oxidizes at the electrode, resulting in an electric current proportional to the analyte concentration. The reaction shown in Figure 6.7 is representative of such an amperometric sensor. Here phage-mediated lysis of the *E. coli* bacteria was used to release the enzyme β-galactosidase. This enzyme, on reaction with the substrate (*p*-aminophenyl β-D-galactopyranoside), produces *p*-aminophenol. The oxidation of *p*-amino-phenol at the electrode produces a current proportional to the initial concentration of the analyte.

In an advanced assay, Neufeld *et al.*[92] used a combination of phagemid (a cloning vector of filamentous phage) and a helper M13 phage. In this technique, rather than lysing the cell, the substrate and the reporter enzyme reaction occurs in the periplasmic region of the bacterial cell. This reaction results in the subsequent release of *p*-aminophenol and its oxidation at the electrode. The advantage of this method is that the bacterial cells remain undamaged at the end of the test. A similar approach was used by Yemini *et al.*[93] for the detection

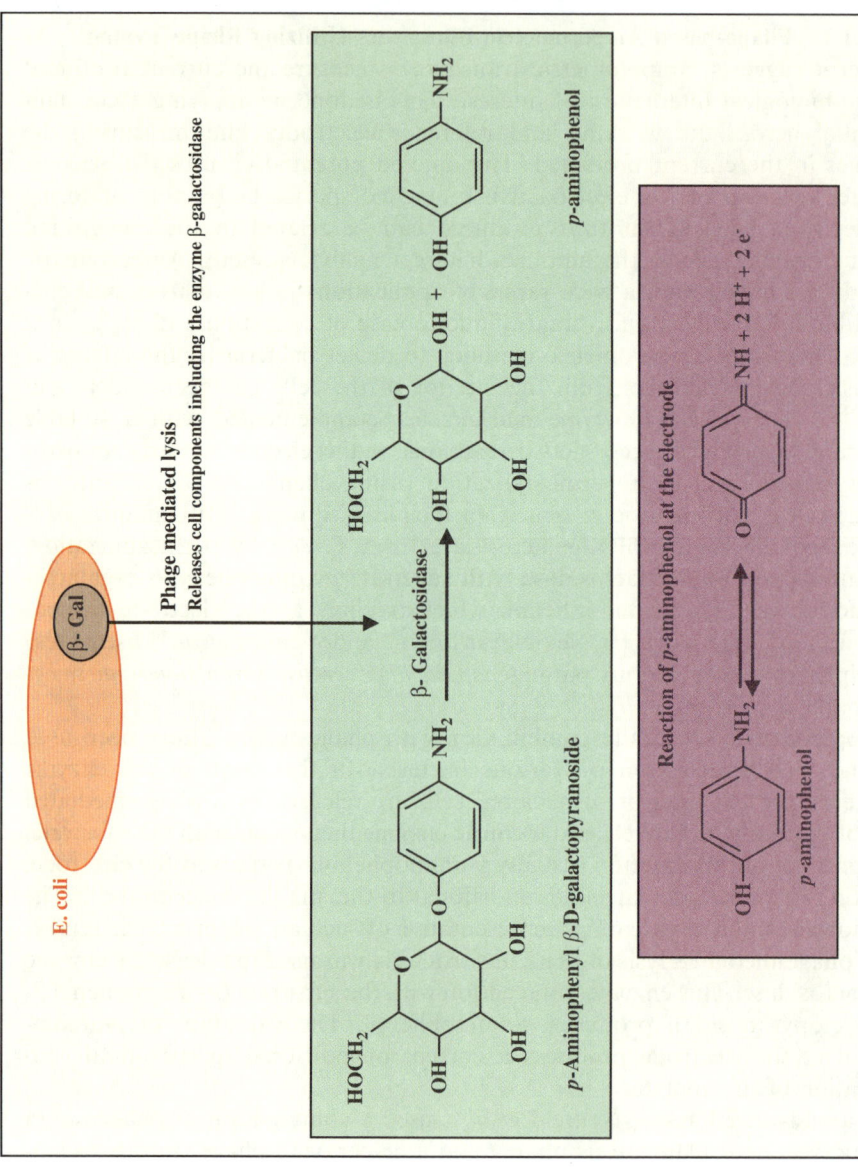

Figure 6.7 Basic reactions involved in amperometric detection of *E. coli*.[90] In this technique, the intrinsic enzyme from the bacteria is released by phage-driven lysis of the cells.

of *B. cereus* and *M. smegmatis* by a phage mediated release of intrinsic enzymes α-glucosidase and β-glucosidase, respectively. These enzymes catalyzed the hydrolysis of their corresponding substrate to yield *p*-aminophenol. A detection limit of 10 cfu ml^{-1} for *E. coli*, *B. cereus* and *M. smegmatis* and a response time of about 10 min were demonstrated. The whole process, including a pre-incubation step, took about 8 h for detection of the lowest concentrations.

Recently, phage typing was used to detect pathogens in nanoscale amperometric biosensors. Utilizing the advances in biochip techniques, Seo and co-workers fabricated a nanowell sensor as a novel potentiometric sensor platform.[94,95] The platform consists of two Ti contact pads and a 150 nm wide Ti nanowell device on an LiNbO$_3$ substrate. When connected to an external preamplifier and spectrum analyzer, the nanowell can probe nanoscale electric field fluctuations. When phage infect bacteria, a transitory ion efflux is generated and can be detected by this sensor. Combined with phage typing, nanowell biochips can provide sensitive, specific and rapid pathogen detection.[94,95]

6.3.2.1.2 Phage-based Electrochemical Biosensors Utilizing Affinity-selected Peptide-displaying Phages.

Electrochemical, impedimetric biosensors using affinity-selected filamentous phage as bioprobes have been developed by Yang and co-workers.[69,72] They covalently immobilized affinity-selected M13 phage particles on gold disk electrodes through a self-assembled monolayer. When the phage particles selectively are bound to target antigens, the impedance of this 'virus electrode' increases. This phage-modified microelectrode has been successfully used to detect prostate-specific membrane antigen (PSMA)[69] (a biomarker for prostate cancer) and antibodies.[72]

Affinity-selected filamentous phage have recently been used in light-addressable potentiometric sensors (LAPS) to detect pathogens. The basic principle of LAPS[96–98] is shown in Figure 6.8. The interaction of the biorecognition element with the target analyte results in a change in ion concentration in the electrolyte. This change in ion concentration is responsible for changes in the surface potential of the electrolyte–sensor interface. The width of the depletion layer changes with changes in the surface potential. The depletion layer appears at the insulator–semiconductor interface when a DC bias is applied. Illuminating the bottom of the semiconductor substrate with modulated light (LED) produces electron–hole pairs and thus an AC photocurrent is generated depending on the width of the depletion layer. The magnitude of the photocurrent generated is then used to determine the surface potential. The primary advantage presented by this type of sensor is ease of miniaturization. Light addressability also permits investigation of localized changes at different regions on a sensor surface. Owing to this unique advantage, LAPS can be used to map profiles of localized chemical changes at different regions on a sensor surface.

In a recent study,[99] affinity-selected filamentous phage were used to detect human phosphatase of regenerating liver-3 (hPRL-3) and the mammary adenocarcinoma cell (MDAMB231) using LAPS. The protein hPRL-3 is a cancer marker and MDAMB231 is a cancer cell. The interaction of the immobilized

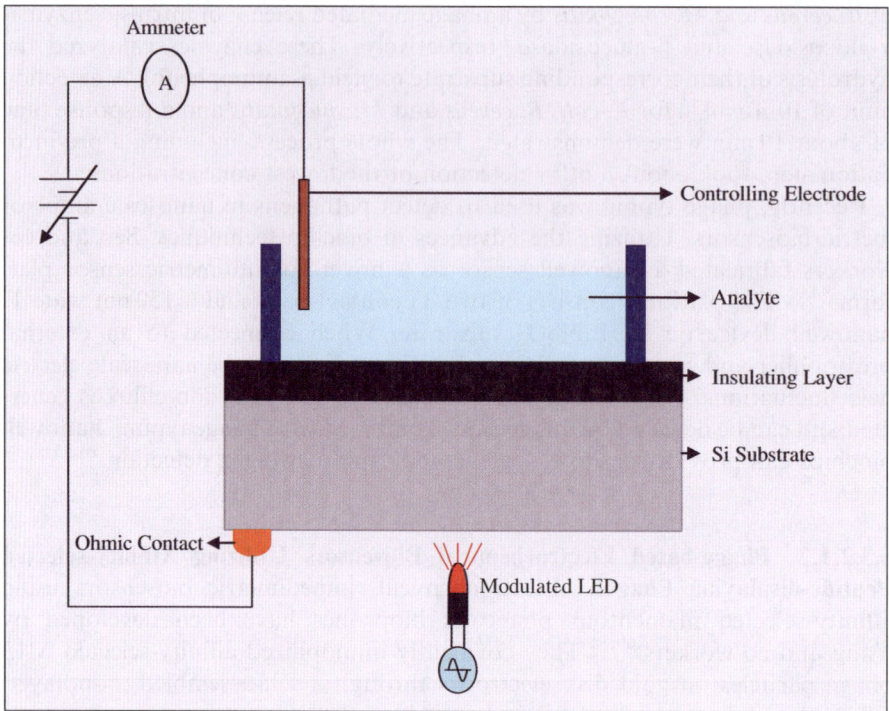

Figure 6.8   Schematic depicting a light-addressable potentiometric sensor (LAPS).

phage and the cancer cells/cancer marker caused a change in the surface potential. Changes in the surface potential were then related to the concentration of the target cells in the analyte. A detection limit of 1×10^3 cells ml^{-1} was reported. However, the authors did not investigate or explain the mechanisms causing the surface potential changes. The performance of the LAPS devices is limited by signal instability and miniaturization and selection of the reference electrodes is a challenge. Another disadvantage of these sensors is the dependence of the sensor response on the pH and ionic strength of the analyte test solutions.

6.3.2.2   Phage-based Optical Biosensors

6.3.2.2.1   Indirect Optical Methods (Fluorescently/Luminescently Labeled Phage).   Affinity-selected phage can be labeled with fluorescent or luminescent markers and then used as a bioprobe to detect various antigens. Fluorescently labeled phage have been used in various bioassays for pathogen detection.[100] Goldman *et al.*[101] utilized phage-displayed peptides selected from a pIII-fused random 12-mers library to detect staphylococcal enterotoxin B (SEB), a causative agent of food poisoning. Selected phage particles were labeled with the dye Cy5 and incorporated into an automated fluorescence-based sensing assay. A detection sensitivity of 1.4 ng per well was achieved with the labeled phage,[101] which is comparable to the detection conduced using antibodies. However, when the fluorescently labeled anti-SEB phage was used in SEB-coated optical-fiber sensors, the signal was much weaker and less specific than that generated by antibodies.[102,103] The reason might lie in the position of the specific peptide on the phage. For the phage used in SEB detection, the binding peptide is fused to the pIII minor coat protein, located on the tip of the phage capsid. This expression format is possibly not optimal for bioprobe applications, since phage with a peptide displayed on pIII protein does not have the advantage of multiple binding sites.[47] In addition to applying phage in biosensing, the same group also used fluorescently labeled phage to detect explosives.[104] They fluorescently labeled M13 phage that was selected to specifically bind to 2,4,6-trinitrobenzene (TNB). The TNB-binding phage was used in a displacement fluorescence assay to detect TNT in artificial seawater. A detection limit of $10 \, mg \, l^{-1}$ was achieved.

Turnbough demonstrated the detection of *B. anthracis* spores using fluorescently labeled phage by fluorescence-activated cell sorting.[105] In this experiment, spores of *B. anthracis* were successfully distinguished from other *Bacillus* species.

Goodridge *et al.*[106,107] used fluorescently labeled phage in combination with antibody-immobilized immunomagnetic beads for the detection of *E. coli* O157:H7 in inoculated ground beef and milk. They quantified the amount of antigen using flow cytometry and epifluorescence microscopy. They reported a detection limit and detection time of $50 \, cfu \, ml^{-1}$ and 7 h (including a pre-enrichment step of 6 h). In this method, phage were tagged with labels and the interaction of phage with the target analyte produced optical signals. The optical signals produced could then be related to the amount of analyte present in the sample.

Some other notable indirect optical techniques that have been reported utilize luciferase-induced bioluminiscence.[108–110] Banaiee *et al.*[109] used luciferase reporter bacteriophage for the detection of *Mycobacterium tuberculosis*. In this work, they demonstrated detection and evaluation of the antibiotic susceptibility of *M. tuberculosis* in clinical samples using luciferase reporter phage. Blasco *et al.*[110] reported the detection of *E. coli* and *Salmonella newport* using a different approach. In this method, phage-mediated lysis was used to release adenylate kinase (AK) and adenosine triphosphate (ATP) from the interior of the cells. When adenosine diphosphate (ADP) is used as a substrate, AK acts as a catalyst to produce more ATP. The ATP then reacts with firefly luciferase to produce luminescence. The authors reported a detection limit of $10^3 \, cfu \, ml^{-1}$ for both *E. coli* and *S. newport* and assay times of 1 and 2 h, respectively.

Recent advances in nanotechnology have resulted in a new class of fluorescence-based assays exploiting new optical labels, such as quantum dots.[111–113] Quantum dots are highly fluorescent semiconductor nanocrystals capable of producing stronger fluorescence signals. They are more resistant to photobleaching than conventional fluorescent dyes.[113–115] Edgar *et al.*[116] utilized this technology to detect *E. coli*. They used filamentous phage with surface peptides that could be biotinylated upon interaction with biotin present in the host cell. The biotinylated phage were then exposed to the streptavidin-conjugated quantum dots. The phage thus bound to the quantum dots and was characterized for the amount of the target analyte using flow cytometry and fluorescence microscopy. Specific detection of about 10 cells ml^{-1} in the experimental sample was achieved.

Utilizing the electrostatic interaction between gold nanoparticles and phage, Souza *et al.* established Au–phage networks.[81] With the phage displaying selected peptides, the network can capture target cells. Meanwhile, the optical properties of such a network can be manipulated and the network can be used as a label in several optical detection systems. They fabricated an Au–RGD-4C network with the phage displaying the peptide CDCRGDCFC (termed RGD-4C) and used this network to detect melanoma cells. The melanoma cells were differentiated due to the enhanced fluorescence signal.[81] However, the need for suitable equipment and microscopes to quantify the amount of analyte would limit the use of such methods for field applications.

6.3.2.2.2 Direct Methods. In direct optical methods, a biological recognition event is measured directly based on changes in the properties of the light used. Currently, most direct optical biosensors utilize antibodies as the biorecognition element. Recently, Balasubramanian *et al.*,[117] Nanduri *et al.*[68] and Liu *et al.*[73] have used a surface plasmon resonance-based (SPR) sensor with phage-based interface to detect various microorganisms. Balasubramanian *et al.*[117] used lytic phage immobilized on the gold surface by physical adsorption to detect different concentrations of *Staphylococcus aureus*. In their work, the SPR platform they used was SPREETA, an integrated device produced by Texas Instruments. They reported a detection limit of 10^4 cfu ml^{-1} in the absence of any amplification or enrichment steps. Using the same transducer (SPREETA), Nanduri *et al.* detected β-galactosidase[68] using a phage-based interface. They used affinity-selected filamentous phage, which were immobilized on an SPR gold surface using simple physical adsorption. A detection limit of about 1 pM was reported. Liu *et al.*[73] reported a phage-displayed SPR biosensor for protein detection. They immobilized M13 phage on an SPR sensor chip through a self-assembled 11-mercaptoundecanoic acid (11-MUA) layer. The M13 phage displayed the 12 amino acid peptide that specifically interacts with protein ET101. The detection of protein ET101 in diluted solutions with different concentrations was performed. A response of 365 ± 8RU corresponding to an ET101 concentration of 4.0 μM was reported.[73] However, the high cost of the apparatus and bulkiness of the equipment again limit the application of these optical biosensors to laboratory measurements.

Recently, Zhu *et al.* developed an optical biosensor platform named the opto-fluidic ring resonator (OFRR).[70] The OFRR utilizes a micro-sized glass capillary with a very thin wall (<4 μm), where the cross-section of the capillary forms a ring resonator. Optical resonance with circulating waveguide modes (WGMs) can be generated in the capillary wall and the WGMs circulate along the circumference of the capillary. Since the capillary wall is extremely thin, the evanescent field generated by the resonant WGMs penetrates into the core by about 100 nm and interacts with the analyte in the capillary core. Any change on the inner surface of the capillary wall, such as binding of a target species to bioprobes immobilized on the wall, causes a change in the resonant wavelength of the WGM that can be detected. Therefore, OFRR provides sensitive, label-free and real-time biomolecule detection. Zhu *et al.* covalently immobilized filamentous R5C2 phage, displaying peptides specifically against streptavidin on the OFRR, and used the phage-based OFRR biosensor to detect strepta-vidin. A detection limit of 100 pM of streptavidin and a K_d of 25 pM were achieved.[70] Zhu *et al.* also developed 'cocktail solutions' to strip bound streptavidin from the phage receptors. By pumping the mixed solution through the capillary, the streptavidin was completely removed from the phage-probes and the biosensor can be reused many times.

6.3.2.3 Phage-based Acoustic Wave Biosensors

Acoustic wave (AW) devices have been widely investigated as biosensor transducers since they provide the capability of real-time detection and several other advantages, such as high sensitivity, simplicity and cost effectiveness.[118–121] In these devices, acoustic waves are generated in the actuating part of the device, causing the oscillation of the device. In combination with the immobilized bioprobes, an AW biosensor detects the target pathogen by monitoring the resonance frequency change caused by the mass of attached target pathogen.

As an important type of AW device, the quartz crystal microbalance (QCM), has been intensively studied as a biosensor platform. Recently, QCM bio-sensors using phage as bioprobes have been reported to detect various patho-gens. Petrenko and Vodyanoy first immobilized phage against β-galactosidase on a QCM surface through biotin-modified phospholipids monolayers that were deposited on the QCM surface using LB coating.[48] The phage-based QCM sensor was then used to detect β-galactosidase. For comparison, β-galactosidase was detected by ELISA using the same phage. Dissociation constants of 0.6 and 30 nM were obtained for the phage-coated QCM sensor and the ELISA method, respectively.[48] The difference in affinities was attrib-uted to the monovalent (ELISA) and divalent (phage-coated QCM sensor) interaction of the phage with β-galactosidase. The study also showed that the immobilized phage was specific and selective towards β-galactosidase. The same group also designed QCM biosensors with immobilized landscape phage for the detection of *S. typhimurium*[56] in addition to β-galactosidase.[74] The immobilization of phage was carried out using physical adsorption. For the detection of *S. typhimurium*, the authors reported a detection limit of 100 cells ml^{-1} and a detection time of 5–10 min.[56] For β-galactosidase detec-tion, a detection limit of a few nanomoles and a response time of 100 s over the range 0.003–210 nM were reported.[74] Yang *et al.*[69] covalently deposited affinity-selected phage on the QCM and used the phage-coated biosensor to detect PSMA and antibody for P8.

Two types of magnetoelastic sensor platforms have been investigated for pathogen detection: magnetoelastic microcantilever (MEMC)[122,123] and magnetoelastic particle resonator.[58–61,67,124–126] Both phage-based magnetoelastic microcantilevers and phage-based magnetoelastic particle resonators have been reported to detect food-borne pathogens, such as *B. anthracis* spores[59,60] and *S. typhimurium*.[58,61]

Similarly to piezoelectric-based microcantilevers, magnetoelastic microcantilevers are composed of one layer of magnetoelastic alloy and one layer of non-magnetoelastic metals. The MEMCs inherit all the advantages of silicon or piezoelectric based microcantilevers, in addition to being wireless. The MEMC exhibits a high quality-merit factor (Q value), which means that the MEMC can function in a liquid environment with relatively high overall sensitivity. Experiments conducted by Li *et al.*[122] showed that the quality factor of MEMCs is as high as 400–500 in air and > 10 in water. The application of the MEMCs as a high-performance biosensor platform has been demonstrated by detecting yeast cells in water.[122] Fu and co-workers reported the detection of *B. anthracis* spores and *S. typhimurium* using phage-based magnetoelastic cantilevers.[123,127]

The phage-based magnetoelastic biosensors possess the advantages of both magnetoelastic resonator platform and phage bioprobe. They have been intensively investigated by Chin's group[58–61,128–138] and are discussed in detail in the next section. Different kinds of phage-based biosensors are summarized in Table 6.3.

6.4 Phage-based Magnetoelastic Particle Resonator Biosensors

6.4.1 Magnetoelastic (ME) Particle Resonator Sensor Platform

Recently, the use of magnetoelastic (ME) materials as transducer platforms for the remote monitoring of food-borne pathogens[58,61,67,124,125,139,140] and other applications in chemical detection and environmental monitoring[141–146] have attracted considerable interest. ME transducers have the unique advantage of detection in the absence of physical wire contacts to the sensor, allowing *in situ* wireless measurements in tubes of flowing liquids or sealed containers. ME sensors are typically amorphous ferromagnetic alloys that are composed of a combination of iron, nickel, molybdenum and boron. These sensors work on the principle of magnetostriction, wherein the material changes in its dimensions in the presence of a magnetic field. Upon application of a magnetic field, the randomly oriented magnetic domains in the material tend to align in the direction of the applied field. The alignment of the magnetic domains in the magnetoelastic material results in a change in the dimensions. By applying a time-varying magnetic field, the magnetoelastic materials can efficiently convert

Table 6.3 Summary of literature reports on the use of phage as a bio-recognition element in various assays.

Transduction	Assay type and mechanism	Target	Ref.
Amperometric electrode	Phage induced cell lysis causing release of components (such as β-galactosidase, α-glucosidase and β-glucosidase)	*E. coli* (K-12, MG 1655) *B. anthracis* *M. smegmatis*	89,92,93
Impedimetric biosensors	Phage display technology to engineer display peptides specific to the target analyte	*PSMA Antibody for P8*	69,72
LAPS	Phage display technology to engineer display peptides specific to the target analyte	*hPRL-3* *MDAMB231*	99
Bioluminescence	Luciferase reporter phage	*M. tuberculosis* *L. monocytogenes*	109
Fluorescence	Fluorescently labeled phage in combination with immuno-magnetic beads	*E. coli* O157:H7	105–107
Quantum dots	Biotinylated phage and strep-tavidin conjugated quantum dot	*E. coli* BL-21	116
Au–phage network	Phage display technology to engineer display peptides specific to the target analyte	*Melanoma cells*	81
SPR	Affinity-selected phage-immo-bilized using physical adsorption/SAMs	*S. aureus* β-galactosidase *L. monocytogenes*	68,73
Opto-fluidic ring resonator	Phage display technology to engineer display peptides specific to the target analyte	*Streptavidin*	70
QCM	Phage display technology to engineer display peptides specific to the target analyte. Phage immobilization: physical adsorption	*S. typhimurium*	48,56
Magnetoelastic cantilever	Phage display technology to engineer display peptides specific to the target analyte. Phage immobilization: physical adsorption	*B. anthracis* *S. typhimurium*	123,127
Magnetoelastic particle resonators	Phage display technology to engineer display peptides specific to the target analyte. Phage immobilization: physical adsorption	*B. anthracis* *S. typhimurium*	58–61, 128–130

the applied magnetic energy into mechanical oscillations. The characteristic fundamental resonance frequency of these oscillations is dependent on the physical properties (elastic modulus, density and Poisson's ratio) and the dimensions of the material. Both the actuation of the sensor and the detection of the response of the sensor can be measured using changes in the impedance of a non-contacting, solenoid, pick-up coil. ME sensors are actuated by the application of an AC magnetic field that causes the sensors to oscillate mechanically. When the frequency of the applied field is in resonance with the natural frequency of the sensor, the conversion of electrical energy to elastic energy is the largest. For a thin, planar, ribbon-shaped sensor of length L, vibrating in its basal plane, the fundamental resonant frequency of longitudinal oscillations is given by[146,147]

$$f = \sqrt{\frac{E}{\rho(1 - \sigma^2)}} \frac{1}{2L} \tag{6.1}$$

where E is Young's modulus of elasticity, ρ is the density of the sensor material and σ is Poisson's ratio.

Any non-magnetoelastic mass added to the sensor surface reduces the mechanical oscillations, causing the resonance frequencies to shift to a lower value. A scan through the range of applied AC frequencies yields a spectrum. The resonance frequency of the sensor can be read from this spectrum and can be tracked for any changes. The changes in the resonance frequency can then be related to the magnitude of non-magnetoelastic mass attached to the sensor surface. Addition of a small mass of Δm [much smaller than the original mass of the sensor $(\Delta m < < M)$] on the sensor surface will result in a change of the natural resonance frequency (f_0) by an amount Δf that can be related to the mass added by

$$\Delta f = -\frac{f_0 \Delta m}{2M} \qquad (\Delta m < < M) \tag{6.2}$$

where f_0 is the initial resonance frequency, M is the initial mass and Δm is the added mass. The negative sign means that the resonance frequency of the ME resonator decreases with an increase in the mass load. Hence the mass load on the magnetoelastic ME resonator can be very easily obtained by simply measuring the shift in the resonance frequency.

The sensitivity (S_m) of a ME resonator sensor can be expressed as follows:[148,149]

$$S_m = \frac{\Delta f}{\Delta m} = -\frac{1}{4L^2 wt} \sqrt{\frac{E}{\rho^3(1 - \nu)}} \tag{6.3}$$

which shows that for a small mass loading, Δm, the resonant frequency decreases and the frequency shift, Δf, is inversely proportional to the size of the sensor platform, that is, smaller sensors have higher sensitivities.

To form a biosensor, a biological recognition element is immobilized on the resonator's surface. It is designed to bind with or capture the target species when the biorecognition element comes into contact with the target species.

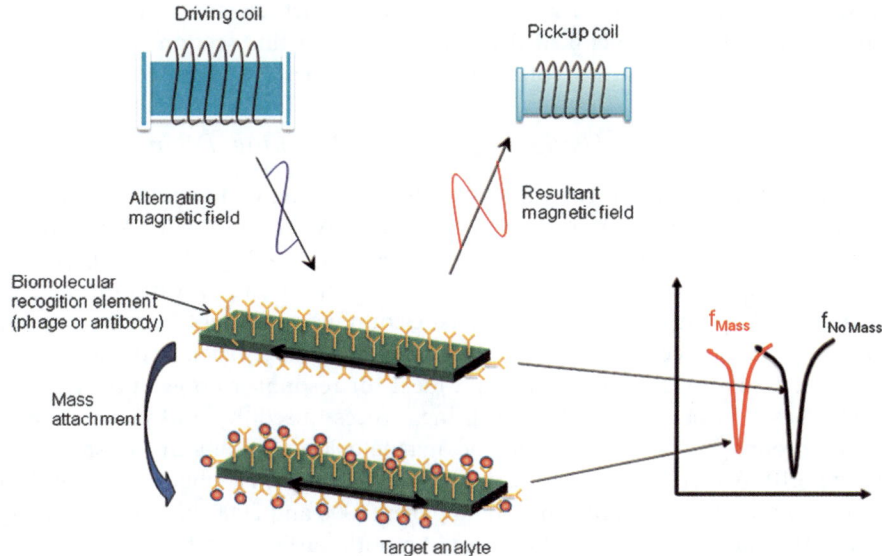

Figure 6.9 Schematic diagram of the working principle of the magnetoelastic biosensor resonator.

This binding leads to an additional mass load on the biosensor. As discussed above, this mass will cause a decrease in the resonance frequency of the biosensor. Therefore, the presence of the target species, such as bacteria/viruses/spores, can be identified by wirelessly monitoring changes in the biosensor's resonance frequency. Figure 6.9 shows schematically the operating principle of the magnetoelastic biosensor.

6.4.2   Fabrication of the Sensor Platform

The simple strip-shaped configuration of the ME platform makes its fabrication process easier than that of other AW sensor platforms, such as QCMs or microcantilevers. The choice of the fabrication method depends on the desired size of the sensor platform. Traditionally, ME sensor platforms have been fabricated by simply mechanically polishing and dicing commercially available amorphous alloy ribbons. Generally, sensor platforms with lengths longer than 1 mm can be fabricated using this method. However, as seen in the previous section, the sensitivity of the sensor will increase as its overall size and characteristic mass are decreased. In order to detect directly a small number of pathogenic bacteria (*e.g.* on the order of 100 or less), the sensor must theoretically be limited to a maximum dimension (length) in the range of 100 μm. It is very difficult to fabricate ME resonators of this size using mechanical polishing and dicing techniques. Sensor platforms with a size down to $3 \times 6 \times 50$ μm have been successfully fabricated by Johnson and co-workers using microelectronic fabrication techniques.[150,151] To obtain a sensor platform with nanoscale dimensions, Li and co-workers fabricated magnetoelastic nanobars using template-based electrochemical deposition.[152,153] In this section, the mechanical polishing/dicing and microfabrication techniques are reviewed.

6.4.2.1 Fabrication Using Mechanical Polishing/Dicing

A biosensor sensor platform was fabricated by dicing commercially available METGLAS 2826MB ribbon (Honeywell, Conway, SC, USA).[154] METGLAS is an iron-rich amorphous alloy of composition $Fe_{40}Ni_{38}Mo_4B_{18}$, prepared using a water-cooled, tape-casting technique. It has a large saturation magnetostriction (12 ppm), a high magnetoelastic coupling coefficient (90%) and good corrosion resistance. These properties result in an efficient conversion of magnetic to elastic energies, making it ideal for resonator applications.

The basic mechanical polishing/dicing process used by Lakshmanan and Wan's groups[58,60,124] to fabricate the magnetoelastic resonators is shown in Figure 6.10. A rectangular strip was cut from the tape-cast ribbon and polished using fine-grit, metallographic polishing paper (1000 and 2000 μm). The polishing allows the thickness to be reduced and smooth surfaces to be obtained. The reduced thickness of the sensor (smaller mass) increases the mass sensitivity of the resonator. The polished strips were then diced to the desired sizes using a computer-controlled automatic micro-dicing saw. Any debris or grease remaining from the dicing process was removed by cleaning the diced sensors ultrasonically in acetone for 20 min. The cleaned resonators were then subjected to a thermal anneal in a vacuum oven at 200 °C for 2 h and oven cooled under a vacuum

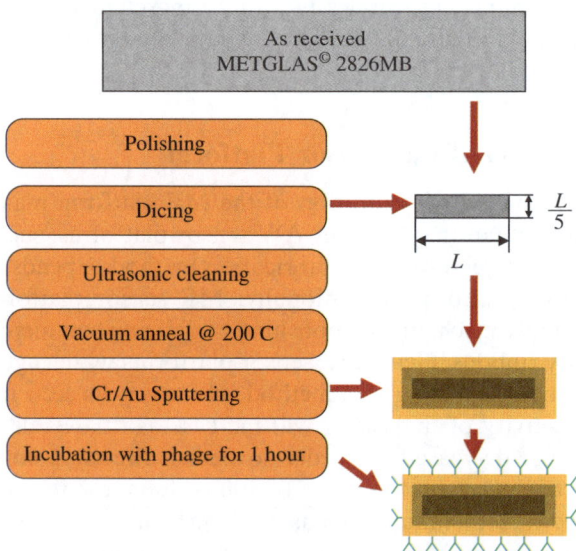

Figure 6.10   Steps in the fabrication of a magnetoelastic resonator using the polishing/dicing technique. This method is limited to fabrication of resonators greater than 500 μm in size.

($> 10^{-3}$ Torr). Research has shown that the annealing process removes residual stresses and improves the morphology and performance of the sensor platform.[151,155] Upon cooling, the sensors were transferred to a Denton Vacuum (Moorestown, NJ, USA) high-vacuum, RF sputtering system. Chromium and then gold were sputtered on to all sides of the resonators. The layer of chromium was sputtered to improve the adhesion of the gold film to the substrate. The gold layer improves the sensors' resistance to corrosion and provides a ready surface for immobilization of the biorecognition elements – antibody or phage. The mechanical dicing method is very simple. However, this method has led to problems with dimensional repeatability (*e.g.* sensors of exactly the same size) and surface quality (*e.g.* poor gold layer adhesion). Additionally, it is extremely difficult to fabricate resonators with dimensions smaller than 500 µm.

6.4.2.2 Microelectronics Fabrication of Magnetoelastic (ME) Resonators

By employing photolithography and physical vapor deposition, Johnson and coworkers fabricated magnetoelastic resonators of micro-scale size with uniform, highly repeatable dimensions and excellent surface quality.[150,151] The microfabricated resonators were made of an amorphous iron–boron binary alloy with a composition near 80:20 at.%. Several factors make this material a good candidate for sensor platform applications. The Fe–B magnetoelastic system has been studied in detail and found to form an amorphous phase with a boron content equal to or greater than the eutectic composition of 16–17 at.%.[156,157] Also, as in other metallic glasses, its soft magnetic properties, moderately high level of magnetostriction (λ_s) and low magneto-crystalline anisotropy combine to result in a very high magneto-mechanical coupling efficiency (k_{33}). The $Fe_{80}B_{20}$ alloy is therefore better suited to high-frequency resonating sensor applications as opposed to the giant magnetostrictive compounds, which are better suited for applications as mechanical stress sensors and high-power acoustic devices.[158] Finally, fabrication of this two-component material is relatively simple and can be achieved using physical vapor deposition (or 'sputtering').

The microfabrication process is shown schematically in Figure 6.11. A wafer was patterned with rectangular shapes of the desired size using a photolithography process. Photoresist SPR-220 and developer 453 from Rohm & Haas Electronic Materials (Philadelphia, PA, USA) were used to form the pattern. Amorphous iron–boron thick films were magnetron sputtered on to silicon wafers using a Discovery-18 sputter system from Denton Vacuum at a base pressure 7×10^{-7} Torr. Thin gold films, which serve as a corrosion protection layer for the resonator and form a biologically compatible surface, were deposited on the wafer before and after the Fe–B film deposition. A lift-off process employing a wash with solvent was used to remove the resonators from the wafer. The resonators were then cleaned with acetone. More details of the fabrication process can be found elsewhere.[150]

Micron-scale magnetoelastic resonators were successfully fabricated using this method and Figure 6.12 shows SEM images.

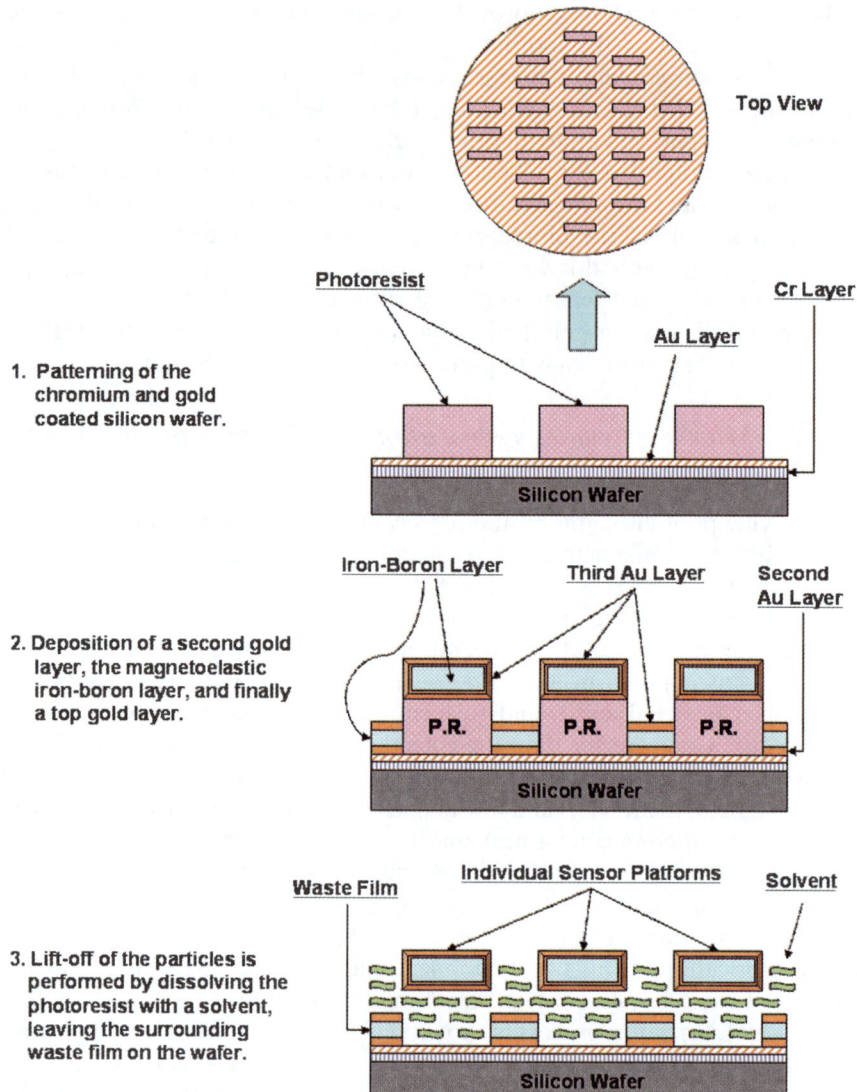

1. Patterning of the chromium and gold coated silicon wafer.

2. Deposition of a second gold layer, the magnetoelastic iron-boron layer, and finally a top gold layer.

3. Lift-off of the particles is performed by dissolving the photoresist with a solvent, leaving the surrounding waste film on the wafer.

Figure 6.11   Diagram of the microelectronics fabrication process used to make micron-scale magnetoelastic resonators.

6.4.3   ME Biosensor Assembly

So far in the investigation of phage-based magnetoelastic biosensors, two kinds of affinity selected phage have been used as bioprobes: JRB7 phage against *B. anthracis* spores and E2 phage against *S. typhimurium*. Both JRB7 and E2 phage were selected from the f8/8 landscape phage library. The selection procedures have been described in detail elsewhere.[51,159] To form phage-based ME

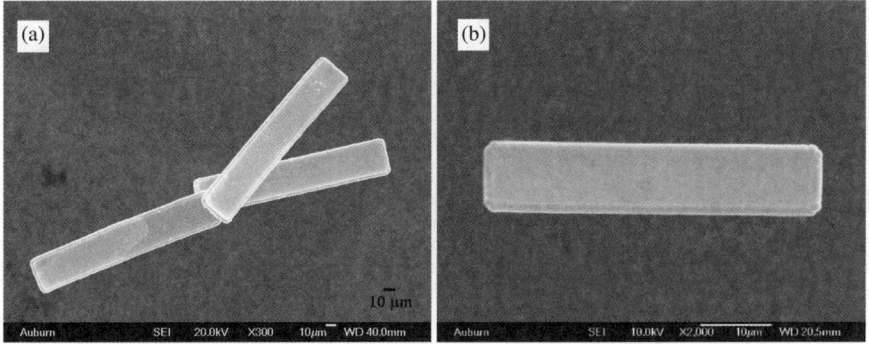

Figure 6.12 SEM images of magnetostrictive particle resonators produced using the techniques described herein: (a) 200 and (b) 50 μm in length.

biosensors, the phages were immobilized on the gold-coated surface of the resonators (sensor platforms) by physical adsorption. Detailed assembly procedures for these biosensors are available in several articles.[58–61]

6.4.4 Performance of Phage-based ME Biosensors

To investigate the performance of the phage-based ME biosensors, Wan, Lakshmanan and co-workers immobilized JRB7 phage[59,60] and E2 phage[58,61] on ME resonators to form biosensors to detect *B. anthracis* spores and *S. typhimurium* bacteria cells, respectively. The performance of the biosensors, including sensitivity, detection limit, specificity and longevity (thermal stability), was systemically characterized. The biosensors were tested in both static and dynamic modes. Static testing involved exposure of the biosensors to static solutions containing the target pathogen in known concentrations. Dynamic testing involved exposure of the biosensors to flowing solutions containing known concentrations of target pathogens in a single flow-through mode. The details of the measurement system and experimental procedures have been described in several papers.[59,61,150] Both JRB7 and E2 ME biosensors exhibit good sensitivity, specificity and selectivity towards the target pathogen, and also excellent longevity. The following sections summarize the research results for JRB7 and E2 phage-based ME biosensors separately.

6.4.4.1 JRB7 Phage-based ME Biosensor for Detection of B. anthracis Spores

6.4.4.1.1 Sensitivity and Detection Limit. Wan and co-workers[59,60] fabricated and compared both mechanically polished/diced ME resonators and microelectronically fabricated resonators with immobilized JRB7 phage. Using the mechanical polishing and dicing technique, they prepared biosensors of size 20 μm × 1 mm × 5 mm.[60] The biosensors were sequentially exposed to static *B.*

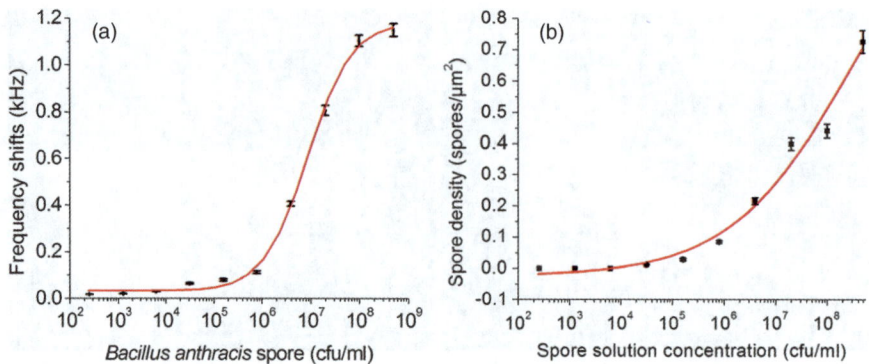

Figure 6.13   The dose response of JRB7 phage-based biosensors with dimensions
$20\,\mu m \times 1\,mm \times 5\,mm$ for detection of *B. anthracis* spores. (a) The mean
values of the resonant frequency shifts as a function of spore solution
concentration from 10^2 to 10^8 cfu ml^{-1}. The smooth line is the sigmoidal
fit to experimental data points ($\chi = 6.06$, $R^2 = 0.97$). (b) The mean values
of bound surface spore density as a function of spore solution concentration from 10^2 to 10^8 cfu ml^{-1}. The smooth line is the sigmoidal fit
to experimental data points ($\chi = 0.043$, $R^2 = 0.98$).[60]

anthracis spore solutions of increasing concentration. Figure 6.13a shows the
measured frequency shifts of the ME biosensors as a function of spore concentration of the analyte.

To confirm that the measured frequency shifts were due to the binding of the
target *B. anthracis* spores to the sensor surface, SEM was used to count the
number of spores attached to the sensor surface. The spore number counting
method used by Wan *et al.*[60] is described as follows: SEM images were taken at
10 randomly chosen regions on the sensor surface. The number of bacterial cells
attached on the sensor surface was counted manually for each of the pictures
taken and an average number of bound cells per unit area was calculated. The
resulting number was multiplied by the entire surface area of the sensor to
obtain the total number of bound cells.

Figure 6.13b shows the measured density of spores attached to the sensor
surface. The results indicate that the measured frequency shifts are due to the
additional mass of the bound *B. anthracis* spores. The biosensor showed a
sigmoidal response curve over the concentration range 10^3–10^7 cfu ml^{-1}.
Experiments were conducted on 50 biosensors to obtain the graphs presented in
Figure 6.13. The sensitivity of the biosensor, measured as the slope of the linear
portion of dose response, was calculated to be 130 Hz per decade or 0.14 cfu
μm^{-2} per decade of spore concentration. The detection limit of the biosensor
was estimated to be 10^3 cfu ml^{-1}, which corresponds to a minimal observable
frequency shift of 25 Hz. The dose dependence curve shows a trend towards
saturation at 10^8 cfu ml^{-1} and above.

Wan and co-workers[59,131] tested biosensors made using microelectronic
fabrication procedures that were also covered with JRB7 phage. Figure 6.14a

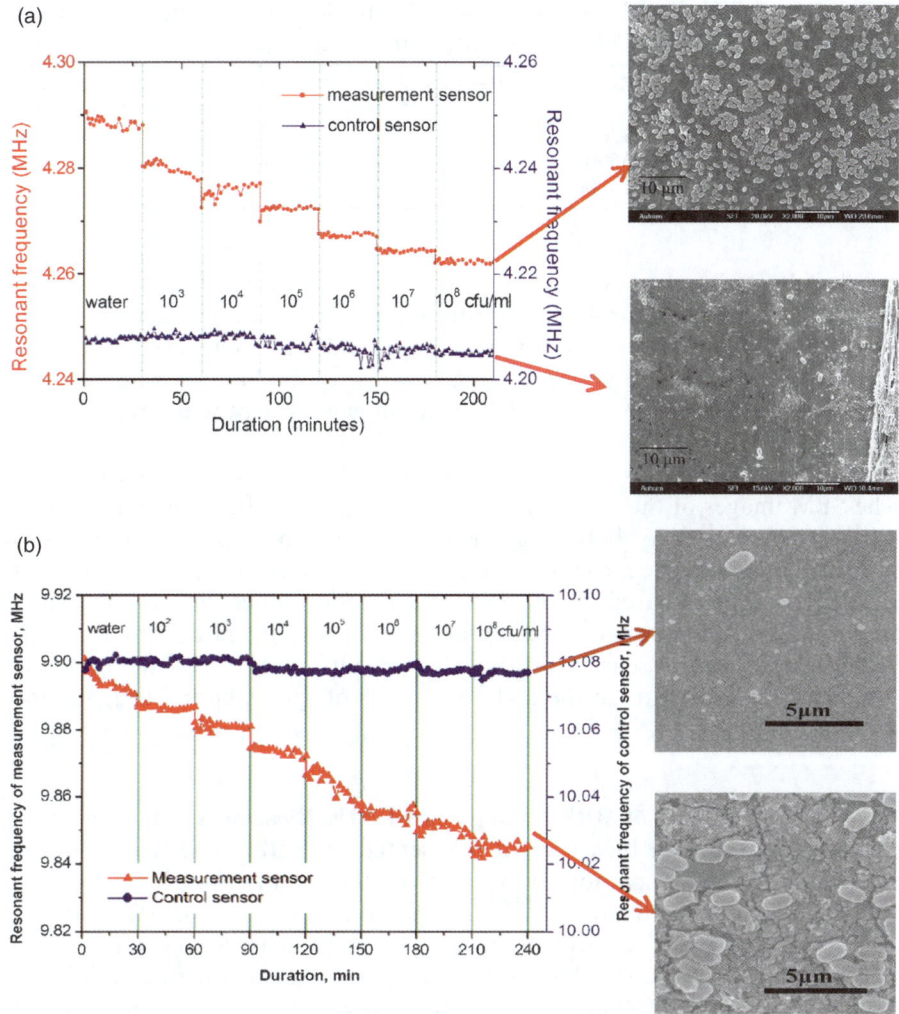

Figure 6.14 The biosensor response curve (frequency shift as a function of time and concentration) of both measurement and control sensors for (a) $4 \times 100 \times 500$ and (b) for $4 \times 40 \times 200 \, \mu m$ size sensors.[59,130]

and b show the resonance frequency responses of the $4 \times 100 \times 500$ and $4 \times 40 \times 200 \, \mu m$ biosensors to *B. anthracis* spore exposure.[59,131] The sensors were exposed to static solutions of increasing spore count every 30 min. The resonance frequencies of both the control and measurement sensors were measured and recorded continuously. Each figure consists of one measurement sensor with JRB7 phage coating and one control sensor without phage coating. Both control and measurement sensors were coated with BSA to block non-specific binding. Based on the experimental results, the $4 \times 100 \times 500 \, \mu m$ sensors were found to have a detection limit of $850 \, cfu \, ml^{-1}$ and a sensitivity of $6.5 \, kHz$

Table 6.4 Summary of the sensitivity and detection limits achieved for JRB7 phage-based biosensors with different dimensions.

Sensor dimensions	Sensitivity (Hz per decade)	Detection limit (cfu ml^{-1})
5.0×1.0×0.02 mm	130	10^3
500×100×4 µm	6500	850
200×40×4 µm	13100	105

per decade. A detection limit of 10^5 cfu ml^{-1} and a sensitivity of 13.1 kHz per decade were achieved for 4×40×200 µm sensors. The size dependence of sensitivity for the ME biosensors is summarized in Table 6.4. The results show that the sensitivity, detection limit and dose response of the biosensor improved as the size of the biosensor decreased. This result is consistent with predictions of equation (6.3).

After the spore exposure tests, the sensor surfaces were observed by SEM. The SEM images of the JRB7 phage biosensor surface after completion of the spore exposure tests for both measurement and control sensors are also shown in Figure 6.14. For the biosensors, spore counts (number of spores attached to the sensor surface) based on the SEM examination agree with the measured frequency shifts of the biosensors. Only a few spores were observed to have bound to the control sensor surface. These results confirm that the measured frequency shifts are due to the additional mass of spores bound to the sensor surfaces.

6.4.4.1.2 Specificity of JRB7 Phage-based ME Biosensors. In evaluating the performance of a biosensor, it is essential to establish selectivity (cross-reactivity with other pathogenic species) of the immobilized biorecognition element. Wan and co-workers[59,60] studied the selectivity of JRB7 phage-based ME biosensors towards *B. anthracis*. The biosensors with immobilized selected phage specific to *B. anthracis* spores were blocked with a solution of Tween-20 and 1% BSA and exposed to suspension (10^8 spores ml^{-1}) of different *Bacillus* spores. The surface density of bound spores as determined by SEM showed that the phage preferentially bound to *B. anthracis* spores in comparison with *B. subtilis*, *B. cereus*, *B. licheniformis* and *B. megaterium* spores. The biosensors showed about a 40-fold higher level of binding to *B. anthracis* than *B. licheniformis* and *B. megaterium* and about 15-fold higher than *B. subtilis* and *B. cereus* spores (Figure 6.15). The above results show that the JRB7 phage does have some cross-reaction with spores of other *Bacillus* species. However, this cross-reaction is much weaker than with *B. anthracis* spores.

6.4.4.1.3 Masking Tests of JRB7 Phage-based ME Biosensors. The selectivity of the JRB7 phage-coated magnetoelastic biosensors was further investigated by measuring the biosensor's response to *B. anthracis* masked with

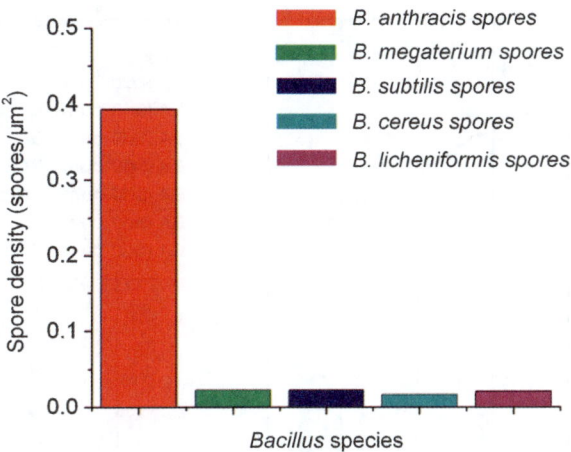

Figure 6.15 Biosensor specificity: phage-coated sensors were exposed to solutions containing 10^8 spores ml^{-1}. Density of attached spores was measured using SEM. Tween-20 with 1% BSA was used for blocking.[60]

B. cereus spores. In this set of experiments conducted by Wan *et al.*,[59] suspension of *B. anthracis* spores at concentrations of 10^3–10^8 CFU ml^{-1} were masked by adding an excess (10^8 CFU ml^{-1}) of *B. cereus* spores. The results are shown in Figure 6.16. At 30 min intervals, each test solution was pipetted off and a new solution of different concentration added. A rapid change in the resonance frequency occurred during the pipetting operation due to the sensor being momentarily exposed to air. A small frequency shift was observed after the sensor was exposed to the initial concentrated *B. cereus* solution (10^8 cfu ml^{-1}). This is consistent with previous observations that JRB7 phage affinity-selected against *B. anthracis* spores cross-react with the *B. cereus* species. The frequency shifts due to the non-specific binding between phage and *B. cereus* spores are small compared with those caused by specific binding between the JRB7 phage and *B. anthracis* spores. The experiment clearly shows that exposure to each concentration of increasing *B. anthracis* spores leads to a further decrease in the resonance frequency. This occurs even in the presence of a background of 10^8 cfu ml^{-1} of *B. cereus* spores. Despite the use of a masking agent, increasing concentrations of *B. anthracis* spores were measured.

6.4.4.1.4 Storage Life and Longevity of JRB7 Phage-based Biosensor. It was found that phage are more robust than other commonly used biorecognition probes such as polyclonal and monoclonal antibodies.[44] To investigate the storage life and longevity of the JRB7 phage-based biosensor, a set of 33 phage-based biosensors and 22 antibody-based biosensors were fabricated by Wan *et al.*[60] Groups of 11 phage-based biosensors were placed in

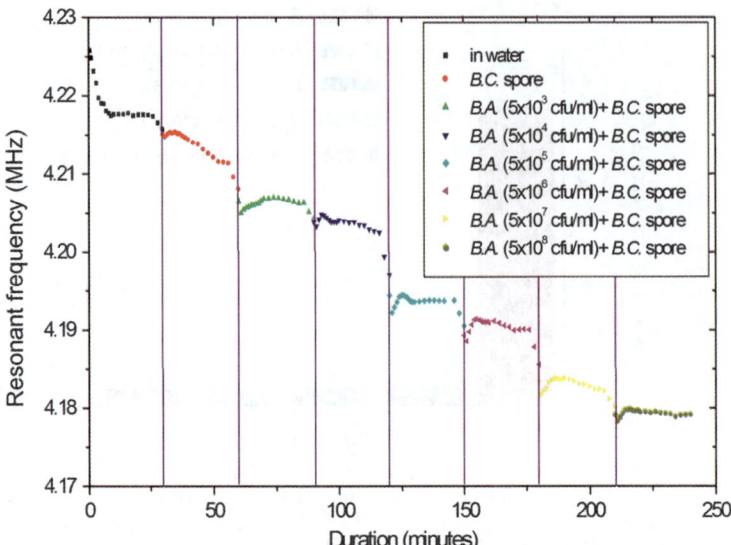

Figure 6.16 The response curve of an MEP ($500 \times 100 \times 4\,\mu m$) as a function of time and spore concentration in a mixed solution of *B. anthracis* and *B. cereus* spores. The concentration of the *B. cereus* spores is 10^8 cfu ml^{-1}.[59]

incubators at temperatures of 25, 45 and 65 °C. Similarly, groups of 11 antibody-based biosensors were place in incubators at temperatures of 45 and 65 °C. After storage, biosensors were removed at specified times and tested at room temperature by exposure to a solution containing *B. anthracis* at a concentration of 10^8 cfu ml^{-1}. These sensors were then examined using SEM. The number of spores bound per unit surface area was determined for each biosensor. Figure 6.17 shows the change in bound spore densities, *i.e.*, the sensor's binding affinity with storage time at different temperatures for both phage-coated and antibody-coated sensors.[60]

A general decrease in the binding affinity with increased storage time and temperature was observed.[60] For phage-coated biosensors, a decline of 50–75% in bound spore density over 100 days was observed. After 100 days of storage, the phage-coated biosensors preserved about 49, 40 and 25% of their original binding affinity for temperatures of 25, 45 and 65 °C, respectively. On the other hand, the antibody-coated sensors showed no binding affinity after only 5 days at 65 and 45 °C. In addition, the initial density of spore binding was less for antibody-coated sensors (about 0.2 spores μm^{-2}) compared to the phage-coated sensors (about 0.5 spores μm^{-2}).

The SEM images of the biosensor's surface for both phage-coated and antibody-coated biosensors exposed to a temperature of 65 °C are shown in Figure 6.18.[60] Phage-coated biosensors showed very high bound spore distributions. Antibody-coated biosensors also have a uniform distribution of spores but start with a lower density than phage-modified sensors. It was clearly

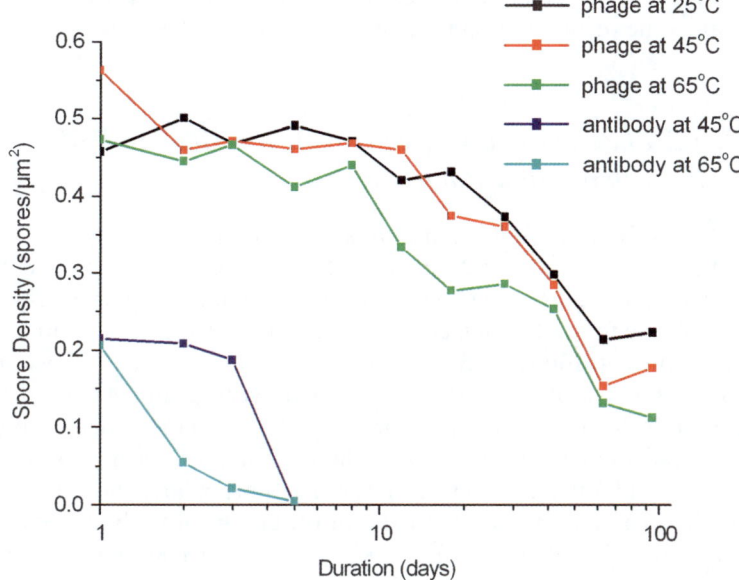

Figure 6.17 The bound surface spore density, which represents the binding affinity of the specific bioprobes, as a function of time at different temperatures for both phage and polyclonal antibody. The antibody-based sensors lost all binding activity after 5 days of storage at 65 and 45 °C.[60]

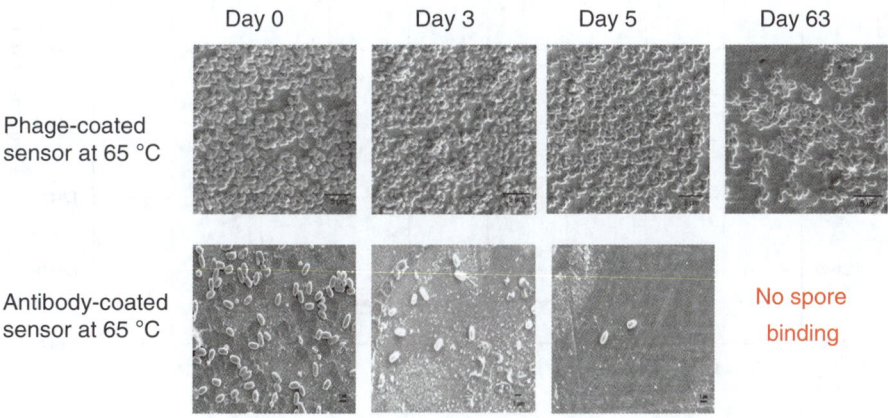

Figure 6.18 SEM images of phage- and antibody-coated sensors after storage at 65 °C. Compared with phage-modified biosensors, the binding affinity of antibody-coated biosensors dropped to zero after being stored at 65 °C for 5 days. Phage-coated biosensors still showed good binding actibity after 2 months.[60]

shown that the JRB7 phage clone used in this research has a better binding ability than some of the best commercially available antibodies.

6.4.4.2   *E2 Phage-based ME Biosensor for Detection of S. typhimurium Bacteria*

6.4.4.2.1   Sensitivity and Detection Limits.   Lakshmanan and co-workers[58,61] fabricated and characterized E2 phage-based ME biosensors for the detection of *S. typhimurium*. Figure 6.19 shows a typical resonance frequency response as a function of time for an E2 phage-based biosensor ($0.015\times0.4\times2$ mm) exposed to different concentrations of *S. typhimurium* bacteria in a flow regime. The biosensor's resonance frequency was recorded every 2 min. At the flow rate used of 50 l min^{-1}, it takes 20 min for 1 ml of the analyte to flow over the biosensor. The resonance frequency at the end of every 20 min was used to calculate the resultant frequency shift for that particular concentration. The resonance frequency decreased with the introduction of each successive concentration (from 5×10^4 to 5×10^8 cfu ml^{-1}) of *S. typhimurium*. The control sensor (no coating of phage or blocking agent) showed a negligible change in resonance frequency, even upon exposure to very high concentrations

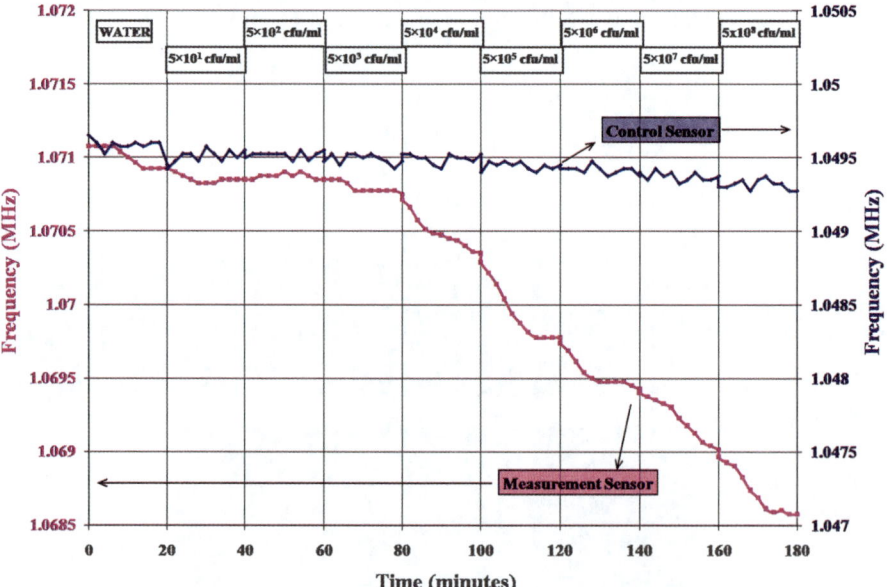

Figure 6.19   Typical frequency response curve for an E2 phage-based biosensor for *S. typhimurium* with dimensions $0.015\times0.4\times2$ mm. A 1 ml volume of each concentration of bacteria-containing solution was passed over the sensor at a flow-rate of 50 µl min^{-1}. The control sensor is not coated with phage.[87]

Table 6.5 Summary of the sensitivity and detection limits achieved for E2 phage biosensors with different dimensions (measurable frequency shift of 50 Hz).

Sensor dimensions (mm)	Sensitivity (Hz per decade)	Detection limit (cfu ml^{-1})
5.0×1.0×0.015	98	10^4
2.0×0.4×0.015	161	950
1.0×0.2×0.015	770	100
0.5×0.1×0.015	1150	60

(5×10^8 cfu ml^{-1}) of bacteria. SEM images were taken after the exposure experiments to confirm the measured frequency shifts were due to binding of *S. typhimurium*.

Similarly to the JRB7 phage-based ME biosensor described above, an increase in sensitivity and reduction in detection limits were obtained for sensors with smaller dimensions.[58,61] The dose–response tests were conducted for E2 phage-based ME biosensors with four different dimensions (0.015×1×5, 0.015×0.4×2, 0.015×0.2×1 and 0.015×0.1×0.5 mm). The measured sensitivity and detection limits for different-sized E2 phage-based biosensors are summarized in Table 6.5. The results show that the detection limits decreased from 10^4 cfu ml^{-1} for a 0.015×1×5 mm E2 biosensor to 60 cfu ml^{-1} for a 0.015×0.1×0.5 mm E2 biosensor. The sensitivity of the response in the linear region increased from 98 Hz per decade for a 0.015×1×5 mm E2 biosensor to 1150 Hz per decade for a 0.015×0.1×0.5 mm sensor. It was found that the number of attached *S. typhimurium* cells on the biosensors agrees well with the measured frequency shifts of the E2 biosensors.[58,61]

6.4.4.2.2 Specificity. To evaluate the specificity of the E2 phage-based ME biosensors, the biosensors were exposed to *S. typhimurium*, *E. coli*, *Salmonella enteritidis* and *Listeria monocytogenes*.[87] The biosensors were individually exposed to static solutions containing known concentrations of each pathogen. After exposure, the frequency shift $\Delta f_{measured}$ was measured and compared with Δf_{SEM} – a theoretical frequency shift obtained by using the SEM to count the total number of bacteria bound to the sensor. This number was then multiplied by the average weight of an *S. typhimurium* bacterial (2 pg) to obtain the change in mass. Equation (6.2) was then used to calculate the shift in resonance frequency due to the attached mass (Δf_{SEM}).

Figure 6.20[87] shows the measured values of the normalized area coverage density, $\Delta f_{measured}$ and Δf_{SEM} [calculated from bacteria counts and equation (6.2)] obtained for the biosensors. Similar values of normalized area coverage density, $\Delta f_{measured}$ and Δf_{SEM} are shown for biosensors exposed to different pathogens. The normalized area coverage density shown in Figure 6.20 was calculated as the ratio of the area coverage density of a certain pathogen (N_P) to the area coverage density of *S. typhimurium* (N_{ST}). The normalized area coverage density of the biosensors exposed to *S. typhimurium*, *S. enteritidis*, *E. coli*

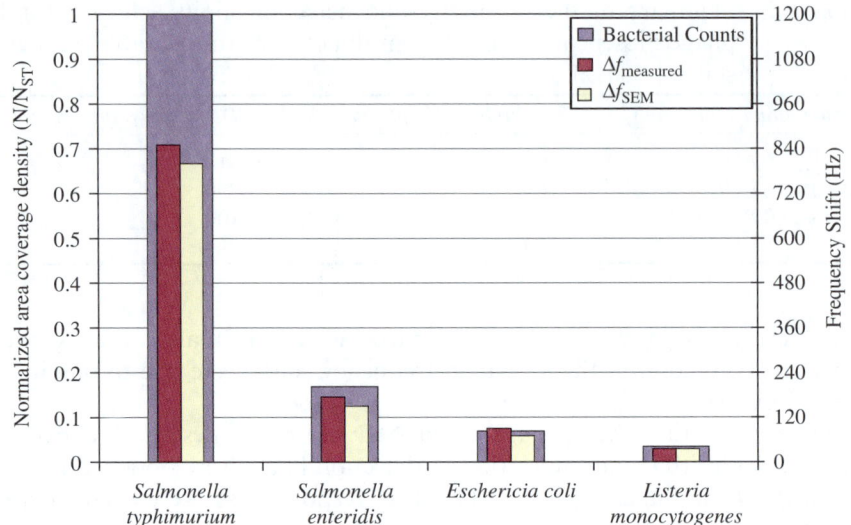

Figure 6.20 Specificity of phage-immobilized sensors exposed to different pathogens at a concentration 5×10^8 cfu ml^{-1}. The normalized area coverage density was calculated from SEM images of the sensor surface (an average of five sensors each). $\Delta f_{measured}$ and Δf_{SEM} are shown on the right.[87]

and *L. monocytogenes* were 1.00, 0.17, 0.06 and 0.03, respectively. These results (Figure 6.20) show a significantly lower affinity of the immobilized phage to pathogens other than *S. typhimurium*. Based on these results, it can be concluded that the E2 phage-based ME biosensors have excellent sensitivity and good selectivity.

6.4.4.2.3 Selectivity in the Presence of High Concentrations of Masking Bacteria.   The effect of masking bacteria on the detection of *S. typhimurium* by the E2 phage-based biosensor was studied by Lakshmanan *et al.*,[87] who exposed the E2 biosensors to three different sets of prepared suspensions: (1) *S. typhimurium*, (2) *S. typhimurium* + *E. coli* and (3) *S. typhimurium* + *E. coli* + *L. monocytogenes*. Based on the resulting dose–response curve, the authors constructed a Hill plot[48,61,74,125] (Figure 6.21) and determined the binding kinetics of the test biosensors. The binding valency was similar for all the three prepared suspensions. The sensitivity, dissociation constant and binding valence of the biosensor in the different mixtures are summarized in Table 6.6.

The binding valencies obtained from the Hill plots were 2.42, 2.79 and 2.91 for suspensions 1, 2 and 3, respectively. This reaffirms the multivalent nature of phage–*Salmonella* interaction on the biosensors. The biosensors exposed to *S. typhimurium* in the presence of masking bacteria had a higher value of the apparent dissociation constant. The lower sensitivity and the higher dissociation constants obtained for the mixtures can be attributed to the presence

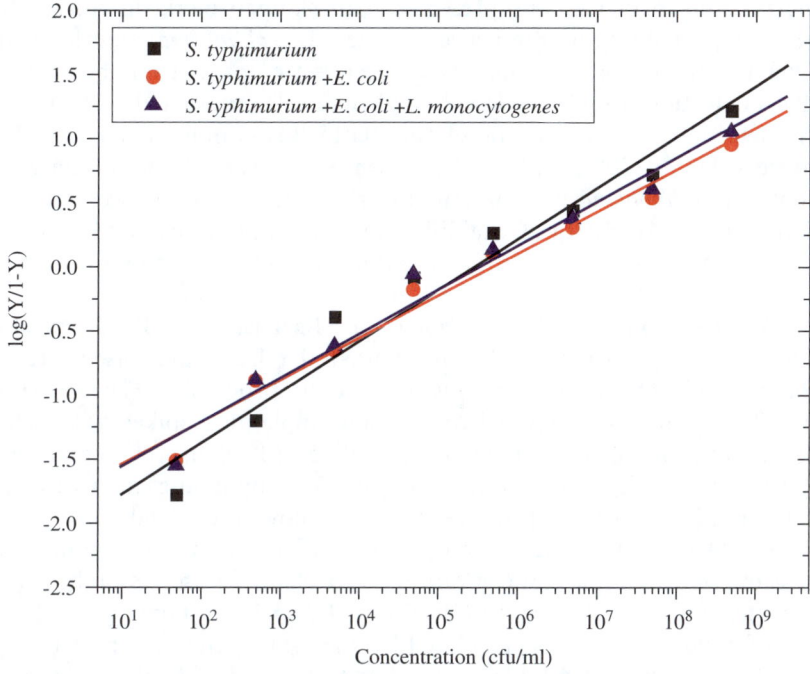

Figure 6.21 Hill plot constructed from the dose–response curves, showing the ratio of occupied (Y) and free phage sites ($1 - Y$) as a function of bacterial concentrations in different mixtures. The straight line is the linear least-squares fit to the data [*S. typhimurium* (■), slope = 0.40 ± 0.03, $R^2 = 0.97$; *S. typhimurium* + *E. coli* (●), slope = 0.33 ± 0.02, $R^2 = 0.98$; and *S. typhimurium* + *E. coli* + *L. monocytogenes* (▲), slope = 0.34 ± 0.02, $R^2 = 0.97$].[87]

Table 6.6 Sensitivity, dissociation constant and binding valence of E2 phage-based ME biosensors in different bacterial mixtures.

Bacterial mixture	S. typhimurium detection sensitivity (Hz per decade)	Binding valence (1/n)	K_d (cfu ml^{-1})	K_d (apparent) = $K_d{}^n$ (cfu ml^{-1})
S. typhimurium	161	2.42	149	1.82×10^5
S. typhimurium + E. coli	131	2.79	82	2.19×10^5
S. typhimurium + E. coli + L. monocytogenes	127	2.91	87	4.41×10^5
Spiked apple juice	155	2.77	89	2.51×10^5
Spiked milk	118	2.5	136	2.16×10^5

of the masking bacteria. The presence of masking bacteria in the mixtures reduces the probability of interaction between *S. typhimurium* and the binding sites on the biosensor. A summary of results obtained from the real food products detection using E2 phage biosensors is also presented in Table 6.6.[61]

The detection limit of the biosensor ($0.015 \times 0.4 \times 2$ mm) in all the above-described tests was 200 cfu ml^{-1}. The biosensor was capable of detecting small amounts of *S. typhimurium*, even in the presence of high concentrations of masking bacteria. Hence it was established that the E2 phage-based ME biosensor could detect *S. typhimurium* with high sensitivity, specificity and selectivity.

6.4.4.2.4 Detection of *S. typhimurium* Bacteria in Food Products. Lakshmanan and co-workers[61,87] have used the E2 phage-based ME biosensors to detect *S. typhimurium* in food products. The biosensors ($0.015 \times 0.4 \times 2$ mm) were exposed to milk and apple juice spiked with increasing concentrations of *S. typhimurium* (5×10^1–5×10^8 cfu ml^{-1}). The biosensor response was studied by flowing food liquids containing increasing concentrations of bacteria over the sensors at a flow-rate of 50 µl min^{-1}. This flow-rate was chosen in order to ensure that laminar flow was maintained. A suspension of a specific concentration of bacteria in the food liquid was allowed to flow over the sensor (20 min at the specified flow-rate). The biosensor's frequency shift was calculated by subtracting the final measured frequency from the initial frequency measured at the introduction of the food liquid. Figure 6.22 represents the average of five different biosensors.

It was found that the dose responses of the biosensors exposed to spiked apple juice and water samples were very similar except at high concentrations.[61,87] The resonance frequency shifts obtained for spiked milk samples were lower than those for spiked water and spiked apple juice samples. The dose response was linear over five aliquots of concentrations (from 5×10^3 to 5×10^7 cfu ml^{-1}) for the three different media. The sensitivity of the biosensor was calculated as the slope of the linear region of the dose–response curve (hertz per decade of concentration change). The sensitivity of biosensors exposed to spiked water, apple juice and milk was 161, 155 and 118 Hz per decade, respectively. The control sensor (Figure 6.22) showed a negligible change in resonance frequency in response to even high concentrations of *S. typhimurium*. The control sensor showed a maximum resonance frequency shift of 50 Hz, whereas a maximum resonance frequency shift of 980 Hz was observed for the phage biosensor. This significant difference in the measured frequency shifts (control *versus* measurement sensor) indicates negligible, non-specific binding of bacteria to the bare gold surface. SEM images of the tested biosensors were used to provide visual verification of bacterial binding to the sensor surfaces.

Based on the Hill plot, constructed from the dose–response data, binding valencies of 2.4, 2.5 and 2.3 were calculated for the biosensors in response to spiked water, milk and apple juice, respectively.[61,87] These values indicate that the binding of *S. typhimurium* to immobilized phage was multivalent in nature. More than two phage binding sites participated in the capture of one *S. typhimurium* cell. The values of K_d (apparent) obtained for biosensors exposed

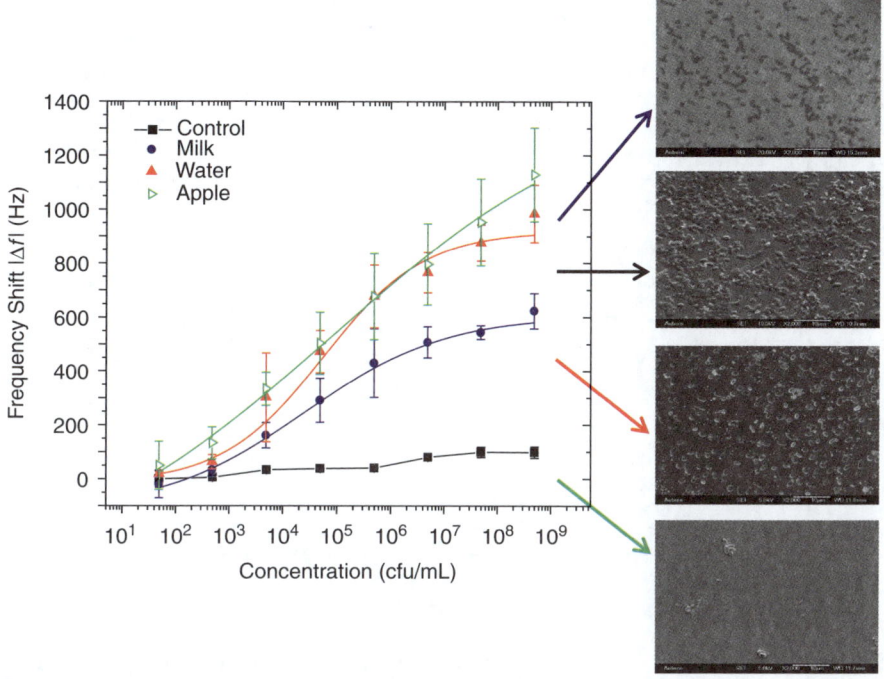

Figure 6.22 Comparison of the dose responses of magnetoelastic biosensors ($0.015 \times 0.4 \times 2$ mm) when exposed to increasing concentrations (from 5×10^1 to 5×10^8 cfu ml^{-1}) of *S. typhimurium* suspensions in water [($\blacktriangle$)$\chi^2 = 0.442$, $R^2 = 0.99$], apple juice [($\blacktriangledown$)$\chi^2 = 0.237$, $R^2 = 0.99$] and fat-free milk [($\bullet$)$\chi^2 = 0.194$, $R^2 = 0.99$]. Control ($\blacksquare$) represents the uncoated (devoid of phage) sensor's response. The curves represent the sigmoid fit of signals obtained.[61,87]

to spiked water, milk and apple juice were 1.82×10^5, 2.51×10^5 and 2.16×10^5 cfu ml^{-1}, respectively. The higher value for K_d (apparent) obtained for the biosensors exposed to milk is also evident from the lower resonance frequency shifts observed. The lower frequency shifts and higher K_d (apparent) values obtained for biosensors exposed to spiked milk samples may result from milk proteins blocking some of the available binding sites.

6.4.4.2.5 Longevity of E2 Phage-based ME Biosensors. The longevity of the E2 phage-based ME biosensors was studied at three different temperatures (25, 45 and 65 °C).[58,61,67,87] A large number of the biosensors were prepared and distributed into three groups that were then incubated in humidified ovens at the three temperatures. At specified intervals of time (in days), five biosensors from each of the three groups were removed from the incubators and equilibrated to room temperature. These 15 biosensors were then exposed to a high concentration (5×10^8 cfu ml^{-1}) of *S. typhimurium* for 45 min, and subsequently dried and analyzed by SEM. The number of individual bacteria that

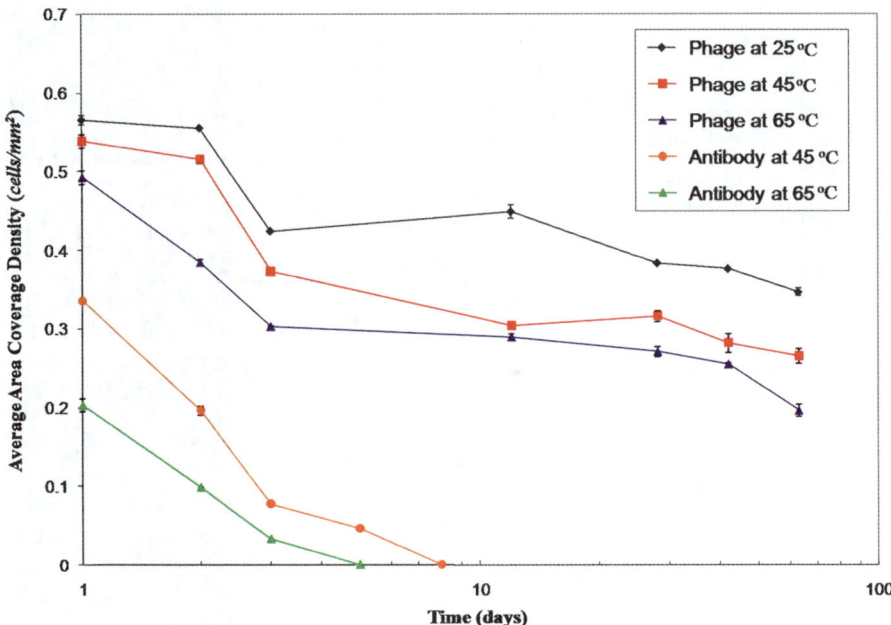

Figure 6.23   Surface coverage densities (average number of cells μm^{-2}) calculated from SEM images of stored magnetoelastic biosensors (25, 45 and 65 °C) after exposure to *S. typhimurium* (5×10^8 cfu ml^{-1}).[67,87]

were bound to the sensor surface were counted and divided by the surface area to obtain surface area coverage densities. Figure 6.23 shows the results of these experiments. The area coverage density (number of bacteria bound to the sensor per unit surface area) was observed to decrease with increasing time and temperature. The phage-based biosensors retained 59, 45 and 33% of their binding affinity at 25, 45 and 65 °C, respectively, after 63 days. The initial binding affinity of the E2 phage-based biosensors was also observed to be much higher than that of the antibody-based biosensors. The magnetoelastic sensors covered with a polyclonal antibody showed a rapid loss of binding affinity, with all binding affinity lost after 8 days at the elevated temperatures of 45 and 65 °C. On the other hand, biosensors with immobilized phage had significant binding even after 2 months of incubation at 65 °C.

6.4.4.3   Sequential Detection of Salmonella typhimurium and Bacillus anthracis Spores Using Multiple Phage-based ME Biosensors

Utilizing multiple phage-based ME biosensors, Huang and co-workers[129,160] demonstrated the simultaneous detection of different pathogens that were sequentially introduced into the measurement system. In their experiments, the detection system included a reference sensor as a control, an E2 phage-coated ME sensor specific to *S. typhimurium* and a JRB7 phage-coated ME sensor specific to *B. anthracis* spores. It was hypothesized that since the JRB7 and E2 biosensors possessed separate characteristic resonance frequencies, two different pathogens would be simultaneously monitored and discriminated. In order to prevent non-specific binding during exposure to multiple analytes, BSA solution was then immobilized on the sensor surfaces to serve as a blocking agent. The authors sequentially exposed this multiple detection system to increasing concentrations of *S. typhimurium* (from 5×10^1 to 5×10^8 cfu ml^{-1}) and *B. anthracis* spores (from 5×10^1 to 5×10^8 cfu ml^{-1}) suspensions in water. The flow-rate was $50\,\mu$l min^{-1} and, for each concentration, a 1 ml suspension was used. Figure 6.24 shows the typical response of the multiple

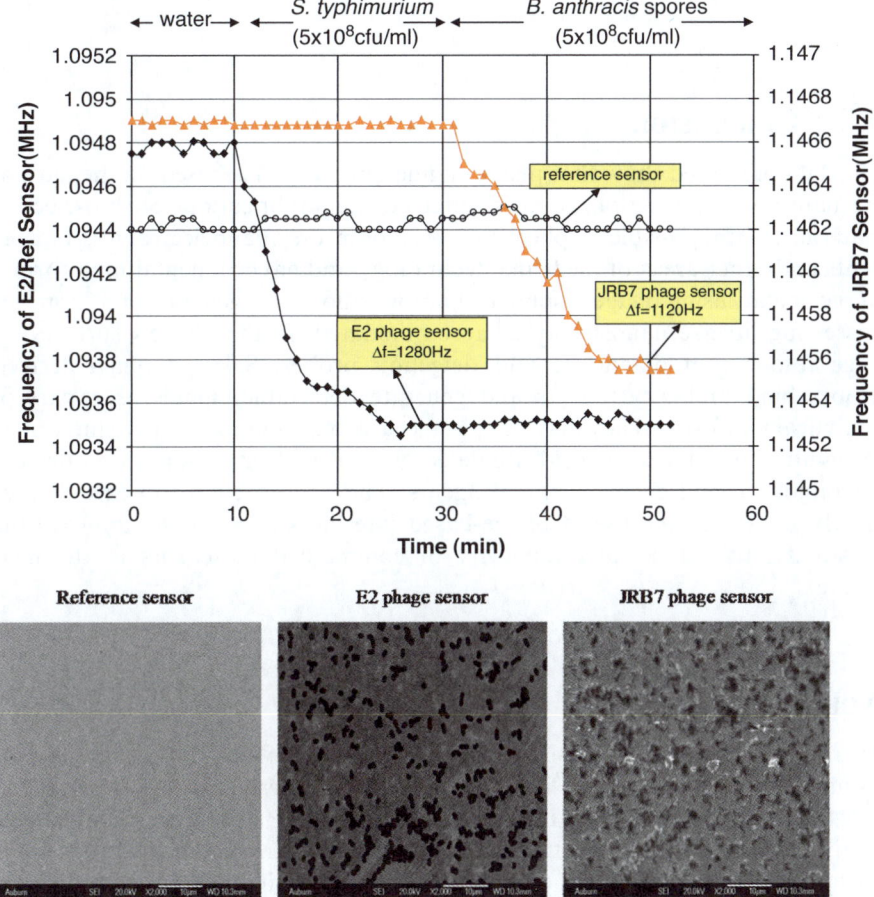

Figure 6.24 Response curves and SEM images for three diced ME biosensors tested simultaneously when exposed to the bacterial/spore suspension with a concentration of 5×10^8 cfu ml^{-1}.[128]

phage-based ME biosensors to water, *S. typhimurium* and *B. anthracis* spore suspensions (5×10^8 cfu ml^{-1} each).[129] The steady-state response of all the sensors in water was observed during the first 10 min of the test. After the introduction of *S. typhimurium*, the E2 phage-coated sensor showed a smooth decrease in resonance frequency due to the binding of these bacteria to the sensor surface. Similarly, subsequent exposure to a 5×10^8 cfu ml^{-1} *B. anthracis* spore solution caused a sudden decrease in the resonance frequency for the JRB7 phage-coated sensor. On exposure to either analyte, the decrease in frequency only occurs when bacteria cells or spores bind to the specific phage on the sensor's surface. Overall, the frequency shift was about 1280 Hz for the E2 phage-coated sensor and about 1120 Hz for the JRB7 phage-coated sensor for 5×10^8 cfu ml^{-1} analyte solutions. The SEM images (also shown in Figure 6.24) of the reference and phage-coated sensors confirmed that the frequency shifts were due to the spores/bacterial cells attached to the corresponding sensors.

6.5 Conclusion

The data presented show that phage engineering, which is based on the natural mechanisms of selection, allows directed nanofabrication of bioselective materials, with possible applications to biosensors, nanoelectronics, biosorbents and other areas of medicine, technology and environmental monitoring. In particular, using phage display technology allows the generation of libraries possessing diverse nanostructures accommodated on the phage's surface – a huge resource of diagnostic and detection probes. Selected phage-derived probes bind biological agents and generate detectable signals as a part of analytical platforms, They may be suitable as robust and inexpensive molecular recognition interfaces for field-use detectors and real-time monitoring devices for biological and chemical threat agents. The data discussed in this chapter clearly show that the use of phage-based interfaces may greatly improve the sensitivity, robustness and longevity of commercial biosensors in the near future.

Acknowledgements

This work was supported by grant NIH-1 R21 AI05564501 'Phage Binding for Continuous Anthrax Spore Detection' and US Army grant (ARO/DARPA) DAAD19-01-10454 'Phage Landscape Libraries as a Source of Substitute Antibodies for Detector Platforms' to Valery A. Petrenko. It also represents part of the ongoing efforts of the Auburn University Detection and Food Safety (AUDFS) Center to improve the safety of the US food supply chain and was funded by CSREES under grants USDA-2005-34394-15674A and 2008-34605-19275I to Bryan A. Chin.

References
1. P. S. Mead, L. Slutsker, V. Dietz, L. F. McCaig, J. S. Bresee, C. Shapiro, P. M. Griffin and R. V. Tauxe, *Emerg. Infect. Dis.*, 1999, **5**, 607.
2. S. F. Altekruse, M. L. Cohen and D. L. Swerdlow, *Emerg. Infect. Dis.*, 1997, **3**, 285.
3. L. Slutsker, S. F. Altekruse and D. L. Swerdlow, *Infect. Dis. Clin. N. Am.*, 1998, **12**, 199.
4. A. R. McClain, *Newsweek*, 2001, **18**, 10.
5. T. J. Torok, R. V. Tauxe, R. P. Wise, J. R. Livengood, R. Sokolow, S. Mauvais, K. A. Birkness, M. R. Skeels, J. M. Horan and L. R. Foster, *JAMA*, 1997, **278**, 389.
6. L. M. Eubanks, T. J. Dickerson and K. D. Janda, *Chem. Soc. Rev.*, 2007, **36**, 458.
7. NATO Organization, NATO Handbook on the Medical Aspects of NBC Defensive Operations. Part II – Biological, NATO, Brussels, 1996.
8. K. R. Klimpel and N. Arora, *CDC Annu. Bioterrorism Rep.*, 1994, **23**, 78.
9. D. V. Kamboj, A. K. Goel and L. Singh, *Def. Sci. J.*, 2006, **56**, 495.
10. P. Leonard, S. Hearty, J. Brennan, L. Dunne, J. Quinn, T. Chakraborty and R. O'Kennedy, *Enzyme Microb. Tech.*, 2003, **32**, 3.
11. M. N. Velasco-Garcia and T. Mottram, *Biosyst. Eng.*, 2003, **84**, 1.
12. D. Ivnitski, I. Abdel-Hamid, P. Atanasov and E. Wilkins, *Biosens. Bioelectron.*, 1999, **14**, 599.
13. A. E. Greenberg, R.R. Trussel, L.S. Clesceri, M.A.H. Franson, (eds), *Standard Methods for the Examination of Water and Wastewater*, American Public Health Association, Washington, DC, 1992.
14. L. D. Mello and L. T. Kubota, *Food Chem.*, 2002, **77**, 237.
15. M. Wainwrigth and J. Lederberg, *History of Microbiology*, Academic Press, New York, 1992.
16. H. Dusch and M. Altwegg, *J. Clin. Microbiol.*, 1995, **33**, 802.
17. J. Ruiz, M. L. Nunez, J. Diaz, I. Lorente, J. Perez and J. Gomez, *J. Clin. Microbiol.*, 1996, **34**, 686.
18. J. D. Perry, M. Ford, J. Taylor, A. L. Jones, R. Freeman and F. K. Gould, *J. Clin. Microbiol.*, 1999, **37**, 766.
19. J. D. Perry, G. Riley, F. K. Gould, J. M. Perez, E. Boissier, R. T. Ouedraogo and A. M. Freydiere, *J. Clin. Microbiol.*, 2002, **40**, 3913.
20. K. B. Mullis and F. A. Faloona, *Methods Enzymol.*, 1987, **155**, 335.
21. J. A. Higgins, S. Nasarabadi, J. S. Karns, D. R. Shelton, M. Cooper, A. Gbakima and R. P. Koopman, *Biosens. Bioelectron.*, 2003, **18**, 1115.
22. J. L. McKillip and M. Drake, *J. Food Protect.*, 2000, **63**, 855.
23. M. Lund, S. Nordentoft, K. Pedersen and M. Madsen, *J. Clin. Microbiol.*, 2004, **42**, 5125.
24. Z. Fu, S. Rogelj and T. L. Kieft, *Int. J. Food Microbiol.*, 2005, **99**, 47.
25. E. R. Castanha, R. R. Swiger, B. Senior, A. Fox, L. N. Waller and K. F. Fox, *J. Microbiol. Methods*, 2006, **64**, 27.

26. G. B. Jensen, N. Fisker, T. Sparso and L. Andrup, *Int. J. Food Microbiol.*, 2005, **104**, 113.
27. R. W. Reiman, D. H. Atchley and K. J. Voorhees, *J. Microbiol. Methods*, 2007, **68**, 651.
28. R. M. Lequin, *Clin. Chem.*, 2005, **51**, 2415.
29. I. Abdel-Hamid, P. Atanasov, A. L. Ghindilis and E. Wilkins, *Sens. Actuators B*, 1998, **49**, 202.
30. I. Abdel-Hamid, D. Ivnitski, P. Atanasov and E. Wilkins, *Anal. Chim. Acta*, 1999, **399**, 99.
31. A. E. G. Cass, *Biosensors: A Practical Approach*, Oxford University Press, Oxford, 1990.
32. C. Jones, A. Patel, S. Griffin, J. Martin, P. Young, K. O'Donnell, C. Silverman, T. Porter and I. Chaiken, Workshop on Chromatography, Electrokinetics and Separations in Porous Media, *J. Chromatogr. A*, 1995, **707**, 3.
33. W. R. Heineman and W. B. Jensen, *Biosens. Bioelectron.*, 2006, **21**, 1403.
34. J. P. Chambers, B. P. Arulanandam, L. L. Matta, A. Weis and J. J. Valdes, *Curr. Issues Mol. Biol.*, 2008, **10**, 1.
35. P. E. Highfield and G. Dougan, *Med. Lab. Sci.*, 1985, **42**, 352.
36. G. Kapperud, T. Vardund, E. Skjerve, E. Hornes and T. E. Michaelsen, *Appl. Environ. Microbiol.*, 1993, **59**, 2938.
37. J. H. Zhai, H. Cui and R. F. Yang, *Biotechnol. Adv.*, 1997, **15**, 43.
38. S. P. Song, L. H. Wang, J. Li, J. L. Zhao and C. H. Fan, *Trends Anal. Chem.*, 2008, **27**, 108.
39. M. Liss, B. Petersen, H. Wolf and E. Prohaska, *Anal. Chem.*, 2002, **74**, 4488.
40. M. Minunni, S. Tombelli, A. Gullotto, E. Luzi and M. Mascini, *Biosens. Bioelectron.*, 2004, **20**, 1149.
41. T. M. A. Gronewold, S. Glass, E. Quandt and A. Famulok, *Biosens. Bioelectron.*, 2005, **20**, 2044.
42. R. Capita, C. Alonso-Calleja, L. Mereghetti, B. Moreno and M. D. Garcia-Fernandez, *J. Appl. Microbiol.*, 2002, **92**, 90.
43. K. Doi, Y. Zhang, Y. Nishizaki, A. Umeda, S. Ohmomo and S. Ogata, *J. Biosci. Bioeng.*, 2003, **95**, 518.
44. J. R. Brigati and V. A. Petrenko, *Anal. Bioanal. Chem.*, 2005, **382**, 1346.
45. V. A. Petrenko and G. P. Smith, *Protein Eng.*, 2000, **13**, 589.
46. V. A. Petrenko, G. P. Smith, X. Gong and T. Quinn, *Protein Eng.*, 1996, **9**, 797.
47. V. A. Petrenko and I. B. Sorokulova, *J. Microbiol. Methods*, 2004, **58**, 147.
48. V. A. Petrenko and V. J. Vodyanoy, *J. Microbiol. Methods*, 2003, **53**, 253.
49. G. P. Smith and V. A. Petrenko, *Chem. Rev.*, 1997, **97**, 391.
50. G. P. Smith, V. A. Petrenko and L. J. Matthews, *J. Immunol. Methods*, 1998, **215**, 151.
51. I. B. Sorokulova, E. V. Olsen, I. H. Chen, B. Fiebor, J. M. Barbaree, V. J. Vodyanoy, B. A. Chin and V. A. Petrenko, *J. Microbiol. Methods*, 2005, **63**, 55.

52. V. A. Petrenko, *Microelectron. J.*, 2008, **39**, 202.
53. C. M. Fauquet, M. A. Mayo, J. Maniloff, U. Desselberger and L. A. Ball, *Virus Taxonomy, 8th Reports of the International Committee on Taxonomy of Viruses*, Elsevier, Amsterdam, 2005.
54. E. V. Olsen, S. T. Pathirana, A. M. Samoylov, J. M. Barbaree, B. A. Chin, W. C. Neely and V. Vodyanoy, *J. Microbiol. Methods*, 2003, **53**, 273.
55. V. Nanduri, A. K. Bhunia, S. I. Tu, G. C. Paoli and J. D. Brewster, *Biosens. Bioelectron.*, 2007, **23**, 248.
56. E. V. Olsen, I. B. Sorokulova, V. A. Petrenko, I. H. Chen, J. M. Barbaree and V. J. Vodyanoy, *Biosens. Bioelectron.*, 2006, **21**, 1434.
57. G. A. Weiss and R. M. Penner, *Anal. Chem.*, 2008, **80**, 3082.
58. R. S. Lakshmanan, R. Guntupalli, J. Hu, D. J. Kim, V. A. Petrenko, J. M. Barbaree and B. A. Chin, *J. Microbiol. Methods*, 2007, **71**, 55.
59. J. H. Wan, M. L. Johnson, R. Guntupalli, V. A. Petrenko and B. A. Chin, *Sens. Actuators B*, 2007, **127**, 559.
60. J. H. Wan, H. H. Shu, S. C. Huang, B. Fiebor, I. H. Chen, V. A. Petrenko and B. A. Chin, *IEEE Sens. J.*, 2007, **7**, 470.
61. R. S. Lakshmanan, R. Guntupalli, J. Hu, V. A. Petrenko, J. M. Barbaree and B. A. Chin, *Sens. Actuators B*, 2007, **126**, 544.
62. L. Olofsson, J. Ankarloo, P. O. Andersson and I. A. Nicholls, *Chem. Biol.*, 2001, **8**, 661.
63. L. Olofsson, J. Ankarloo and I. A. Nicholls, *J. Mol. Recognit.*, 1998, **11**, 91.
64. R. H. J. van der Linden, L. G. J. Frenken, B. de Geus, M. M. Harmsen, R. C. Ruuls, W. Stok, L. de Ron, S. Wilson, P. Davis and C. T. Verrips, *BBA–Protein Struct. M.*, 1999, **1431**, 37.
65. A. Usami, A. Ohtsu, S. Takahama and T. Fujii, *J. Pharm. Biomed. Anal.*, 1996, **14**, 1133.
66. H. Dooley, S. D. Grant, W. J. Harris and A. J. Porter, *Biotechnol. Appl. Biochem.*, 1998, **28**, 77.
67. R. Guntupalli, R. Lakshmanan, J. Wan, D. J. Kim, T. Huang, V. Vodyanoy and B. Chin, *Sens. Instrum. Food Qual. Safety*, 2008, **2**, 27.
68. V. Nanduri, S. Balasubramanian, S. Sista, V. J. Vodyanoy and A. L. Simonian, *Anal. Chim. Acta*, 2007, **589**, 166.
69. L. M. C. Yang, P. Y. Tam, B. J. Murray, T. M. McIntire, C. M. Overstreet, G. A. Weiss and R. M. Penner, *Anal. Chem.*, 2006, **78**, 3265.
70. H. Y. Zhu, I. M. White, J. D. Suter and X. D. Fan, *Biosens. Bioelectron.*, 2008, **24**, 461.
71. F. N. Dultsev, R. E. Speight, M. T. Florini, J. M. Blackburn, C. Abell, V. P. Ostanin and D. Klenerman, *Anal. Chem.*, 2001, **73**, 3935.
72. L. M. C. Yang, J. E. Diaz, T. M. McIntire, G. A. Weiss and R. M. Penner, *Anal. Chem.*, 2008, **80**, 5695.
73. F. F. Liu, Z. F. Luo, X. Ding, S. G. Zhu and X. L. Yu, *Sens. Actuators B*, 2009, **136**, 133.
74. V. Nanduri, I. B. Sorokulova, A. M. Samoylov, A. L. Simonian, V. A. Petrenko and V. Vodyanoy, *Biosens. Bioelectron.*, 2007, **22**, 986.
75. Q. Wen and J. X. Tang, *Phys. Rev. Lett.*, 2006, **97**.

76. A. P. Lyubartsev, J. X. Tang, P. A. Janmey and L. Nordenskiold, *Phys. Rev. Lett.*, 1998, **81**, 5465.
77. J. X. Tang, P. A. Janmey, A. Lyubartsev and L. Nordenskiold, *Biophys. J.*, 2002, **83**, 566.
78. T. T. Nguyen, I. Rouzina and B. I. Shklovskii, *J. Chem. Phys.*, 2000, **112**, 2562.
79. T. E. Angelini, H. Liang, W. Wriggers and G. C. L. Wong, *Proc. Natl. Acad. Sci. USA*, 2003, **100**, 8634.
80. T. E. Angelini, H. Liang, W. Wriggers and G. C. L. Wong, *Eur. Phys. J. E*, 2005, **16**, 389.
81. G. R. Souza, D. R. Christianson, F. I. Staquicini, M. G. Ozawa, E. Y. Snyder, R. L. Sidman, J. H. Miller, W. Arap and R. Pasqualini, *Proc. Natl. Acad. Sci. USA*, 2006, **103**, 1215.
82. S. Huang, H. Yang, R. Lakshmanan, M. L. Johnson, I. Chen, J. Wan, H. C. Wikle, V. A. Petrenko, J. Barbaree, Z. Y. Cheng and B. A. Chin, *Biotechnol. Bioeng.*, 2008, **101**, 1014.
83. J. X. Tang, T. Ito, T. Tao, P. Traub and P. A. Janmey, *Biochemistry*, 1997, **36**, 12600.
84. B. Y. Ha and A. J. Liu, *Phys. Rev. Lett.*, 1998, **81**, 1011.
85. S. A. Overman, D. M. Kristensen, P. Bondre, B. Hewitt and G. J. Thomas, *Biochemistry*, 2004, **43**, 13129.
86. M. T. Record, W. T. Zhang and C. F. Anderson, *Adv. Protein Chem.*, 1998, **51**, 281.
87. R. Lakshmanan, PhD dissertation, Materials Research and Education Center, Auburn University, Auburn, AL, 2008.
88. A. S. Mittelmann, E. Z. Ron and J. Rishpon, *Anal. Chem.*, 2002, **74**, 903.
89. T. Neufeld, A. Schwartz-Mittelmann, D. Biran, E. Z. Ron and J. Rishpon, *Anal. Chem.*, 2003, **75**, 580.
90. F. Perez, I. Tryland, M. Mascini and L. Fiksdal, *Anal. Chim. Acta*, 2001, **427**, 149.
91. I. Benhar, I. Eshkenazi, T. Neufeld, J. Opatowsky, S. Shaky and J. Rishpon, *Talanta*, 2001, **55**, 899.
92. T. Neufeld, A. S. Mittelman, V. Buchner and J. Rishpon, *Anal. Chem.*, 2005, **77**, 652.
93. M. Yemini, Y. Levi, E. Yagil and J. Rishpon, *Bioelectrochemistry*, 2007, **70**, 180.
94. S. Seo, M. Dobozi-King, R. F. Young, L. B. Kish and M. S. Cheng, *Microelectron. Eng.*, 2008, **85**, 1484.
95. S. Seo, H. C. Kim, M. S. Cheng, X. C. Ruan and W. Ruan, *J. Vac. Sci. Technol. B*, 2006, **24**, 3133.
96. J. C. Owicki, L. J. Bousse, D. G. Hafeman, G. L. Kirk, J. D. Olson, H. G. Wada and J. W. Parce, *Annu. Rev. Bioph. Biomol. Struct.*, 1994, **23**, 87.
97. D. G. Hafeman, J. W. Parce and H. M. McConnell, *Science*, 1988, **240**, 1182.
98. T. Yoshinobu, M. J. Schoning, R. Otto, K. Furuichi, Y. Mourizina, Y. Ermolenko and H. Iwasaki, *Sens. Actuators B*, 2003, **95**, 352.

99. Y. F. Jia, M. Qin, H. K. Zhang, W. C. Niu, X. Li, L. K. Wang, X. Li, Y. P. Bai, Y. J. Cao and X. Z. Feng, *Biosens. Bioelectron.*, 2007, **22**, 3261.
100. N. K. Petty, T. J. Evans, P. C. Fineran and G. P. C. Salmond, *Trends Biotechnol.*, 2007, **25**, 7.
101. E. R. Goldman, M. P. Pazirandeh, J. M. Mauro, K. D. King, J. C. Frey and G. P. Anderson, *J. Mol. Recognit.*, 2000, **13**, 382.
102. G. P. Anderson, K. D. King, K. L. Gaffney and L. H. Johnson, *Biosens. Bioelectron.*, 2000, **14**, 771.
103. K. D. King, J. M. Vanniere, J. L. Leblanc, K. E. Bullock and G. P. Anderson, *Environ. Sci. Technol.*, 2000, **34**, 2845.
104. E. R. Goldman, M. P. Pazirandeh, P. T. Charles, E. D. Balighian and G. P. Anderson, *Anal. Chim. Acta*, 2002, **457**, 13.
105. C. L. Turnbough, *J. Microbiol. Methods*, 2003, **53**, 263.
106. L. Goodridge, J. R. Chen and M. Griffiths, *Int. J. Food Microbiol.*, 1999, **47**, 43.
107. L. Goodridge, J. R. Chen and M. Griffiths, *Appl. Environ. Microb.*, 1999, **65**, 1397.
108. P. Billard and M. S. DuBow, *Clin. Biochem.*, 1998, **31**, 1.
109. N. Banaiee, M. Bodadilla-del-Valle, S. Bardarov, P. F. Riska, P. M. Small, A. Ponce-De-Leon, W. R. Jacobs, G. F. Hatfull and J. Sifuentes-Osornio, *J. Clin. Microbiol.*, 2001, **39**, 3883.
110. R. Blasco, M. J. Murphy, M. F. Sanders and D. J. Squirrell, *J. Appl. Microbiol.*, 1998, **84**, 661.
111. B. Dubertret, P. Skourides, D. J. Norris, V. Noireaux, A. H. Brivanlou and A. Libchaber, *Science*, 2002, **298**, 1759.
112. A. Sukhanova, M. Devy, L. Venteo, H. Kaplan, M. Artemyev, V. Oleinikov, D. Klinov, M. Pluot, J. H. M. Cohen and I. Nabiev, *Anal. Biochem.*, 2004, **324**, 60.
113. W. C. W. Chan and S. M. Nie, *Science*, 1998, **281**, 2016.
114. M. Bruchez, M. Moronne, P. Gin, S. Weiss and A. P. Alivisatos, *Science*, 1998, **281**, 2013.
115. X. Michalet, F. Pinaud, T. D. Lacoste, M. Dahan, M. P. Bruchez, A. P. Alivisatos and S. Weiss, *Single Mol.*, 2001, **2**, 261.
116. R. Edgar, M. McKinstry, J. Hwang, A. B. Oppenheim, R. A. Fekete, G. Giulian, C. Merril, K. Nagashima and S. Adhya, *Proc. Natl. Acad. Sci. USA*, 2006, **103**, 4841.
117. S. Balasubramanian, I. B. Sorokulova, V. J. Vodyanoy and A. L. Simonian, *Biosens. Bioelectron.*, 2007, **22**, 948.
118. D. S. Ballantine, R. M. White, S. J. Martin, A. J. Ricco, G. C. Frye, E. T. Zellers and H. Wohltjen, *Acoustic Wave Sensors: Theory, Design and Physico-chemical Applications*, Academic Press, New York, 1997.
119. O. Tamarin, C. Dejous, D. Rebiere, J. Pistre, S. Comeau, D. Moynet and J. Bezian, *Sens. Actuators B*, 2003, **91**, 275.
120. R. Raiteri, M. Grattarola, H. J. Butt and P. Skladal, *Sens. Actuators B*, 2001, **79**, 115.

121. C. Bailey, B. Fiebor, W. Yan, V. Vodyanoy, R. Cernosek and B. A. Chin, *Proc. SPIE*, 2002, **4575**, 138.
122. S. Li, L. Orona, Z. Li and Z. -Y. Cheng, *Appl. Phys. Lett.*, 2006, **88**, 073507.
123. L. Fu, S. Li, K. Zhang, I. H. Chen, V. A. Petrenko and Z. -Y. Cheng, *Sensors*, 2007, **7**, 2929.
124. R. Guntupalli, J. Hu, R. S. Lakshmanan, T. S. Huang, J. M. Barbaree and B. A. Chin, *Biosens. Bioelectron.*, 2007, **22**, 1474.
125. R. Guntupalli, R. S. Lakshmanan, J. Hu, T. S. Huang, J. M. Barbaree, V. Vodyanoy and B. A. Chin, *J. Microbiol. Methods*, 2007, **70**, 112.
126. R. Guntupalli, R. Lakshmanan, M. L. Johnson, J. Hu, T.-S. Huang, J. Barbaree, V. Vodyanoy and B. A. Chin, *Sens. Instrum. Food Qual. Saf.*, 2007, **1**, 3.
127. L. Fu, S. Li, K. Zhang, Z. -Y. Cheng and J. Barbaree, *Proc. SPIE*, 2007, **6556**, 655619.
128. S. T. Ge, S. J. Huang, F. Y. Zhang, Q. Y. Cai and C. A. Grimes, *Sensor Lett.*, 2009, **7**, 79.
129. S. Huang, H. Yang, R. S. Lakshmanan, M. L. Johnson, J. Wan, I. H. Chen, H. C. Wikle, V. A. Petrenko, J. M. Barbaree and B. A. Chin, *Biosens. Bioelectron.*, 2009, **24**, 1730.
130. W. Shen, R. S. Lakshmanan, L. C. Mathison, V. A. Petrenko and B. A. Chin, *Sens. Actuators B*, 2009, **173**, 501.
131. F. Xie, H. Yang, S. Li, W. Shen, J. Wan, M. L. Johnson, H. C. Wikle, D.-J. Kim and B. A. Chin, *Intermetallics*, 2009, **17**, 270.
132. S. Li, M. L. Johnson, J. Wan, V. A. Petrenko and B. A. Chin, *ECS Trans.*, 2008, **16**, 177.
133. P. Pang, X. L. Xiao, Q. Y. Cai, S. Z. Yao and C. A. Grimes, *Sens. Actuators B*, 2008, **133**, 473.
134. S. C. Roy, J. R. Werner, D. Kouzoudis and C. A. Grimes, *Sensor Lett.*, 2008, **6**, 280.
135. S. C. Roy, J. R. Werner, G. Mambrini and C. A. Grimes, *Sensor Lett.*, 2008, **6**, 285.
136. X. L. Xiao, M. L. Guo, Q. X. Li, Q. Y. Cai, S. Z. Yao and C. A. Grimes, *Biosens. Bioelectron.*, 2008, **24**, 247.
137. K. G. Ong, K. F. Zeng, X. P. Yang, K. Shankar, C. M. Ruan and C. A. Grimes, *IEEE Sens. J.*, 2006, **6**, 514.
138. Q. Z. Lu, H. L. Lin, S. T. Ge, S. L. Luo, Q. Y. Cai and C. A. Grimes, *Anal. Chem.*, 2009, **81**, 5846.
139. K. G. Ong, J. S. Bitler, C. A. Grimes, L. G. Puckett and L. G. Bachas, *Sensors*, 2002, **2**, 219.
140. K. G. Ong, J. M. Leland, K. F. Zeng, G. Barrett, M. Zourob and C. A. Grimes, *Biosens. Bioelectron.*, 2006, **21**, 2270.
141. C. A. Grimes and D. Kouzoudis, *Sens. Actuators A*, 2000, **84**, 205.
142. M. K. Jain, S. Schmidt, K. G. Ong, C. Mungle and C. A. Grimes, *Smart Mater. Struct.*, 2000, **9**, 502.

143. M. K. Jain, S. Schmidt, C. Mungle, K. Loiselle and C. A. Grimes, *IEEE. Trans. Magn.*, 2001, **37**, 2767.
144. C. M. Ruan, K. F. Zeng, O. K. Varghese and C. A. Grimes, *Anal. Chem.*, 2003, **75**, 6494.
145. K. Shankar, K. F. Zeng, C. M. Ruan and C. A. Grimes, *Sens. Actuators B*, 2005, **107**, 640.
146. P. G. Stoyanov and C. A. Grimes, *Sens. Actuators A*, 2000, **80**, 8.
147. L. D. Landau and E. M. Lifshitz, *Theory of Elasticity*, Pergamon Press, Oxford, 1986.
148. C. A. Grimes, P. G. Stoyanov, D. Kouzoudis and K. G. Ong, *Rev. Sci. Instrum.*, 1999, **70**, 4711.
149. C. Liang, S. Morshed and B. C. Prorok, *Appl. Phys. Lett.*, 2007, **90**.
150. M. L. Johnson, J. H. Wan, S. C. Huang, Z. Y. Cheng, V. A. Petrenko, D. J. Kim, I. H. Chen, J. M. Barbaree, J. W. Hong and B. A. Chin, *Sens. Actuators A*, 2008, **144**, 38.
151. M. L. Johnson, O. LeVar, S. H. Yoon, J. -H. Park, S. Huang, D. -J. Kim, Z. Cheng and B. A. Chin, *Vacuum*, 2009, **83**, 958.
152. S. Li, L. Fu, C. Wang, S. Lea, B. Arey, M. Engelhard and Z. -Y. Cheng, *MRS Proc.*, 2006, **962E**, 0962.
153. Z. Y. Cheng, S. Li, K. Zhang, L. Fu and B. A. Chin, *Adv. Sci. Technol.*, 2008, **54**, 19.
154. http://www.metglas.com/PRODUCTS/page5_1_2_7_1.htm.
155. S. Huang, J. Hu, J. Wan, M. L. Johnson, H. Shu and B. A. Chin, *Mater. Sci. Eng. C*, 2008, **28**, 380.
156. W. G. Moffatt, *Handbook of Binary Phase Diagrams*, General Electric, Schenectady CT, 1976.
157. T. Van Rompaey, K. C. H. Kumar and P. Wollants, *J. Alloys Compd.*, 2002, **334**, 173.
158. G. Engdahl, *Handbook of Giant Magnetostrictive Materials*, Academic Press, San Diego, CA, 2000.
159. J. Brigati, D. D. Williams, I. B. Sorokulova, V. Nanduri, I. H. Chen, C. L. Turnbough and V. A. Petrenko, *Clin. Chem.*, 2004, **50**, 1899.
160. S. Huang, S. Li, H. Yang, M. L. Johnson, J. Wan, I. Chen, V. A. Petrenko, J. Barbaree and B. A. Chin, *J. Microsyst. Technol. Appl.*, 2008, **3**, 87.

CHAPTER 7

Phage-mediated Detection of Biological Threats

STEVEN RIPP

University of Tennessee, 676 Dabney Hall, Knoxville, TN 37996, USA

7.1 Introduction

Bacteriophage are capable of specific recognition of their bacterial host cells and this innate identification attribute has been exploited for decades in phage typing schemes to identify food-borne and clinical bacterial isolates. More recently, this basic premise of phage/host specificity has been advanced via phage-mediated signaling elements that indicate when a phage/host attachment, infection or lysis event has occurred, thereby providing a very simple and rapid means for bacterial recognition. Phage-based sensors can take many forms, ranging from straightforward visualization of tagged phage as they attach to their appropriate host bacterial cells to actual incorporation of trackable labels inside the phage for delivery and subsequent expression within the host cell itself to those that bypass labeled phage altogether and rely rather on the end product amplification and measurement of progeny phage released from target infected cells. This chapter discusses these novel phage-mediated reporter assays and their application towards the detection of priority biological agents.

RSC Nanoscience & Nanotechnology No. 17
Phage Nanobiotechnology
Edited by Valery A. Petrenko and George P. Smith
© Royal Society of Chemistry 2011
Published by the Royal Society of Chemistry, www.rsc.org

7.2 Phage Typing Schemes

Over 5000 bacteriophage have been morphologically characterized and grouped as to susceptible host.[1] This diversity and often absolute specificity displayed by phage has long been exploited in diagnostic 'phage typing' schemes to identify bacterial targets uniquely. After spreading a solution of the unknown bacterial culture on a solid growth medium, small drops of several different phage solutions are added. Wherever a certain phage is capable of infecting and killing (lysing) the bacterium, a cleared area called a plaque is formed (Figure 7.1). Based on which phage form plaques and which do not, in other words, the pattern of specificity of the phage for infection of that particular bacterial host (or the phage's host range), the unknown bacterium can be epidemiologically identified. Phage typing schemes are widely available for nearly all pathogenic microorganisms and their successful clinical use in microbial identification down to the serotype or serogroup levels has been well documented.[2–5] This includes diagnostic phage sets for virtually all National Institute of Allergy and Infectious Diseases (NIAID) Category A, B and C bacterial pathogens (*i.e.*, *Yersinia enterocolitica*, *Y. pestis*, *Bacillus anthracis*, *Campylobacter jejuni*, diarrheagenic *Escherichia coli*, *Listeria monocytogenes*, *Staphylococcus*, *Clostridium botulinum*, *C. perfringens*, *Vibrio cholerae*, *Salmonella*, *Mycobacterium tuberculosis*, *Brucella* spp., *Burkholderia* spp. and *Shigella* spp.).

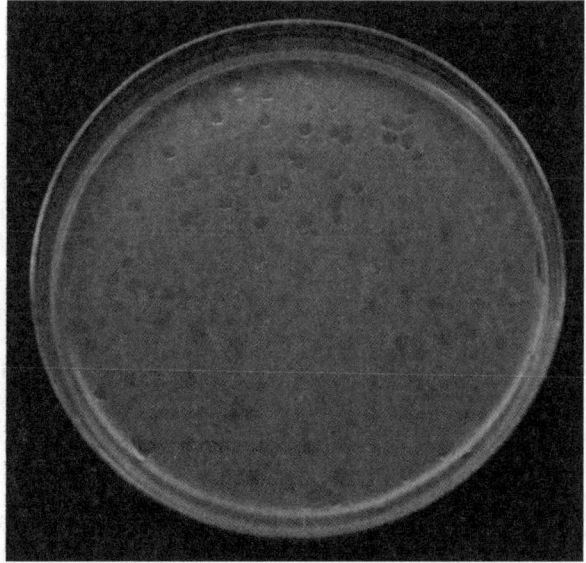

Figure 7.1   Phage typing takes advantage of a phage's host range to identify unknown bacteria specifically based on phage/bacterium infection patterns as visualized by the formation of zones of bacterial lysis (plaques).

7.3 Exploiting Phage Specificity for Bacterial Detection

Phage typing, although specific and sensitive, is not rapid. Typically an overnight incubation is required prior to visualization of plaques. In addition, the unknown bacterium needs to be cultured. Recognizing these disadvantages, several groups have devised novel phage/host recognition assays that are faster and easier to perform. However, no matter what the assay, the fundamentals of phage typing and phage host ranges stand and it is only innovative 'visualization' methods apart from seeing plaque formation on a growth medium that distinguishes these assays. The visualization signals so far incorporated into these assays are either colorimetric, fluorescent, bioluminescent or electrochemical, and their emissions can be linked to either the initial attachment of a phage to its host cell, the actual phage/host infection event or the ultimate lysing of the bacterial host cell and release of progeny phage. Assays are rapid, typically achieving detection within a few hours, and maintain exceptional detection limits, often down to less than $10\,\mathrm{cfu\,ml}^{-1}$. Further, in the majority of assays, only living bacteria are detected since phage require a viable host to sustain their infection and propagation ritual. Analogous bacterial detection methods based on the polymerase chain reaction (PCR) or immunological antibody–antigen reactions are more prone to false-positive signals due to their inability to distinguish the hazardous living from the harmless dead, as occurs, for example, when dead cells and cellular constituents remain after pasteurization or disinfection procedures. Phage can also be propagated to high titers easily and inexpensively, making assays very cost effective, and exhibit sufficient resilience to be storable for months to years.

7.3.1 Labeled Phage

Perhaps the simplest phage-mediated detection scheme associates a visible label with the phage itself and monitors when the phage has attached to its host cell. Goodridge *et al.*,[6,7] in what is referred to as the fluorescent-bacteriophage assay (FBA), stained the DNA of the *E. coli* O157:H7 specific phage LG1 with the fluorescent dye YOYO-1. In a 30 min procedure, the *E. coli* cells are first immunomagnetically separated from the mixed culture using paramagnetic beads coated with polyclonal antibodies against *E. coli* O157:H7 (Dynal Dynabeads) and then specifically labeled with the fluorescent LG1 phage (Figure 7.2), analogous to an enzyme-linked immunosorbent assay (ELISA) where a labeled antibody targets a specific cellular antigen. Using flow cytometry, the now fluorescently labeled *E. coli* could be rapidly enumerated. Detection limits in artificially contaminated ground beef approached approximately $2\,\mathrm{cfu\,g}^{-1}$ after a 6 h enrichment and $10\,\mathrm{cfu\,ml}^{-1}$ in raw milk after a 10 h enrichment. Kenzaka *et al.*[8] developed a similar 30 min FBA procedure using DAPI (4′,6-diamidino-2-phenylindole)-labeled phage T4 to detect *E. coli* and, using epifluorescence microscopy, successfully applied it towards the enumeration of *E. coli* in fecally contaminated canal water samples in Thailand.

Figure 7.2 YOYO-1-labeled phage LG1 attaches to and forms a 'halo' of fluorescence around *E. coli* O157:H7 cells. Bar = 10 μm, 1000× magnification. Reproduced from Goodridge *et al.*,[7] with permission.

Mosier-Boss *et al.*[9] labeled P22 phage with fluorescent SYBR gold dye. In this case, the labeled DNA was actually injected into the target *Salmonella* Typhimurium (*Salmonella enterica* subspecies *enterica* serovar Typhimurium) host cells where it could be visualized internally via epifluorescent microscopy.

Taking a very unique approach to using labeled phage as biological probes, Edgar *et al.*[10] nanoengineered phage-conjugated quantum dots to detect *E. coli*. Quantum dots are fluorescent probes consisting of colloidal semiconductor nanocrystals. With high quantum yield and excellent photostability, their application in biological probing and labeling has surged. Phage T7 was engineered to display a biotinylation peptide on its major capsid protein. Upon infection, progeny phage synthesized within the *E. coli* cell became biotinylated. Once released from the cell during lysis, these masses of biotinylated phage could be detected by a streptavidin-functionalized quantum dot designed to attach to and label only biotinylated phage. If no suitable host cell was present, then the infection and release of biotinylated phage did not occur and the quantum dot labels were simply washed away. Using fluorescence microscopy, a single quantum dot conjugated phage could be visualized. Flow cytometry permitted high-throughput quantification. Detection limits approached 10 cells ml^{-1} within an assay time of 1 h. With the availability of a wide variety of quantum dots with different emission spectra, the potential for evolving this assay towards a multiplexed format for simultaneous detection of multiple bacterial targets is highly feasible.

7.3.2 Reporter Phage

Reporter phage are genetically designed to carry a reporter gene that is then transferred to the bacterial host upon infection. Originally developed for the fundamental analysis of factors affecting gene expression, reporter genes simply encode for a protein that is easy to detect and assay. Applicable examples include the bioluminescent detection of bacterial (*lux*) or firefly (*luc*) luciferase,

the fluorescent detection of green fluorescent protein (GFP), the chemiluminescent detection of β-galactosidase (*lacZ*) or ice crystal formation by the ice
nucleation gene (*inaW*).

7.3.2.1 lux Reporter Phage

Ulitzur and Kuhn[11] first demonstrated the use of reporter phage for the
detection of bacteria with their incorporation of the *Vibrio fischeri*-derived *luxA*
and *luxB* (*luxAB*) luciferase genes into phage λ Charon 30. The reporter phage
by itself cannot express the *luxAB* genes and therefore does not bioluminesce.
However, upon infection of *E. coli*, the phage genome, along with its accom-
panying *luxAB* genes, enters the host cell where it is transcribed and translated
to yield ultimately progeny phage and, in concert, the LuxAB protein. By
adding a solution of *n*-decanal, the LuxAB protein generates an immediate
burst of bioluminescent light that can be easily detected and measured using
any variety of photomultiplier tube (PMT)- or charge-coupled device (CCD)-
based instruments. Thus, simply by assaying for the emission of a light signal,
one can conclude that an infection event has occurred and, therefore, that a
suitable host for that reporter phage was present. Using this *luxAB* reporter
phage, it was possible to detect between 10 and 100 *E. coli* cells ml^{-1} in milk or
urine within 1 h. Kodikara *et al.*[12] continued to demonstrate the utility of this
reporter phage under real-world conditions using swab samples from swine
slaughterhouse surfaces and carcasses. Swabs were stomached in small volumes
of nutrient media and combined with *luxAB* reporter phage and *n*-decanal
substrate. Bioluminescence could be detected in these samples within 1 h at a
detection limit of 10^4 cells cm^{-2} or g^{-1}. With a 4 h enrichment, detection limits
increased to 10 cells cm^{-2} or g^{-1}.

Several other *luxAB*-incorporated phage reporters have since been developed
for a variety of pathogens. Using transposon mutagenesis, Waddell and
Poppe[13] inserted the *luxAB* genes into phage ΦV10 for the specific detection of
E. coli O157:H7 within 1 h. Chen and Griffiths[14] constructed several P22-based
lux reporter phage specific for the A, B and D$_1$ *Salmonella enterica* serotypes.
Salmonella isolates in pure culture were detected down to 10 cfu ml^{-1} after a 6 h
enrichment. Thouand *et al.*[15] further evolved the assay to a more commercially
viable format consisting of four steps: (1) addition of the *luxAB* reporter phage
to a 12–14 h enriched culture, (2) a 1 h incubation sufficient to allow contact
between phage and host *Salmonella*, (3) addition of fresh growth medium
followed by a 2–4 h incubation and (4) addition of the *n*-decanal reagent and
measurement of light. The assay was performed on various poultry feed, feces
and litter samples artificially inoculated with *Salmonella* Typhimurium. Within
the approximate 16 h assay format, bacterial concentrations enriched above 10^6
cfu ml^{-1} could be detected *via* bioluminescent light emission. Although likely
not to be considered a rapid assay due to the requisite enrichment step, it does
surpass the normal 24 h or greater incubation times required by standard
selective plating methods and would ultimately be a very straightforward

hands-off assay if adapted to robotic liquid handling integration. In another use of this reporter phage, Chen and Griffiths[14] injected *Salmonella* Enteritidis directly into poultry eggs at a lower inoculum of 63 cfu ml^{-1}. After a 24 h incubation, *luxAB* reporter phage and the *n*-decanal substrate were injected and the entire egg was imaged under a BIQ Bioview Image Quantifier (Cambridge Imaging, Cambridge, UK). This not only permitted *in situ* detection of *Salmonella*, but also visually defined its location within the egg. Referred to as bioluminescent imaging (BLI), this detection technology has its main focus in medical diagnostic imaging, where luciferase expressing reporter cell lines can be visualized and tracked in real time directly within a living organism to monitor, for example, the spread of a bacterial or viral infection or the response of that infection to different treatment regimens.[16] Since natural biolumines-cence in living tissue is virtually non-existent, there is minimal background noise to contend with and optical signals can be detected very precisely. The vast majority of environmental, food and water samples similarly yield very low or no background bioluminescence, making BLI techniques highly applicable towards luciferase reporter phage technologies and the real-time sensing, monitoring and tracking of pathogens in a variety of sample matrices, although expensive imaging instrumentation currently limits these applications.

A *luxAB* reporter phage for *Listeria monocytogenes* was developed by Loessner *et al.*[17] in phage A511, which infects approximately 95% of the ser-ovars responsible for human listeriosis. In this case, the *luxAB* genes were derived from *Vibrio harveyi* and were linked to a strong promoter for the phage's major capsid protein. *L. monocytogenes* was artificially inoculated into several food sources (various cheeses, chocolate pudding, cabbage, lettuce, ground beef, liverwurst, milk and shrimp) from 0.1 to 1000 cfu g^{-1} and, after a 20 h enrichment, combined with A511::*luxAB* reporter phage and *n*-decanal substrate. In most foods tested, detection limits approached one to less than one cell g^{-1}. Foods with more complex microbial background flora, such as ground beef, yielded detection limits of 10 cells g^{-1}. A total of 348 naturally contaminated meats, poultry, dairy products and other environmental samples were also assayed and positive results corresponded to positives obtained with traditional plating methods. Considering that plating requires 72–96 h, the less than 24 h A511::*luxAB* reporter phage assay demonstrates significant efficiency with comparable results.

Using *lux*, but in a uniquely different manner, Ripp *et al.*[18] developed a phage reporter assay capable of continuous real-time light emission. The *luxAB* reporters so far described do not generate bioluminescence until the requisite *n*-decanal substrate is added, and, once added, produce only a single time point burst of light. In addition to *luxAB*, the *lux* operon additionally consists of the *luxC*, *luxD* and *luxE* genes that in the wild-type *V. fischeri* or *V. harveyi* cells are responsible for synthesizing the *n*-decanal (aldehyde) substrate that in con-junction with cellular FMNH$_2$ produces self-generated bioluminescence in these cells.[19] Reporters that carry this full operon, referred to as *luxCDABE*, therefore do not require *n*-decanal addition to produce bioluminescence, but rather produce bioluminescence naturally and continuously. With proper genetic manipulation, the bioluminescent response can be controlled to turn on and off based on a user-defined target. Due to the small head sizes of phage and the limited size of DNA that can be packaged within these heads, it has not yet been possible to insert the entire *luxCDABE* gene cassette into a phage genome, which is why *lux*-based phage reporters carry only the smaller *luxAB* segment. Bacterial cells, however, can more easily be integrated with the complete *lux-CDABE* operon, resulting in the creation of numerous bacterial bioluminescent bioreporters for environmental sensing and gene expression studies.[20] Ripp's group first constructed a *luxCDABE*-based bacterial bioluminescent bioreporter, referred to as OHHLux, that bioluminesced in the presence of the quorum sensing signaling compound *N*-3-(oxohexanoyl)-L-homoserine lactone (OHHL). OHHL is generated by another *lux* gene, *luxI*, which was inserted into the genome of phage lambda behind a strong left arm promoter. When this recombinant lambda reporter phage infects an *E. coli* cell, the host *E. coli* cell expresses the *luxI* gene and begins synthesizing OHHL. OHHL diffuses out of the *E. coli* cell into the surrounding medium where it interacts with the OHHLux bioluminescent bioreporter and triggers it to autonomously generate bioluminescence (Figure 7.3A). Therefore, the final bioluminescent signal is linked to and indicates the initial phage/host infection event. When tested in pure culture, the assay could detect *E. coli* at $1\,\mathrm{cfu\,ml^{-1}}$ within $\sim 10\,\mathrm{h}$. In washings from artificially contaminated leaf lettuce, a detection limit of $\sim 100\,\mathrm{cfu\,ml^{-1}}$ was

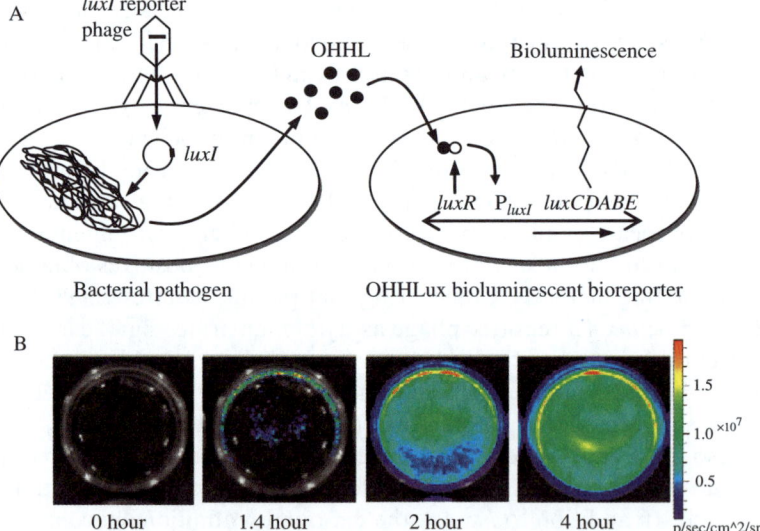

Figure 7.3   (A) The *luxI* reporter phage detection scheme uses diffusible OHHL quorum signaling molecules and a responsive *luxCDABE*-based bioluminescent bioreporter to identify the phage/host infection event. This permits real-time *in situ* sensing of bioluminescence, as depicted in (B), where a spinach leaf rinsate artificially contaminated with *E. coli* O157:H7 is imaged over time to yield escalating light emissions.

achieved in 22 h. Since no user manipulation is required, besides combining the phage reporter and bioluminescent bioreporter with the sample, these assays are extremely straightforward. The assay was later defined to the specific detection of *E. coli* O157:H7 through similar *luxI* insertion into phage PP01.[21] Detection limits were obtained in artificially contaminated apple juice (1 cfu ml^{-1} within 22 h) and tap water (1 cfu ml^{-1} within 12.5 h). In addition, bioluminescent imaging was used to monitor light emission continuously from artificially contaminated spinach leaf rinsate placed under a Xenogen IVIS CCD camera, where *in situ* real-time detection of *E. coli* O157:H7 at 1 cfu ml^{-1} was obtained within 4 h after a 2 h enrichment (Figure 7.3B).

7.3.2.2 *luc Reporter Phage*

The *luc* genes were first isolated from *Photinus pyralis* (firefly) and encode for the eukaryotic luciferase that catalyzes the two-step conversion of D-luciferin (which, like the *luxAB n*-decanal substrate, must be added to the reaction) to oxyluciferin with bioluminescent emission at 560 nm. (Due to a mode of action similar to that of bacterial *lux*, the firefly *luc* gene is sometimes denoted *fflux*.) The application of *luc* as a reporter phage signaling element has primarily been associated with *Mycobacterium*-specific phages (TM4, L5 and D29) and the detection and assessment of drug susceptibility in *M. tuberculosis*.[22,23] Susceptibility to anti-mycobacterial drugs can be determined by comparing the bioluminescence output kinetics from reporter phage added to antibiotic-free or antibiotic-amended *Mycobacterium* cultures. If the drug is effective, fewer host *Mycobacterium* cells are available for phage infection and less light is emitted compared with the antibiotic-free control. Conventional antibiotic susceptibility tests require several weeks to complete, whereas *luc* reporter phage can obtain results within several days. Using *luc* reporter phage phAE142, Bardarov *et al.*[24] assayed 20 sputum samples both to detect the presence of *M. tuberculosis* and to assess antibiotic susceptibility. For detection, reporter phage phAE142 exhibited earlier detection at inocula greater than 10^4 cfu ml^{-1} (median 7 days) whereas standard Mycobacterial Growth Indicator Tube (MGIT) assays performed better at lower concentrations (median 9 days). For antibiotic susceptibility testing, results could be obtained in 3 days using reporter phage whereas 12 days were required using standard tests. Recognizing the simplicity of these phage reporter assays and the prevalence of tuberculosis in developing and under-developed countries, Riska *et al.*[25] developed a cost-effective Polaroid film-based device for recording bioluminescence output. Referred to as the Bronx Box, the device simply integrates photographic film with a multi-well microtiter sample plate. Exposed spots of light on the film correspond to wells containing *luc* reporter phage and viable, infectable *M. tuberculosis* cells (*i.e.*, cells unaffected by the antibiotic). Using clinical isolates, there was a 100% correspondence in antibiotic susceptibility profiles between the Bronx Box and standard laboratory methods, with the Bronx Box providing results in 94 h compared with 3 weeks for the standard methods.[26]

7.3.2.3   GFP Reporter Phage

The green fluorescent protein (GFP) gene, derived from the jellyfish *Aequorea victoria*, encodes a 238 amino acid protein that, when activated by cyclization of a tyrosine (Tyr66) residue, fluoresces at approximately 508 nm after excitation by ultraviolet or blue light. Although used extensively as a reporter gene in prokaryotic and eukaryotic cells, its exploitation in phage reporter systems is surprisingly sparse. Funatsu *et al.*[27] first demonstrated GFP in phage lambda for the general detection of *E. coli*. Reporter phage, designated GFP-λ TriplEx, were combined with a mixed culture of *E. coli* and *Mycobacterium smegmatis* for 4 h and then observed under an epifluorescent microscope. The reporter phage-specific *E. coli* cells fluoresced whereas the non-phage-specific *M. smegmatis* cells did not, signifying that this reporter assay could distinctly identify target *E. coli* cells within a mixed culture environment (Figure 7.4).

Tanji *et al.*[28] constructed another *E. coli*-specific GFP reporter phage using phage T4. Since phage T4 is lytic, it lysis or destroys its host cell after infection, which is not a desired outcome when using host-derived fluorescence as a signal for host cell presence. They therefore inactivated the lytic activity of phage T4 by eliminating its ability to produce the lysozyme enzyme responsible for host cell wall degradation and cell lysis. Thus, this T4e⁻/GFP reporter phage does not kill the cells that it is designed to detect, thereby optimizing its detection limits. When added to a mixed culture of target *E. coli* cells and non-target *Pseudomonas aeruginosa* cells, T4e⁻/GFP phage-infected fluorescent *E. coli* cells could be located and identified within a 1 h assay. Miyanaga *et al.*[29] later used T4e⁻/GFP reporter phage to detect *E. coli* directly in sewage influent. Since the host range

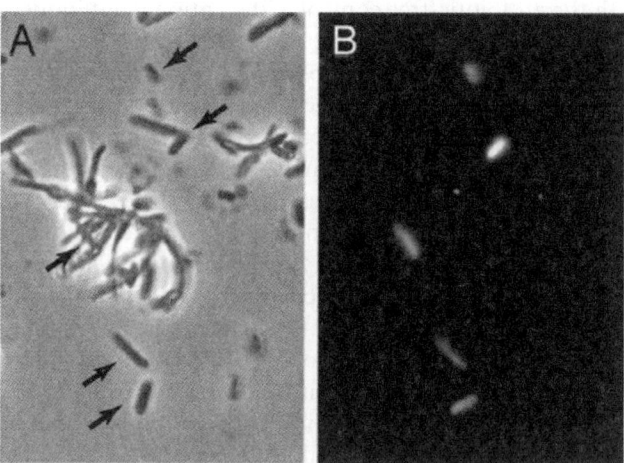

Figure 7.4   (A) Phase-contrast micrograph showing a mixed culture of *E. coli* and *M. smegmatis* cells (arrows indicate *E. coli* cells). (B) Fluorescent micrograph showing fluorescence emission only from *E. coli* cells after infection by reporter phage GFP-λ TriplEx. Reproduced from Funatsu *et al.*,[27] with permission.

and infectivity of the T4e⁻/GFP reporter phage were not inclusive of all *E. coli* encountered in the sewage samples, only a varied percentage ($\sim 8\%$) of the total *E. coli* population was detectable throughout the year-long study, which signifies the typical predicament that one encounters when relying on a single phage reporter to detect a wide range of bacterial targets. It is likely impossible to find a single phage with a host range sufficiently broad to infect the many targets desired and in most cases a suite of phage reporters with intersecting host ranges will have to be constructed. Namura *et al.*[30] went on to do this by isolating two other phage from sewage and endowing them with the GFP label to create lysozyme-inactivated reporter phage IP008e⁻/GFP and IP052e⁻/GFP. However, neither of these reporters generated sufficient fluorescence within infected *E. coli* cells, requiring a second insertion of GFP to create the phage reporters IP008e⁻/2×GFP and IP052e⁻/2×GFP. Together, these reporters demonstrated a host range covering nearly 50% of the *E. coli* sewage isolates. Actual environmental testing of these reporters has not yet been reported.

A much narrower host range GFP reporter phage was created by Oda *et al.*[31] using the *E. coli* O157:H7-specific phage PP01. The GFP gene was inserted into the phage genome by homologous recombination and used in mixed culture assays (*E. coli* O157:H7 and *E. coli* K12) where *E. coli* O157:H7 cells could be readily discriminated within 10 min based on fluorescence emission. To improve its sensitivity, phage PP01's lysozyme activity was also inactivated to create reporter phage PP01e⁻/GFP,[32] which showed clearer and sharper epifluorescent images that translated into easier and more direct identification of *E. coli* O157:H7 in mixed cultures. It was also possible to segregate healthy cells from stressed cells based on fluorescence output, with healthy cells generating bright green fluorescence due to high phage reporter replication rates and maximal expression of GFP whereas stressed cells yielded faded fluorescence due to the inability of reporter phage to propagate within these metabolically encumbered cells. This is an important attribute since it permits the simultaneous detection of healthy cells and cells within potential viable but non-culturable (VBNC) states. Conventional plating methods fail to enumerate VBNC cells and require supplemental time-consuming methods to do so.

7.3.2.4   *lacZ* Reporter Phage

The *lacZ* gene encodes a β-galactosidase enzyme that catalyzes the hydrolysis of β-galactosides. Traditionally, *lacZ* bioreporters were assayed colorimetrically by adding the substrate *o*-nitrophenyl-β-D-galactoside (ONPG) to permeabilized cells to generate a yellow by-product. Due to low sensitivities and narrow dynamic ranges, however, ONPG has largely been replaced with superior fluorescent, luminescent or chemiluminescent substrates. Goodridge and Griffiths[33] successfully inserted the *lacZ* reporter gene into phage T4 for biosensing of *E. coli*. *Via* addition of a chemiluminescent substrate to phage-infected *E. coli* cells, they were able to achieve detection down to 100 cfu ml⁻¹ in pure culture within 12 h.

7.3.2.5   Ice Nucleation Reporter Phage

The *inaW* gene from *Pseudomonas fluorescens* encodes for ice nucleation activity. When *inaW* is expressed, the InaW protein is incorporated into the outer cellular membrane and catalyzes ice formation at temperatures between –2 and –10 °C. As a reporter gene, therefore, the expression of *inaW* can be linked to ice formation at supercooled temperatures. Wolber and Green[34] inserted the *inaW* gene into *Salmonella* phage P22 to create the BIND (Bacterial Ice Nucleation Diagnostic) assay. When this reporter phage infects *Salmonella*, the *inaW* gene is expressed. Samples are cooled to –9.5 °C and examined for freezing, with those samples containing host *Salmonella* cells freezing whereas samples devoid of *Salmonella* do not freeze. An indicator dye turns orange if freezing occurs and fluorescent green if it does not, thereby providing a simple colorimetric or fluorescent endpoint to the assay. Testing in artificially contaminated food matrices such as raw egg and milk indicated detection limits of less than 10 cells ml^{-1}. The assay was further improved by inclusion of a modified most probable number statistical protocol and immunomagnetic separation to enhance detection limits to less than 5 cells ml^{-1} in a 3 h assay.[35] The BIND assay was sold as a commercial kit but is no longer marketed.

7.3.3   Phage Amplification

Rather than relying on the phage/host infection event essential for the above-described reporter phage assays, phage amplification exploits the ultimate outcome of phage infection, that is, cell lysis and the release of progeny phage. In its simplest application, wild-type lytic phage are added to a bacterial culture. If appropriate bacterial hosts are present, the phage will infect, reproduce within the cell and lyse the cell to release amplified numbers of new progeny phage. This burst of new phage synthesis or 'phage amplification' therefore indicates that host cells compatible with the phage's host range were present. Hirsh and co-workers[36,37] first demonstrated this technique in the early 1980s using phage Felix-01 to detect *Salmonella*. Addition of Felix-01 to a sample followed by an incubation period to allow for phage/host infection to occur resulted in increased numbers of phage if appropriate *Salmonella* hosts were present. This amplification in phage numbers was detected by high-performance liquid chromatography (HPLC). With an unpretentious detection limit of 10^6 *Salmonella* g^{-1} or ml^{-1} sample, in combination with the complexities of HPLC analysis, these assays never became popular. Over a decade later, Stewart *et al.*[38] demonstrated a much simpler and more sensitive phage amplification assay, again using phage Felix-01 for detecting *Salmonella* and also phages NCIMB 10116 and 10884 for detecting *Pseudomonas aeruginosa*. Phage were combined with bacterial cultures and incubated for up to 25 min to allow for infection to occur. Residual phage, *i.e.*, those that had not participated in an active infection, were then destroyed *via* the addition of a virucide, in this case, pomegranate rind. After 3 min, virucidal activity was neutralized and 'helper cells' were added to provide fresh hosts for the new progeny phage to infect. A standard top agar plaque plate was then prepared, where each plaque ideally represented a single phage/host infection event. By enumerating plaques, an estimate of the number of phage present could be calculated. Within 4 h, this assay could detect as few as 600 *S. typhimurium* or 40 *P. aeruginosa* cells ml^{-1}. The advantage of phage amplification assays over reporter phage assays is that wild type phage can be used, so no up-front genetic engineering is required, which saves considerable time and expense. The disadvantage is that there is no universal virucide. Each phage/host combination must be pre-tested with any number of virucidal agents to ensure that the virucide inactivates free phage and not host bacterial cells. Additionally, as was later discovered with pomegranate rind, seasonal growth variations caused variability in virucidal efficacy. de Siqueira *et al.*[39] suggested the use of loose-leaf tea infusions as a more stable virucide in their assays with the *Salmonella* phages Felix-01 and P22.

Favrin *et al.*[40] bypassed the use of virucides altogether in their assays using phage SJ2 for the detection of *Salmonella* Enteritidis. *Salmonella* cells were first concentrated using immunomagnetic separation and then incubated with phage SJ2 for 10 min. Use of the paramagnetic beads consequently allowed the *Salmonella* cells to be easily washed and recovered while simultaneously removing free phage from the sample. Upon continued incubation, *Salmonella* cells infected with phage now lysed and released their phage progenies. Healthy *Salmonella* 'signal-amplifying cells' (SACs) were then added to serve as new hosts for the progeny phage and their subsequent infection could be monitored simply by measuring optical density, where a decrease in optical density indicated that signal-amplifying cell concentrations were declining due to infection and lysing by phage while an increase in optical density indicated an unaffected and growing population of signal-amplifying cells (Figure 7.5). The assay was performed on artificially contaminated skimmed milk powder, chicken rinses and ground beef, with an average detection limit of 3 cfu g^{-1} or ml^{-1} within a total assay time of 20 h, inclusive of requisite pre-enrichment incubations.[41] It has also been applied towards the detection of *E. coli* O157:H7 using phage LG1 and anti-*E. coli* paramagnetic beads, with a detection limit of 2 cfu g^{-1} ground beef within an assay timeframe of 23 h.[41]

Although paramagnetic beads negate the use of virucides, there is an associated cost to their application that makes them to some extent not cost effective. Jassim and Griffiths[42] developed a phage amplification assay that did not require their use, but rather relied on fluorochromic stains to detect target host cell viability associated with phage infection and lysis. The *Bac*Light Live/Dead assay (Molecular Probes) uses SYTO9 dye to stain viable cells fluorescent green and propidium iodide to stain dead cells fluorescent red. The propidium iodide will penetrate only those cells with damaged cell membranes, as would occur due to a phage infection event. Therefore, after performing the phage amplification assay, each dye is added to the sample and the ratio of green to red fluorescent cells is used to indicate the occurrence of phage infection and, therefore, the presence of target bacterial cells. Using phage NCIMB 10116,

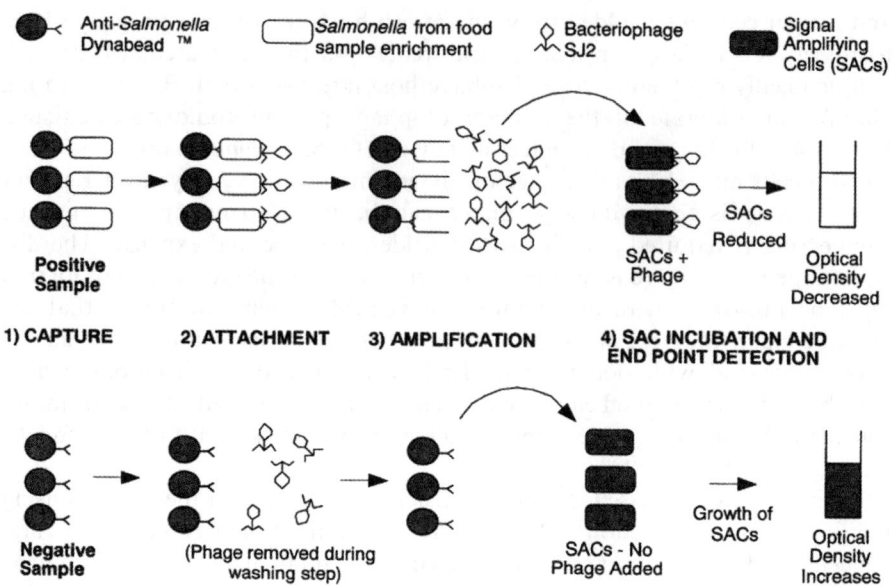

Figure 7.5   Phage amplification assay incorporating immunomagnetic separation. *Salmonella* cells, if present in the sample, are captured with polyclonal antibody-coated paramagnetic beads. The addition of *Salmonella*-specific phage leads to infection and cell lysis and the amplified release of progeny phage. The addition of healthy signal-amplifying cells provides new hosts for these phage and subsequent infection (or non-infection if *Salmonella* was not initially present) can be monitored by optical density. Reproduced from Favrin et al.,[41] with permission.

target *P. aeruginosa* cells could be detected down to 10 cfu ml^{-1} in pure culture within 4 h with no requisite pre-enrichment.

Ulitzur and Ulitzur[43] avoided altogether the removal of non-infecting phage in their assays by using novel phage mutant repair mechanisms to ensure that endpoint plaque formation was due only to infected target bacteria. Phage carrying amber mutations (phage Felix-01 for *Salmonella*), ultraviolet light-irradiated mutations (phage OE for *E. coli*) or temperature-sensitive mutations (phage AR1 for *E. coli* O157:H7) were engineered. These phage could not form plaques on their host cells unless their mutations were repaired by recombination or complementation, thereby negating the need to wash and/or centrifuge the assay samples to remove free phage. For example, two temperature-sensitive phage mutants were allowed to co-infect their *E. coli* O157:H7 targets at the permissive temperature (37 °C), then incubated at their restrictive temperature (42 °C) to prevent further infection cycles. Subsequent plaque formation was therefore only possible if the mutation had been repaired since any remaining mutant phage, due to their temperature sensitivity, could not plaque at 42 °C and the number of plaques thus reflected the number of *E. coli* host cells in the sample. Detection was achieved down to one cell ml^{-1} in a 3.5 h assay. Similar co-infection strategies with the other phage mutants yielded detection limits of 10 or less target cells ml^{-1} in 3–5 h assay formats.

Perhaps the most applied phage amplification assay is for the clinical detection of *Mycobacterium tuberculosis* using the commercially available *FASTPlaque*TB kit.[44] The phage (Actiphage) is added to the sample for 1 h followed by the addition of virucide (Virusol). Since these tests are highly standardized for this specific phage/host combination, the virucidal approach works exceedingly well. After a 5 min incubation, the virucide is neutralized and a rapid growing mycobacterial cell suspension is added (Sensor cells) to promote additional phage infection and amplification. Resulting phage are enumerated on plaque plates. Two large-scale studies verified the detection of 65–83% of confirmed *M. tuberculosis* infections in sputum samples within 2 days using this assay.[45] Although direct microscopic methods can achieve results in 2 h with corresponding sensitivity, the simplicity of phage amplification, especially when packaged in kit form, and the ability to perform these assays with no ancillary costly equipment and minimal user training, afford significant benefits, particularly in economically disadvantaged countries and low-resource environments. A FASTPlaque-Response kit is also available for establishing rifampicin resistant *M. tuberculosis*. The sample is pre-incubated in the presence or absence of the antibiotic rifampicin and then subjected to the phage amplification assay. If the cells are resistant to rifampicin, the number of plaques enumerated will be similar in both samples. If the cells are sensitive to rifampicin, the number of plaques in the rifampicin-treated sample will be smaller than that in the rifampicin-free sample. Susceptibility testing for various other anti-tuberculosis drugs (isoniazid, ethambutol, streptomycin, pyrazinamide, ciprofloxacin) can additionally be achieved.[44]

The amplified progeny phage end product of phage amplification assays has been merged with several atypical detection technologies. Madonna *et al.*,[46] for example, used matrix-assisted laser desorption/ionization mass spectrometry (MALDI-MS) to identify the molecular weight signature of the phage capsid protein. *E. coli* in pure culture was concentrated by immunomagnetic separation and then infected with phage MS2. Analysis of 1 µl of this sample by MALDI-MS was sufficient to detect the MS2 capsid protein, providing a detection limit of approximately 10^4 *E. coli* cells ml^{-1} in an assay time of 2 h. Guan *et al.*[47] combined phage amplification with a competitive enzyme-linked immunosorbent assay (cELISA) to detect *Salmonella* Typhimurium using phage BP1 and a biotinylated version of BP1. *Salmonella* cultures were incubated with wild-type BP1 phage and resulting phage supernatants added to ELISA microtiter plates coated with *Salmonella* Typhimurium smooth lipopolysaccharide (LPS) to which the phage attached. The biotinylated version of the phage was additionally added, which could be detected by the colorimetric substrate 3,3′,5,5′-tetramethylbenzidine peroxidase (TMB). If excess wild-type BP1 phage were present, due to the availability of suitable *Salmonella* host cells, then few biotinylated phage would attach and a weak colorimetric signal would be detected. If no target *Salmonella* were present, then BP1 replication would not occur and excess biotinylated phage would bind to the smooth LPS to yield an intense yellow color.

7.3.4   Electrochemical-based Sensing Assays

Electrochemistry measures the potential (potentiometric), current (ampero-metric) or conductivity (conductometric) of a solution. Phage-based assays that use electrochemistry typically take advantage of phage-mediated host cell lysis and the subsequent release of cellular constituents conducive to electrochemical measurements. The extracellular release of enzymes serves as a well applied example. Neufeld *et al.*[48] used a lytic version of phage lambda (λ_{vir}) to infect specifically *E. coli* cells. The cellular constituents released by *E. coli* after cell lysis include, among many other components, the enzyme β-D-galactosidase. The presence of β-D-galactosidase can be measured amperometrically with a potentiostat via the addition of the substrate *p*-aminophenyl-β-D-galactopyr-anoside (β-PAPG) to yield the product *p*-aminophenol, which is oxidized at the carbon anode. *E. coli* could be detected within 6–8 h at a detection limit of 1 cfu per 100 ml. Theoretically, any phage/host combination could be detected in this manner provided that the phage is lytic and an appropriate electrochemically detectable enzymatic marker is released by the target cell. *Bacillus cereus*, for example, was detected with phage B1-7064 using α-glucosidase as an intrinsic cellular enzyme, as was *Mycobacterium smegmatis* using the enzyme β-gluco-sidase and the phage D29.[49] However, in these assays, a cocktail of phage would likely be needed to infect adequately across the target bacterial genera, species or strains desired, and false negatives from cross-infections or naturally lysing cells would need to be accounted for. In later work, this group addressed some of these specificity problems by inserting the marker enzyme (DNA encoding for alkaline phosphatase) into the phage.[50] Now, only when the phage actually participates in a host cell infection event will the enzyme be synthe-sized. The exogenously added substrate *p*-aminophenyl phosphate enters the phage-infected cell and reacts with alkaline phosphatase to form the electro-chemically detectable product *p*-aminophenol. Thus, minimal assay steps are required beyond simply mixing the alkaline phosphatase incorporated phage with the bacterial cell culture. This assay detected one *E. coli* cfu ml^{-1} in less than 3 h in pure and mixed cultures.

The SEPTIC (SEnsing of Phage-Triggered Ion Cascade) assay uses electro-chemistry to measure microscopic voltage fluctuations occurring after a phage injects its nucleic acid into its host cell. Immediately after injection, the bacterial host will emit approximately 10^8 ions into the surrounding medium and this release can be measured with two thin metal film microelectrodes. Since wild-type phage are used, no *a priori* phage manipulations are needed and, given that infection is one of the earliest phage/host signals available, detection can occur seconds after combining the phage with its host. Dobozi-King *et al.*[51] demonstrated the assay using *E. coli* as the target cell in a 5 µl nanowell sensor chip.[52] Although in these experiments *E. coli* was detected at 10^7 cfu ml^{-1}, a theoretical detection limit of 1 cfu ml^{-1} was hypothesized based on potential improvements in fluid conductivity and/or reductions in thermal noise.

Growth in a microbial culture can be monitored electrochemically by mea-suring changes in electrical parameters occurring as complex growth media substrates are broken down into smaller highly charged molecules such as acids. Chang *et al.*[53] hypothesized that the presence of phage within a bacterial culture, provided that suitable host cells were present, would impede culture growth, and by comparing conductance measurements between phage-supplemented and phage-free samples or between phage-specific and non-phage-specific bacterial cultures, one could straightforwardly screen samples for the presence of phage-specific pathogens. Phage AR1 and its *E. coli* O157:H7 host were used to demonstrate the technique. Pure cultures of *E. coli* O157:H7 or non-O157:H7cells at 10^6 cfu ml^{-1} with or without phage AR1 addition were placed in test-tubes fitted with platinum electrodes and conductance measurements were taken every 6 min over 24 h. The resulting conductance curves could discriminate between *E. coli* O157:H7 and non-O157:H7 cultures within the 24 h period. Samples other than pure culture will need to be tested to verify fully the functionality of this assay and, again, a cocktail of phage would likely be required, but in terms of simplicity, the assay does show promise.

7.3.5 Surface Plasmon Resonance-based Sensing Assays

Spreeta is a miniaturized biosensor marketed by Texas Instruments that uses surface plasmon resonance (SPR) to detect changes in refractive index due to receptor–ligand binding interactions. Generally, the receptor consists of an antibody designed to bind to the antigenic ligand of a bacterial pathogen. Recognizing the similar capacity of phage to capture specific bacterial targets, Balasubramanian *et al.*[54] physically adsorbed phage 12600 on the gold surface of a Spreeta sensor for SPR-based detection of *Staphylococcus aureus*. In a flow-through format, *S. aureus* cells were pumped across the Spreeta channel, where they contacted and attached to the immobilized phage. Non-phage-specific bacterial cells, in this case *Salmonella* Typhimurium, simply passed through the device and were deposited as waste. Integrated software monitors the changes in refractive index, corrects for signal noise and displays the results. Using this set-up, detection of 10^4 *S. aureus* cells ml^{-1} could be achieved in near-real time. Although this compares poorly with the detection limits of other assays, this SPR-based phage recognition platform is still in its initial developmental stages and will certainly improve as it is refined.

7.3.6 The Phage-mediated Adenylate Kinase Assay

The adenylate kinase enzyme is present in virtually all cells and reversibly moderates the conversion of ADP to ATP and AMP ($2ADP \leftrightarrow ATP + AMP$). Its release from a cell can be used in conjunction with bacteriophage for identification purposes. The *E. coli*-specific phage NCIMB 10359 and the *Salmonella*-specific Newport phage were used by Blasco *et al.*[55] to detect these bacterial species. Upon infection, cells lysed and released adenylate kinase. ADP was then added to drive the above reaction towards the generation of ATP and subsequent ATP pools were detected with a firefly luciferase assay. Approximately 10^4 cfu ml^{-1} of *E. coli* or *Salmonella* could be detected in less than 2 h. Wu *et al.*[56] later optimized the assay incubation times and phage concentrations to increase detection limits to 10^3 cfu ml^{-1}.

7.4 Conclusion

Phage-mediated identification of bacterial targets offers a rapid, sensitive and specific monitoring scheme that can be integrated with a wide assortment of detector and measurement technologies. Although this diversity provides numerous assay formats, every assay nonetheless relies on the same fundamental characteristics of phage/host specificity to achieve their ultimate detection endpoint. Considering the global abundance of phage with which we share this planet, appropriate phage/host specificity ranges should theoretically exist for any bacterial host desired. Although only a handful of bacterial hosts have actually been incorporated into phage-based assays, the infancy of these methods, many of which remain at the proof-of-concept stage, should not imply any practical bacterial detection limits. However, expecting a single phage to converge sufficiently across all bacterial genera, species, strains or serotypes required is likely unrealistic and it is clear that in most instances a cocktail of phage reporters will be needed. For phage requiring genetic manipulation, this may seem a daunting task, but the escalating accessibility of whole genome phage sequences provides the essential platform to do so reasonably straightforwardly. In addition, advancements in landscape phage engineering, random mutagenesis and DNA shuffling permit the synthesis of phage libraries tailored to a specific host bacterium, thus allowing virtually *any* bacterial target to be detectable by a customized phage. With applications ranging from the detection of pathogens in food and water sources to homeland defense and military-related biological threat identification to the *in vivo* diagnostic tracking of bacterial infections and disease states, the innovative simplicity, specificity and sensitivity of phage-mediated sensors cannot be overlooked.

References

1. H. W. Ackerman, *Arch. Virol.*, 2001, **146**, 843.
2. H. W. Ackerman, R. R. Azizbekyan and R. L. Bernier, *et al.*, *Res. Microbiol.*, 1995, **146**, 643.
3. M. J. Corbel, *Ann. Inst. Pasteur Microbiol.*, 1987, **138**, 70.
4. M. Heyndrickx, N. Rijpens and L. Herman, *Focus Biotechnol.*, 2001, **2**, 193.
5. P. A. Mackowiak, *Am. J. Med.*, 1979, **67**, 293.
6. L. Goodridge, J. Chen and M. W. Griffiths, *Int. J. Food Microbiol.*, 1999, **47**, 43.

7. L. Goodridge, J. Chen and M. W. Griffiths, *Appl. Environ. Microbiol.*, 1999, **65**, 1397.
8. T. Kenzaka, F. Utrarachkij, O. Suthienkul and M. Nasu, *J. Health Sci.*, 2006, **52**, 666.
9. P. A. Mosier-Boss, S. H. Lieberman, J. M. Andrews, F. L. Rohwer, L. E. Wegley and M. Breitbart, *Appl. Spectrosc.*, 2003, **57**, 1138.
10. R. Edgar, M. McKinstry and J. Hwang, *et al.*, *Proc. Natl. Acad. Sci. USA*, 2006, **103**, 4841.
11. S. Ulitzur and J. Kuhn, in *Bioluminescence and Chemiluminescence: New Perspectives*, ed. J. Scholmerich, R. Andreesen, A. Kapp, M. Ernst and W. G. Woods, Wiley, New York, 1987, p. 463.
12. C. P. Kodikara, H. H. Crew and G. S. A. B. Stewart, *FEMS Microbiol. Lett.*, 1991, **83**, 261.
13. T. E. Waddell and C. Poppe, *FEMS Microbiol. Lett.*, 2000, **182**, 285.
14. J. Chen and M. W. Griffiths, *J. Food Protect.*, 1996, **59**, 908.
15. G. Thouand, P. Vachon, S. Liu, M. Dayre and M. W. Griffiths, *J. Food Protect.*, 2008, **71**, 380.
16. S. H. Thorne and C. H. Contag, *Proc. IEEE*, 2005, **93**, 750.
17. M. J. Loessner, M. Rudolf and S. Scherer, *Appl. Environ. Microbiol.*, 1997, **63**, 2961.
18. S. Ripp, P. Jegier and M. Birmele *et al.*, *J. Appl. Microbiol.*, 2006, **100**, 488.
19. E. A. Meighen, *Annu. Rev. Genet.*, 1994, **28**, 117.
20. H. Harms, M. C. Wells and J. R. van der Meer, *Appl. Microbiol. Biotechnol.*, 2006, **70**, 273.
21. S. Ripp, P. Jegier, C. M. Johnson, J. Brigati and G. S. Sayler, *Anal. Bioanal. Chem.*, 2008, **391**, 507.
22. C. Carriere, P. F. Riska and O. Zimhony, *et al.*, *J. Clin. Microbiol.*, 1997, **35**, 3232.
23. G. J. Sarkis, W. R. Jacobs and G. F. Hatfull, *Mol. Microbiol.*, 1995, **15**, 1055.
24. S. Bardarov, H. Dou and K. Eisenach, *et al.*, *Diagn. Microbiol. Infect. Dis.*, 2003, **45**, 53.
25. P. F. Riska, Y. Su and S. Bardarov, *et al.*, *J. Clin. Microbiol.*, 1999, **37**, 1144.
26. M. H. Hazbon, N. Guarin and B. E. Ferro, *et al.*, *J. Clin. Microbiol.*, 2003, **41**, 4865.
27. T. Funatsu, T. Taniyama, T. Tajima, H. Tadakuma and H. Namiki, *Microbiol. Immunol.*, 2002, **46**, 365.
28. Y. Tanji, C. Furukawa, S. H. Na, T. Hijikata, K. Miyanaga and H. Unno, *J. Biotechnol.*, 2004, **114**, 11.
29. K. Miyanaga, T. F. Hijikata, C. Furukawa, H. Unno and Y. Tanji, *Biochem. Eng. J.*, 2006, **29**, 119.
30. M. Namura, T. Hijikata, K. Miyanaga and Y. Tanji, *Biotechnol. Prog.*, 2008, **24**, 481.
31. M. Oda, M. Morita, H. Unno and Y. Tanji, *Appl. Environ. Microbiol.*, 2004, **70**, 527.

32. R. Awais, H. Fukudomi, K. Miyanaga, H. Unno and Y. Tanji, *Biotechnol. Prog.*, 2006, **22**, 853.
33. L. Goodridge and M. W. Griffiths, *Food Res. Int.*, 2002, **35**, 863.
34. P. K. Wolber and R. L. Green, *Trends Biotechnol.*, 1990, **8**, 276.
35. P. Irwin, A. Gehring, S. I. Tu, J. Brewster, J. Fanelli and E. Ehrenfeld, *J. AOAC Int.*, 2000, **83**, 1087.
36. D. D. Crane, L. D. Martin and D. C. Hirsh, *J. Microbiol. Methods*, 1984, **2**, 251.
37. D. C. Hirsh and L. D. Martin, *Appl. Environ. Microbiol.*, 1983, **46**, 1243.
38. G. S. A. B. Stewart, S. A. A. Jassim, S. P. Denyer, P. Newby, K. Linley and V. K. Dhir, *J. Appl. Microbiol*, 1998, **84**, 777.
39. R. S. de Siqueira, C. E. R. Dodd and C. E. D. Rees, *Int. J. Food Microbiol.*, 2006, **111**, 259.
40. S. J. Favrin, S. A. Jassim and M. W. Griffiths, *Appl. Environ. Microbiol.*, 2001, **67**, 217.
41. S. J. Favrin, S. A. Jassim and M. W. Griffiths, *Int. J. Food Microbiol.*, 2003, **85**, 63.
42. S. A. A. Jassim and M. W. Griffiths, *Lett. Appl. Microbiol.*, 2007, **44**, 673.
43. N. Ulitzur and S. Ulitzur, *Appl. Environ. Microbiol.*, 2006, **72**, 7455.
44. R. J. Mole and W. O. C. Maskell, *J. Chem. Technol. Biotechnol.*, 2001, **76**, 683.
45. M. Pai, S. Kalantri, L. Pascopella, L. W. Riley and A. L. Reingold, *J. Infect.*, 2005, **51**, 175.
46. A. J. Madonna, S. Van Cuyk and K. J. Voorhees, *Rapid Commun. Mass Spectrom.*, 2003, **17**, 257.
47. J. W. Guan, M. Chan, B. Allain, R. Mandeville and B. W. Brooks, *J. Food Protect.*, 2006, **69**, 739.
48. T. Neufeld, A. Schwartz-Mittelmann, D. Biran, E. Z. Ron and J. Rishpon, *Anal. Chem.*, 2003, **75**, 580.
49. M. Yemini, Y. Levi, E. Yagil and J. Rishpon, *Bioelectrochemistry*, 2007, **70**, 180.
50. T. Neufeld, A. S. Mittelman, V. Buchner and J. Rishpon, *Anal. Chem.*, 2005, **77**, 652.
51. M. Dobozi-King, S. Seo, K. JU, R. Young, M. Cheng and L. B. Kish, *J. Biol. Phys. Chem.*, 2005, **5**, 3.
52. S. Seo, M. Dobozi-King, R. F. Young, L. B. Kish and M. S. Cheng, *Microelectron. Eng.*, 2008, **85**, 1484.
53. T. C. Chang, H. C. Ding and S. W. Chen, *J. Food Protect.*, 2002, **65**, 12.
54. S. Balasubramanian, I. B. Sorokulova, V. J. Vodyanoy and A. L. Simonian, *Biosens. Bioelectron.*, 2007, **22**, 948.
55. R. Blasco, M. J. Murphy, M. F. Sanders and D. J. Squirrell, *J. Appl. Microbiol.*, 1998, **84**, 661.
56. Y. Wu, L. Brovko and M. W. Griffiths, *Lett. Appl. Microbiol.*, 2001, **33**, 311.

CHAPTER 8

Genetically Engineered Virulent Phage Banks for the Detection and Control of Bacterial Biosecurity Threats

FRANÇOIS IRIS,[a] FLAVIE POUILLOT,[b] HÉLÈNE BLOIS,[b] MANUEL GEA[a] AND PAUL-HENRI LAMPE[a]

[a] BIO-MODELING SYSTEMS SAS, 26 rue Saint Lambert, 75015, Paris, France; [b] Pherecydes Pharma, Biocitech, Bâtiment lavoisier, 102 avenue Gaston Roussel, 93230, Romainville, France

8.1 Introduction

The therapeutic potential of obligate lytic (virulent) phage in the treatment of pathological bacterial infections has long been known and successfully utilized. However, in Western societies, the development of highly effective antibiotics, easy to produce and administer, that can quickly and efficiently control many life-threatening bacterial infections, rapidly made obsolete the more cumbersome phage therapies. With the advent of mass resistance to most antibiotics in a plethora of pathogenic bacterial strains, phage therapy is now regaining attention.[1] T4, the archetype of virulent phage, has evolved very efficient strategies to subvert host functions and impose the expression of its genome.[2] T4 has a myriad of relatives in Nature that differ significantly in their host range.[3] These phage detect and bind to outer membrane proteins of their hosts,

RSC Nanoscience & Nanotechnology No. 17
Phage Nanobiotechnology
Edited by Valery A. Petrenko and George P. Smith
© Royal Society of Chemistry 2011
Published by the Royal Society of Chemistry, www.rsc.org

such as the Omp transporters, via tail fibers bearing host-specificity determinants (adhesins) encoded by the genes *gp37* and *gp38*.[4] The amino acid sequences of the receptor recognition proteins encoded by these genes are critical for phage–host interactions, largely defining the host range of a phage. Thus mutations in these genes restrict or increase the range of bacterial hosts that a phage can detect and infect.[5,6] These genes appear as mosaics with parts derived from a common gene pool and different segments within each gene constitute host range cassettes consisting of two constant domains and one contiguous array of four hypervariable regions.[5–7] Exchanges involving these constant domains can replace the endogenous specificity determinants, thus mediating acquisition of determinants allowing a phage to cross species boundaries and infect taxonomically distant hosts.[8] Mutations affecting the host range of tailed -phage map to these regions and specific mutagenesis within the hypervariable domains has, in the past, been used on T4 phage active against *Escherichia coli* to generate host range mutants redirected towards *Yersinia pseudotuberculosis*.[9]

Bacteriophage are probably the most abundant biological entities on the planet[10] with a global population estimated at 10^{31}. It is simply not know when bacterial viruses evolved, that is, whether or not they pre-date modern bacteria and represent remnants of former cellular life-forms that lost the competition against the modern forms and persisted only as dependent parasites of modern bacterial life. However, there is ample evidence for continued exchange of genetic elements between phage, bacterial genomes and various other mobile genetic elements. This may explain the sometimes fuzzy distinction between phage, plasmids and pathogenicity islands and the chimeric nature of some phage.[11]

Phage are known in over 140 bacterial genera. They occur in archaea and eubacteria, in cyanobacteria, in exospore and endospore formers, spirochetes, mycoplasmas and chlamydias, in aerobes, anaerobes, budding, gliding, ramified, sheathed and stalked bacteria. Podovirus particles have even been found within bacterial endosymbionts of paramecia.[12] Tailed phage appear to be monophyletic and constitute the oldest known virus group, representing at least 96% (4950 phage) of all known bacteriophage. Of these phage, about 80% possess dsDNA genomes packaged within an icosahedral head connected to a long (~ 50–$200\,$nm) tail.[13]

As a result of their continuous co-evolution with phage, bacteria have evolved numerous strategies allowing them to resist phage predation efficiently, even when several different phage are used concurrently.[14] Among these strategies, the accumulation of mutations affecting the structure and the expression levels of outer membrane transporters figures prominently.[15] Furthermore, the majority of bacterial pathogens contain prophage or phage remnants integrated into the bacterial DNA, many of which encode virulence factors that contribute to protecting the cell from phage predation.[16] Indeed, toxins of *Corynebacterium diphtheriae* (diphtheria), *Clostridium botulinum* (botulism), *Streptococcus pyogenes* (scarlet fever), *Staphylococcus aureus* (food poisoning) and *E. coli* (Shiga toxin), are all phage encoded, although, as far as is known, these genes do not play a role in the lifecycle of the phage.[11,12] The list of phage-encoded fitness factors is growing rapidly and involves a wide range of different genes that include ADP–ribosyl transferase toxins, superantigens, lipopoly-saccharide-modifying enzymes, type III effector proteins, detoxifying enzymes, hydrolytic enzymes and proteins conferring serum resistance. In exceptional cases, phage tail genes appear to have developed dual functions and also serve as adhesion proteins for bacterial host attachment (*Streptococcus mitis pblA* and *pblB* genes).[11]

Considering the vast numbers of phage and bacteria coexisting in environments such as ocean water and sediments, it has been estimated that, in these niches alone, some 10^{25} phage infections are initiated every second worldwide,[17] and this has probably occurred for the last 3 billion years.

Thus, just as with antibiotics, the rapid emergence of bacterial resistance in response to recurrent predation by a same group of phage remains a major biosecurity issue.[18] The counter-measures most frequently adopted have consisted in attempting to isolate, from natural environments, new phage capable of infecting and destroying resistant bacterial populations.[14] However, this takes considerable time and effort (generally of the order of at least 1 year). Other avenues envisaged have been the use of filamentous and also obligate lytic phage engineered to express enzymes that repress the host's DNA repair machinery (SOS DNA repair system), hence inhibiting the development of resistance to antibiotics while enhancing the lethal effects of these drugs,[19] or that degrade biofilms, thus locally enhancing bacterial lysis.[20] However, although certainly presenting undeniable merits, these approaches perpetuate a situation where counter-measures always lag well behind the appearance of resistance, since these approaches can only be implemented once the occurrence of the undesirable phenomenon, known to be inevitable, has become evident. Indeed, in each case, the bacteria targeted must first have been characterized at both structural and molecular levels. Filamentous phage can only address flagellate cells and bacteria possess numerous mechanisms that rapidly counteract SOS inhibition to maintain highly efficient DNA repair and recombination systems accommodating horizontal genetic transfers,[21] while the option of using lytic phage must be initiated from a virus that can adsorb on and replicate in the target cells. This not only leaves the emergence of bacterial resistance to phage predation unaffected, but also recent studies have demonstrated that a high adsorption rate is detrimental to phage effectiveness in biofilm environments,[22] hence favoring passive effects (relying on self-replication to bring the phage concentration above the bacterial 'inundation threshold' at the site of infection), which can only be effective if both target bacteria and infective phage concentrations are low.[23]

An alternative would be the production of large, genetically engineered, virulent phage populations from which particles capable of defeating any bacterial resistance strategy could be rapidly isolated. Moreover, provided that such phage banks can be produced, they need not be limited to addressing Gram-negative bacteria only. Phages targeting Gram-positive cells could also be utilized. Hence, after thorough screening against currently known bacterial pathogens (be they natural or engineered), the availability of such phage banks could provide the means both to detect rapidly the presence of and to treat topical infections by any of the known pathogens and also by previously uncharacterized emergent strains, provided that a sample can be obtained for rapid screening against the banks.

We utilized systems biology to model both the mechanisms indispensable for virulent T4 phage infection and replication concurrently with those the host needs to implement to resist T4 phage predation. The models demonstrated that the bacterial host will try anything to escape predation and we have no idea what will be the successful strategy. Furthermore, this strategy is likely to vary between locations (populations) for the same host. The best-fit solution required the use a stochastic approach in the genetic engineering of the phage's host recognition mechanisms (Figure 8.1). As a result, three different novel technological approaches would be required for the production of genetically engineered virulent phage banks. First, a technology (TAPE) allowing one to modify concurrently and stochastically multiple coding domains within any

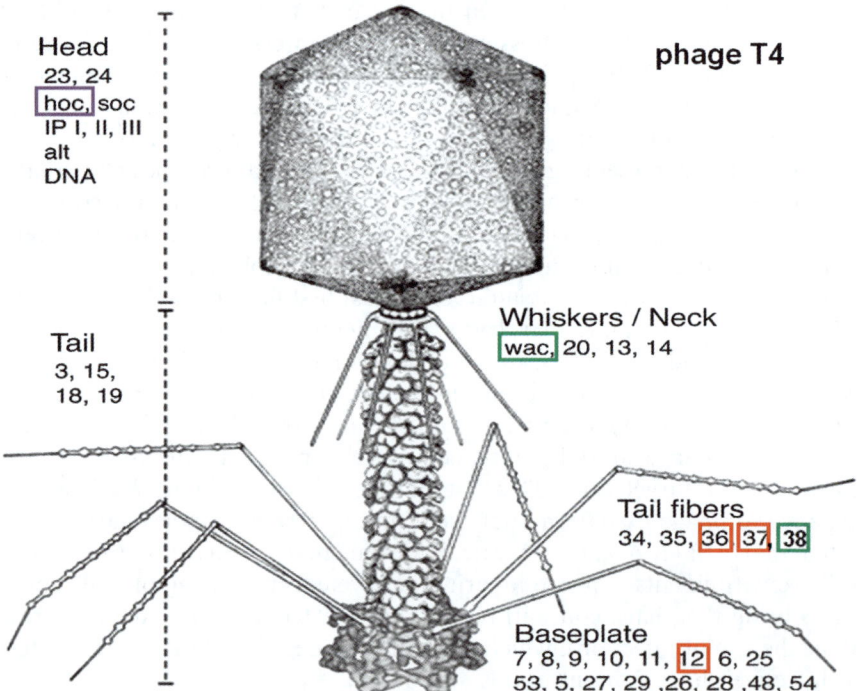

Figure 8.1   Schematic representation of the T4 bacteriophage. The genes encoding structures directly and indirectly involved with host range properties are boxed in red and green, respectively. The hoc gene (boxed in purple) encodes a capsid-specific protein that can be modified for detection purposes, thus allowing real-time monitoring of target detection and phage progeny bursts sizes (detector–killer principle).

gene, such as those critical for phage–host interactions, while preserving intact multiple coding regions within the same gene would be needed to generate a multitude of domain-targeted variants. Then, another technology (AB-Accus) allowing one to recombine efficiently and stochastically, at a population level, the multitude of variants generated for at least two genes into the genomes of the members of a wild-type phage population in replacement of the native sequences would be necessary. Finally, a third technology (RIPh) allowing one to interrupt reversibly the lytic cycle of an obligate virulent phage within its host would be indispensable to allow the above homologous recombinations to be performed at high efficiency so that, upon reinstatement and completion of the lytic cycle, each recombinant parental particle would produce a progeny different from that of most other recombinant particles.

8.2 Host Range engineering

To produce the widest possible range of functional host range variants, large-scale stochastic modifications will have to be introduced into the variable regions that constitute the host range cassettes in the *gp37* and *gp38* genes. This implies that, in each gene, X regions must be simultaneously targeted and stochastically modified, at Y different sites, in Z different manners, while conserving intact N different domains, all this independently and concurrently. To achieve this efficiently, we developed a nested polymerase chain reaction (PCR)-based technology, named TAPE (for Targeted Accelerated Protein Evolution), that utilizes the interplay between error-prone and high-fidelity reactions to introduce random point mutations into a sequence while preserving the identity of predefined inner segments. The modifications were introduced by (1) decreasing the fidelity of the Taq polymerase during DNA synthesis without significantly decreasing the level of amplification achieved in the PCR reaction, and, (2) the random incorporation of nucleotide analogs. Different regions can be subjected to different mutation rates depending on the application. This is followed by a series of selective high-fidelity reactions that reconstruct a wide spectrum of variant forms of the genetic region of interest while automatically eliminating the products containing mutations within the domains selected to remain constant. The T4 genes *gp37* and *gp38* were used to develop and validate the method described in Figure 8.2.

The DNA sequence to be modified is first analyzed to define the domains that must be conserved (Blast and CD searches: http://www.ncbi.nlm.nih.gov/sutils/static/blinkhelp.html and http://www.ncbi.nlm.nih.gov/Structure/cdd/cdd.shtml). These are then used to construct overlapping PCR primers. The gene or genetic region to be modified is then isolated from the corresponding genome using high-fidelity PCR amplification. This PCR product is subsequently utilized as a template in a series of error-prone PCR reactions (Figure 8.2A) addressing different variable domains bracketed by primers corresponding to constant domains or parts thereof. Each PCR reaction can be independently adjusted to different theoretical mutation rates (1.3–15%; Table 8.1a and b).

A   **Step 1:** gp38 High fidelity amplification and mutagenesis

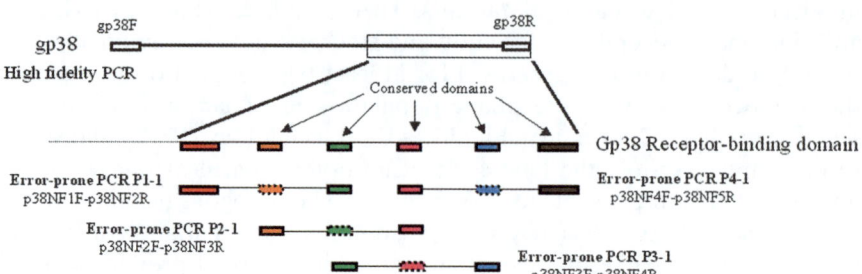

B   **Step 2A:** Selective high fidelity amplification of desired fragments

This sub-step selectively amplifies the fragments where the external priming sites ▬ , ▬ , ▬ , ▬ and ▬ have been largely preserved (mutation-free).

C   **Step 2B:** Selective high fidelity reconstruction of desired fragments.

In this sub-step, fragments that terminate with a priming site corresponding to the internal conserved domain found in other fragments serve as primers for the selective amplification of the fragments where the internal conserved domains ▬ , ▬ , ▬ and ▬ have been preserved (mutation-free).

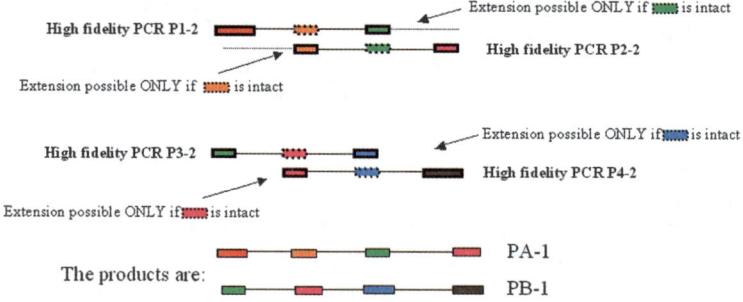

D   **Step 3:** Reconstruction of pg38 RBD via high fidelity PCR.

In this step, the PA-1 and PB-1 fragments serve as extension primers for each other and result in multiple reconstituted Gp38 RBD variants in which all the « conserved » sequences have been preserved while the variable regions have been extensively mutated.

The resulting PCR products are then recovered and individually re-amplified using high-fidelity reactions (Figures 8.2B and 8.3A and C1), the purposes of which are to 'lock in' the mutations introduced while ensuring that the fragments extremities (priming sites) have been conserved intact. Only the templates in which the priming sites have been conserved largely mutation free will be efficiently amplified. The PCR products are then purified and fragments with overlapping extremities are utilized for selective reconstruction of the entire genetic region consisting of randomly variable domains interspersed between constant domains (Figures 8.2C and 8.3B and C2). In this high-fidelity PCR reaction, each fragment serves as the primer for its partly overlapping counterpart and dual extension (partial reconstruction) will be possible only if the overlapping segments (conserved domains or parts thereof) have nearly intact complementary sequences. Different reconstruction schemes (Figure 8.2D) were tested. For both

---

Figure 8.2   Schematic representation of the 'TAPE' procedure (see also text). (A) Step 1. The gene to be stochastically modified is first isolated by PCR amplification. Comparative sequence analysis with other isoforms is used to identify the positions of conserved domains and/or variable regions. PCR primers corresponding to entire conserved domains or the extremities thereof are then used in nested, error-prone PCR reactions generating fragments embodying two inner variable regions and two outer conserved domains. This approach allows to compensate for the fact that stochastic mutagenesis is not very efficient over very short regions. With each successive PCR cycle, mutations will be stochastically and incrementally introduced anywhere over the entire length of each fragments. (B) Step 2A. It is now necessary to eliminate the products carrying mutations in the conserved domains while starting reconstruction, in the correct order, of the entire coding sequence. This is achieved in two steps. A high-fidelity PCR reaction, using the outer conserved domains as primers, is carried out to amplify selectively, in each of the mutagenesis PCR reactions products, the fragments where the outer priming sites have been largely preserved mutation free. This high-stringency reaction uses at least 25 PCR cycles, leading to the elimination by dilution of most undesirable fragments generated by each mutagenesis reaction. (C) Step 2B. The amplified products are recovered and overlapping fragments, in which one outer conserved domain in one fragment corresponds to the inner conserved domain in its overlapping counterpart, are then used in high-fidelity PCR reactions where each fragment serves as the primer for its overlapping counterpart. In this reaction, where the participating fragments all have largely intact outer conserved domains, only fragments where the inner conserved domain has been preserved largely mutation free will allow effective priming and subsequent exponential amplification of the merged products. This constitutes a negative selection step where undesirable fragments gradually and automatically exclude themselves from inclusion into partly reconstructed and amplified gene segments containing four conserved domains interspersed between three variable regions. (D) Step 3. Overlapping partly reconstructed fragments generated in the previous step are now sequentially utilized for the final reconstruction of the entire coding sequence, generating a large variety of constructs (Step 4), all presenting identical structures and successions of conserved domains interspersed between a large spectrum of stochastically modified variable regions.

Table 8.1 Comparative efficiencies of the two mutagenesis methods used. The mutagenesis efficiencies arising from the random incorporation of nucleotide analogues (0.5 mM dPTP; method 2) were compared with those achieved by decreasing the fidelity of the Taq polymerase (method 1). For this comparison, the five overlapping hypervariable domains of *gp38* fragments (P1–P5) encoding the C-terminus receptor-recognition domain of this tail fiber protein (a) and the first two (P1 and P2) of the 15 overlapping fragments constituting the *gp37* gene (b) were independently subjected to a PCR reaction made error-prone by either of the two methods. The resulting products were then sequenced ($n = 4$) and the frequencies of mutated sites recorded in terms of percentages. Using 15 error-prone PCR cycles, the random incorporation of nucleotide analogues produces much higher levels of mutagenesis than the destabilization of Taq fidelity [averages of $7.7 \pm 2.6\%$ *versus* $1.6 \pm 1.5\%$, respectively (a)]. However, the results obtained with the latter approach appeared considerably more stochastic (standard error as large as the mean) and therefore difficult to control. Although neither method appeared particularly dependent upon the lengths of DNA fragments (a), mutagenesis by random incorporation of nucleotide analogues is highly sensitive to the number of error-prone PCR cycles (b), thus allowing control over the levels of mutagenesis through the modulation of both the concentrations of nucleotide analogs in the reaction and the number of error-prone PCR cycles.

(a) Comparative efficiencies (%) of the two mutagenesis methods used

Method	P1 (60 bp)	P2 (75 bp)	P3 (27 bp)	P4 (258 bp)	P5 (33 bp)
1	1.7 ± 2.3	1.3 ± 0.0	1.9 ± 2.6	1.6 ± 0.5	1.6 ± 2.2
2	7.5 ± 1.1	5.3 ± 0.0	9.3 ± 2.6	8.8 ± 3	7.6 ± 6.4

(b) Influence of PCR cycle numbers upon the rates (%) of mutagenesis using the nucleotide analogs method

Number of cycle	P1 (225 bp)	P2 (255 bp)
5	3.8 ± 0.3	3.7 ± 0.3
10	7.8 ± 0.4	7.7 ± 1.3
15	8.2 ± 0.3	7.8 ± 0.5
20	10.3 ± 1.9	9.0 ± 0.6
25	13.8 ± 1.3	11.8 ± 2.8
30	16.9 ± 0.6	13.2 ± 1.3

gp37 (2897 bp) and *gp38* (C-terminus, 583 bp) genes, batch reconstructions were successfully achieved using up to six fragments simultaneously (data not shown). However, we did not encounter limits using sequential schemes and versions of the *gp37* gene made to contain up to 13 conserved domains interspersed between variable regions were successfully obtained (Figure 8.3B). Reconstructed genes were then analyzed by batch sequencing to define the frequencies, positions and

types of mutations introduced in the variable regions and to verify the levels of sequence conservation within the constant domains. Peaks superpositions in sequencing chromatograms, indicating the presence of a mixture of different sequences, revealed the positions and frequencies of alternative nucleotides stochastically introduced in variable domains (Figure 8.4B and C). Peak superpositions were rarely observed in the conserved domains, their sequences remaining essentially identical with those of the corresponding regions in the wild-type *gp37* (Figure 8.3A, B and C) and *gp38* T4 genes, indicating that all had been largely preserved, irrespective of their numbers and lengths (Figure 8.3C). Hence TAPE appears to be a robust, trustworthy, easy to implement and to manipulate technology that can be applied for the large-scale stochastic diversification of multiple targeted domains within any known gene and protein coding sequence (enzymes, receptors, antibodies, *etc.*).

8.3 Production of a Genetically Engineered T4 Phage Bank with Vastly Increased Host Range

The very large diversification of T4 coding sequences obtained through TAPE implies that homologous recombinations into the phage's genome need to be carried out at population levels if this diversity is to be preserved and harnessed. Plasmid-encoded molecular systems allowing very efficient *in vivo* homologous recombination reactions using electroporated PCR products[24,25] can easily be introduced into bacteria permissive to T4 replication, such as the *E. coli* K12 strain DK8. However, the T4 genome is very rapidly activated upon its entry into a host, resulting in the quasi-immediate appropriation of the bacterial transcriptional machinery and the conditional inhibition of host-dependent gene activation,[26,27] including the expression of plasmid-encoded functions. This phenomenon effectively abolishes the efficacy of plasmid-borne recombination systems.

However, our models indicated how the mechanisms utilized by the phage to impose step-by-step its genetic program upon the host cell could be utilized to inhibit reversibly the production of the essential T4 early proteins, which, through mechanisms such as ADP–ribosylation of RNA polymerase and of other host proteins, allow initial phage-directed mRNA synthesis reactions to escape from host control.[28] Hence, reversibly blocking T4 transcriptional takeover together with the entire phage replication program would preserve the efficacy of plasmid-encoded recombination systems while maintaining, at least for a few hours, the integrity of the phage's genome within the host, thereby allowing homologous recombination to be efficiently performed.

8.4 Reversible Inhibition of the T4 Lytic Cycle Within the Bacterial Host

The T4 early genes essential for transcriptional takeover (such as *motA*, *asiA*, *alt* and *mod*)[27–29] are apparently transcribed as concatenated, run-through

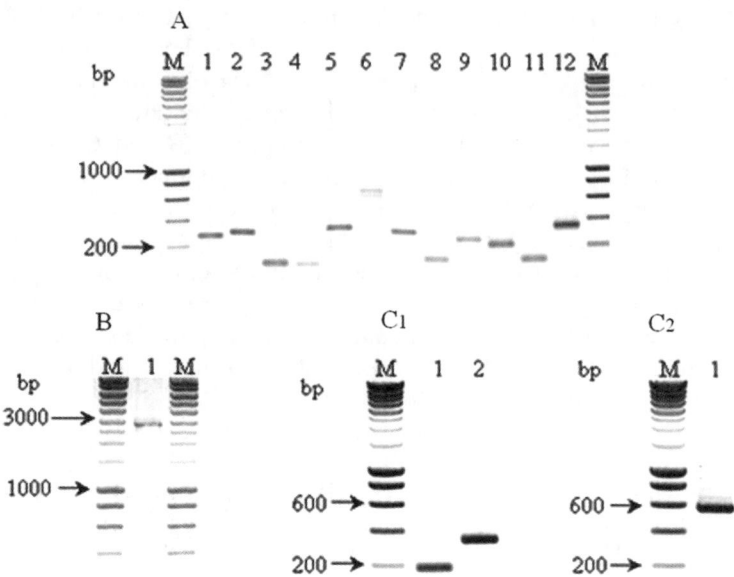

Figure 8.3 Mutagenesis and reconstruction of *gp37* and *gp38*. The two genes encoding the host recognition receptor functions in the T4 genome were subjected to the TAPE procedure. The *gp37* gene (2897 bp) was organized into 13 conserved domains interspersed between 12 variable regions (designated P1–P12) and the coding region corresponding to the C-terminus of the gp38 protein (583 bp) was organized into six conserved domains interspersed between five variable regions (designated P1–P5) and subjected to PCR mutagenesis by random incorporation of nucleotide analogues [0.5 mM dPTP and 20 PCR-cycles (10% nominal mutation rate; see Table 8.2)]. The resulting randomly mutated fragments were then subjected to the entire high-fidelity reconstruction procedure described in Figuer 8.2. The products generated at each step were visualized by gel electrophoresis. (A) The results obtained following high-fidelity amplification of each post-mutagenesis fragment (Step 2) generated for *gp37* (P1–P12), each lane (1–12) showing the corresponding PCR fragment (P1 = lane 1, expected size of 275 bp; P2 = lane 2, expected size of 300 bp; P3 = lane 3, expected size of 111 bp; P4 = lane 4, expected size of 102 bp; P5 = lane 5, expected size of 326 bp; P6 = lane 6, expected size of 690 bp; P7 = lane 7, expected size of 302 bp; P8 = lane 8, expected size of 130 bp; P9 = lane 9, expected size of 254 bp; P10 = lane 10, expected size of 224 bp; P11 = lane 11, expected size of 132 bp; and P12 = lane 12, expected size of 341 bp; M = molecular size marker). (B) The results obtained at completion of sequential reconstruction of *gp37* variants (lane 1 = construct of P1–P12, expected size of 2897 bp) containing 13 conserved domains and 12 variable regions (Step 4). (C) The results obtained during partial [C1; lane 1 = P1 + P2, expected size of 203 bp; lane 2 = P3 + P4, expected size of 345 bp (Step 3)] and complete sequential reconstruction of the C-terminus-encoding *gp38* region [C2; lane 1 = construct of P1–P5, expected size of 583 bp (Step 4)] from randomly mutated PCR products representing six conserved domains and five variable regions.

RNAs and our models clearly indicated that host-encoded transcription termination factors, such as the homo-hexameric Rho[30] terminator, should be required for the production of the corresponding very early phage proteins. Our models further suggested that conditional over-expression of a mutated, non-functional Rho protein within the host would interfere with the formation of a functional Rho complex, thereby reversibly inhibiting production of the essential T4 early proteins while minimally affecting host viability under controlled laboratory conditions.

Two pathways for transcription termination have been identified in *E. coli* and termination events are divided roughly equally between them.[31] Intrinsic termination is promoted by a sequence at the end of the mRNA transcript, which forms a stem–loop structure followed by an A–U-rich region. The weak stability of the A–U-rich region, combined with the RNA stem–loop structure in the exit channel of the RNA polymerase, forces mRNA to disengage from the RNA polymerase and DNA. The other pathway requires a homo-hexameric helicase called Rho (Figure 8.5). The Rho factor attaches to the mRNA transcript and uses its helicase function to track along the transcript towards the moving mRNA polymerase. Upon catching up with the polymerase, Rho catalyzes the disassociation of mRNA from both genomic DNA and the RNA polymerase. The Rho hexamer is a broken circle, resembling a 'lock washer' in appearance. The gap is approximately 12 Å wide, sufficient to accept an mRNA strand into the center of the hexamer. The Rho monomer unit contains two domains, each of which binds RNA. The N-terminal domain, which contains the primary mRNA binding site, loads on to the mRNA target sequence (referred to as '*rut*'). The C-terminal domain, which contains the secondary mRNA binding site, binds mRNA for helicase/translocation activity. mRNA bound to the primary site is located towards the center of the ring rather than the periphery and the mRNA is directed inwards towards the central cavity. Thus, mRNA is directed from the primary sites to the secondary sites that can be accessed through the notched opening in the hexamer. Upon mRNA binding to the secondary site(s), the ring closes. If the Rho hexamer is closed at the time secondary interaction should occur, the mRNA will remain proximal *via* interaction with the primary site until the open state is presented. It could also be that mRNA binding to the primary sites might promote the open form of Rho. In the open form, the catalytic elements of the ATP-binding sites, located at subunit interfaces, are not in the proper configuration for ATP hydrolysis. Ring closure may thus bring the catalytic elements into register to commence ATP hydrolysis, thereby propelling Rho in pursuit of the RNA polymerase. In the final step, Rho unwinds RNA from genomic DNA and the RNA polymerase disengages.[32–35]

We used TAPE to produce non-functional variants of the *E. coli rho* gene (Rho*) that were then cloned into IPTG-inducible expression vectors and introduced into DK8 cells harboring the heat shock-inducible Red recombinase system.[36] The cells were then infected with wild-type T4 phage with and without prior induction of Rho* over-expression. Induction of Rho* practically coincident with phage infection had no effects upon phage replication and massive

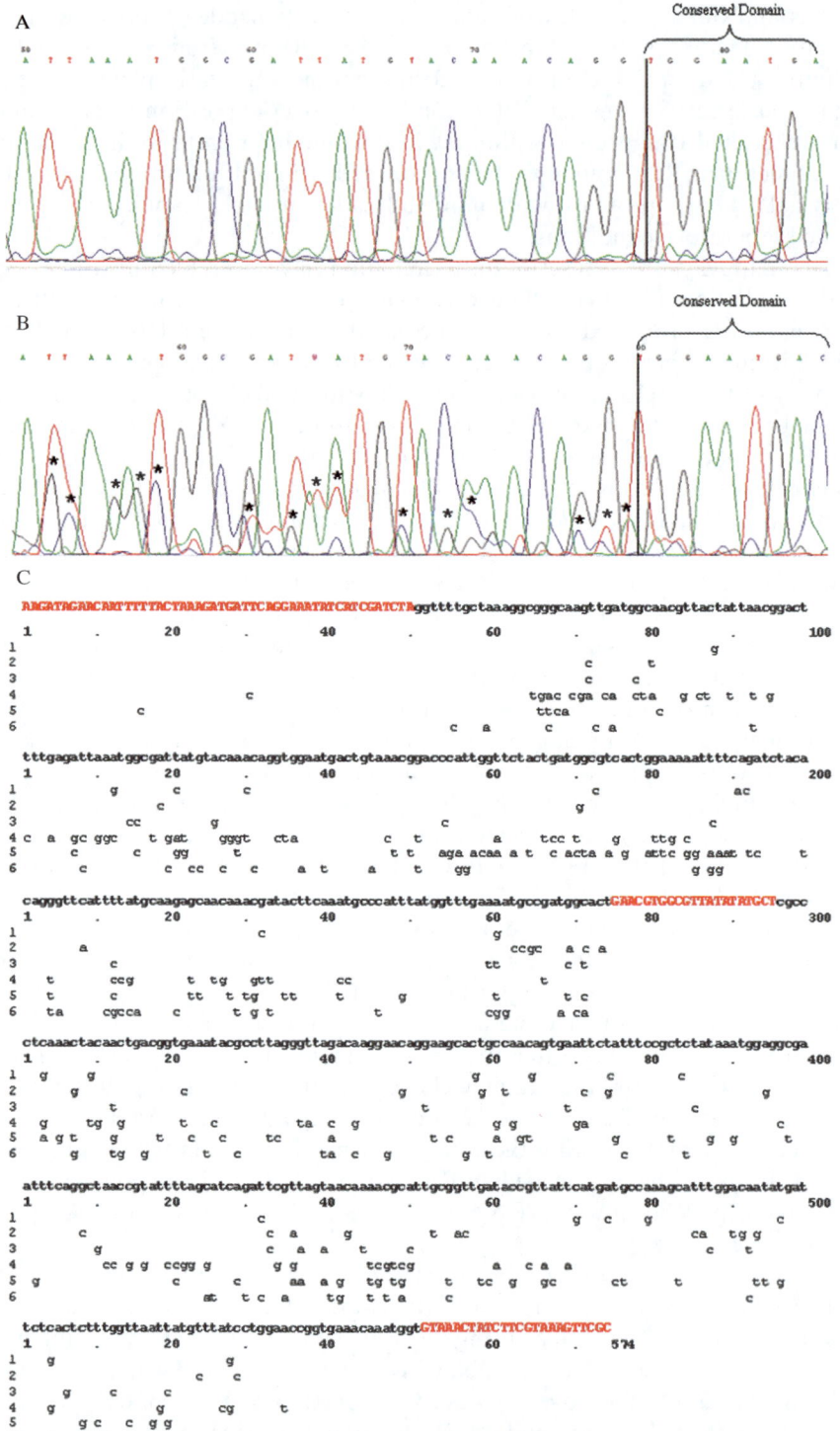

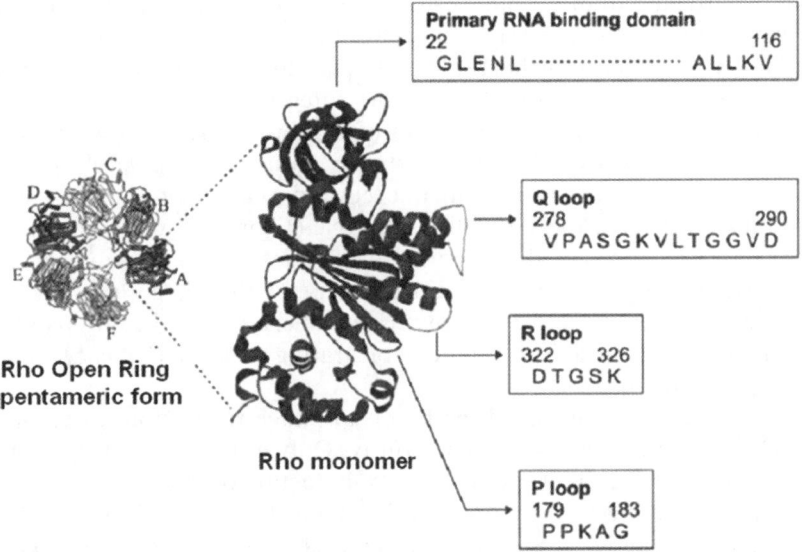

Figure 8.5 Monomeric functional domains and polymeric architecture of *E. coli* Rho. The figure shows an open 'lock washer'-like pentameric form of Rho. Each monomer has four defined functional domains: the primary RNA binding domain that can bind to a single-stranded DNA molecule and also a single-stranded RNA molecule extending from residues 22 to 116; the P-loop, for ATP binding and ATPase activity, includes residues from 179–183; the Q-loop, from residues 278 to 290, and R-loop, from residues 322 to 326, constitute the secondary RNA binding site. Adapted from Banerjee *et al.*[34]

Figure 8.4 Sequencing analyses of stochastically mutated and partly reconstructed *gp37*. Fragments of the wild-type *gp37* gene corresponding to the P1 and P2 segments were generated by PCR and subjected to the PCR-mediated reconstruction procedure without prior mutagenesis. Both the wild-type (A) and mutated (B; nominal mutation rate of 6%) P1 + P2 reconstructed fragments were then subjected to batch DNA sequencing analyses ($N = 4$; 200 ng of PCR products each). This approach allows the chromatograms to reveal the positions and densities of mutations generated (peak superpositions marked *). In addition, comparative sequence analyses (C) were preformed upon individual P1 + P2 fragments reconstructed from material subjected to different nominal mutation rates (from 3%, line 1, to 18%, line 4). These analyses demonstrate that (1) the reconstruction protocol does not introduce sequence shuffling and (2) the TAPE procedure efficiently performs selective stochastic mutagenesis of multiple domains while concurrently preserving the sequence integrity of multiple selected regions within the same gene, even when very high nominal mutation rates are being implemented (line 4 = 18%, 5 = 13% and 6 = 9%). These analyses were independently repeated three times and gave highly comparable results.

host lysis ensued within 30 min post-infection. However, induction of Rho* over-expression 5–10 min prior to exposure to T4 particles, although without effects upon the efficacy of host infection, entirely blocked phage replication without causing the loss or affecting the integrity of the phage's genome within the host, even after several days of continuous Rho* over-expression (Figure 8.6A). Following 30 min of strong induction, the levels of Rho* expression can be reduced to minimal levels and maintained there for several days without restoring the T4 replication cycle (Figure 8.6A). Interruption of Rho* expression through the removal of all traces of ITPG from bacterial cultures results in the systematic reinstatement of the phage lytic cycle and the production of an infective progeny (Figure 8.6A and C), irrespective of the duration of prior inhibition, thus demonstrating both the efficacy and the reversibility of the process.

Furthermore, Rho*-mediated inhibition of phage replication (a technology we labeled 'RIPh') is very stable and sturdy. Only a few cells in large popula-tions (10^9 cells ml^{-1}) exhibit spontaneous re-induction of the T4 lytic cycle and, most importantly, electroporation of PCR fragments (0.6–3 kb) into bacterial cells over-expressing Rho* and harboring a 'deactivated' T4 genome is without significant effects upon the inhibition of phage replication (Figure 8.6A and C). However, several entirely unexpected phenomena associated with Rho*-mediated inhibition of phage replication rapidly became apparent. In DK8 cells, the 'deactivated' T4 genome behaves as a phagemid and is replicated concurrently with the host cell's genome, resulting in its propagation within the growing, Rho*-expressing bacterial population (Figure 8.6B). Rho*-expressing bacteria harboring a 'deactivated' T4 genome are resistant to subsequent T4 super-infection (pseudo-lysogeny), irrespective of the duration of Rho* expression (Figure 8.6C), hence protecting the bacterial population from the effects of spontaneous re-induction of the T4 lytic cycle in some of the cells (Figure 8.6A). Finally, Rho*-expressing cells harboring a 'deactivated' T4 genome can be glycerolated prior to or after induction of homologous recombination into the T4 genome, stored at $-80\,^{\circ}$C for weeks and reutilized without liberating the 'deactivated' phage (data not shown). Nevertheless, the duration of storage has adverse effects since, with increasing time beyond 4–5 weeks, we observed rapidly increasing levels of spontaneous re-induction of the T4 lytic cycle in the revived bacterial cultures (data not shown).

8.5   Large-scale Recombinations into the Genomes of an Infective Wild-type T4 Population

The T4 *hoc* gene was utilized as a target for the insertion, via Red-driven homologous recombination,[24] of a fused Hoc–GFP synthetic construct into Rho* 'deactivated' T4 genomes.

To this effect, the PCR synthetic constructs were electroporated into Rho*-expressing DK8 cells harboring both the heat shock-inducible Red recombinase system together with a T4 genome and, following a 30 min recovery period, the

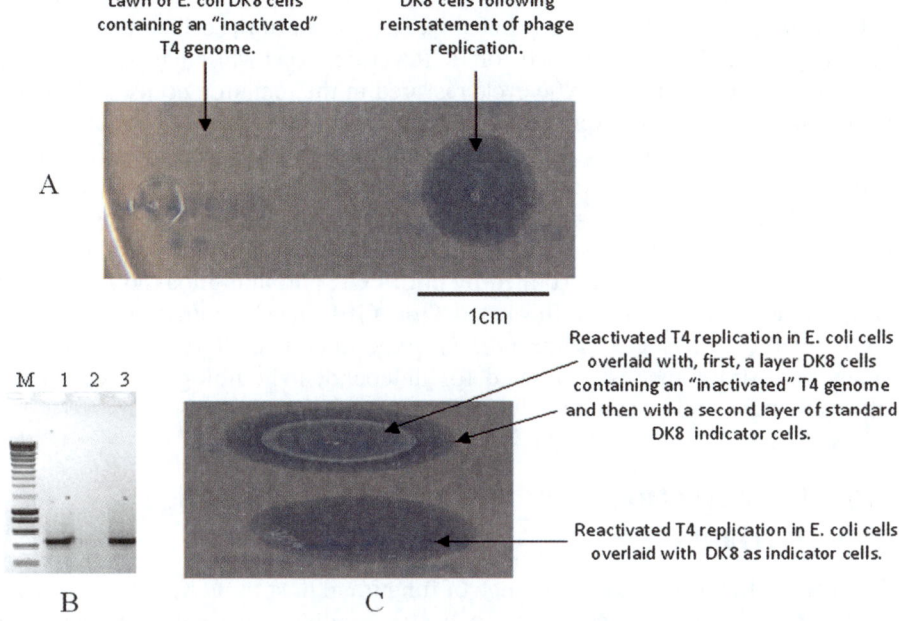

Figure 8.6 The Rho*-mediated inhibition of the T4 lytic cycle is reversible, cells harboring a 'deactivated' T4 genome are immune to subsequent T4 super-infection while the 'deactivated' T4 genome behaves as a phagemid. Over-expression of the non-functional variant of the *rho* gene (Rho*) in DK8 *E. coli* hosts harboring a wild-type T4 phage genome was inhibited (see text) and, following a latent period of 10 min, the hosts were mixed with non-infected cells, aliquoted on absorbent discs and overlaid with a lawn of Rho* over-expressing DK8 cells harboring a wild-type T4 phage genome (A) or with a second layer of standard indicator DK8 cells (C). In both cases, inhibition of Rho* over-expression prior to plating resulted in reactivation of the phage's lytic cycle and the production of a T4 progeny capable of infecting and lysing standard indicator DK8 cells [dark discs in (A) and (C)]. However, this treatment could not cause the lysis of DK8 cells harboring a 'deactivated' T4 genome [white halo in (B) and lysis zone strictly limited to the surface of the absorbent discs in (A) and (C)], hence demonstrating that these cells are resistant to T4 super-infection and present the behavior of lysogenic cells. This is further demonstrated by the non-expanding faint blue blotches visible on the lower left in (A). These correspond to DK8 cells in which Rho*-mediated phage inactivation failed, leading to spontaneous reinstatement of phage replication and host lysis that did not spread through neighboring cells. Interestingly, Rho* expression inhibits preserves the phage genome which now propagates concurrently the host. DK8 cells over-expressing Rho* and harboring a wild-type T4 phage genome were collected 3 and 96 h following phage infections, lysed and the total cellular DNA subjected to PCR analysis directed towards the T4-specific *gp38* gene followed by gel electrophoresis. (B) shows that, in spite of a near complete absence of host lysis, PCR analyses revealed the presence at high frequency of T4 *gp38* genes in host cells 3 h (lane 1) and 96 h (lane 3) following infection, while failing to give a product from non-infected host cells (lane 2).

cells were maintained at 34 °C for 10 min to induce the recombinase system and subsequently returned to 25 °C for 60 min. An aliquot of the cells was then lysed and the total cellular DNA used for PCR verifications while Rho* expression was suppressed and the T4 lytic cycle restored in the remaining cells. Following lysis, the phage progeny were recovered and an aliquot was used for PCR and immunoassay verifications while the remaining particles were used to infect standard DK8 cells. The subsequent second-generation progeny was recovered and used for PCR and immunoassay verifications. The results (Figure 8.7) indicated remarkable levels of homologous recombination efficiency (mean of $10^{-4} \pm 1 \times 10^3$, $n = 4$), as ascertained by both PCR and immunoassays, together with stable transmission of the fused Hoc–GFP gene within the descent of recombinant T4 particles (Figure 8.7A). Experiments in which up to three T4 genes were simultaneously targeted for independent homologous recombination gave very similar results (Figure 8.7B).

8.6 Construction of a T4 Bank of Host Range Variants

To produce the widest possible range of functional host range variants, we used TAPE to introduce large-scale stochastic modifications into the variable regions that constitute the host range cassettes in the *gp37* and *gp38* genes. The pools of TAPE-generated *gp37* and *gp38* variants where then sequentially recombined as above into the genomes of a wild-type T4 population. The lytic cycle was restored and allowed to proceed for one round of lysis only (25–30 min) in order to preserve the recombinant progeny produced in this first lytic cycle which would otherwise be rapidly swamped out by non-recombinant, wild-type progeny with each successive round of host infection and lysis.

The ensuing bank of T4 recombinant progeny was screened against bacterial strains evolutionarily close to (*Yersinia ruckeri* ATCC 29908) and also far removed from (*Pseudomonas aeruginosa* ATCC 47053) the original *E. coli* K12 strain used for production of the bank.

On the first screening pass, recombinant particles capable of infecting and destroying both bacterial species tested were isolated (Figure 8.8).

8.7 Discussion

The mutability of bacteriophage offers particular advantages in the treatment of bacterial infections not afforded by other antimicrobial therapies. Indeed, antibiotics are static molecules, incapable of selectively responding to resistant bacteria *in vivo*. However, although mutations at genome level can generate potentially beneficial variants, they also produce deleterious genetic loads. Natural phage types with a historical beneficial-to-deleterious mutation ratio (b/d ratio) of approximately 0.1 per genome per generation have been reported to offer a reasonable balance between beneficial diversity and deleterious mutational load.[37] Recent studies, addressing situations where phage and hosts

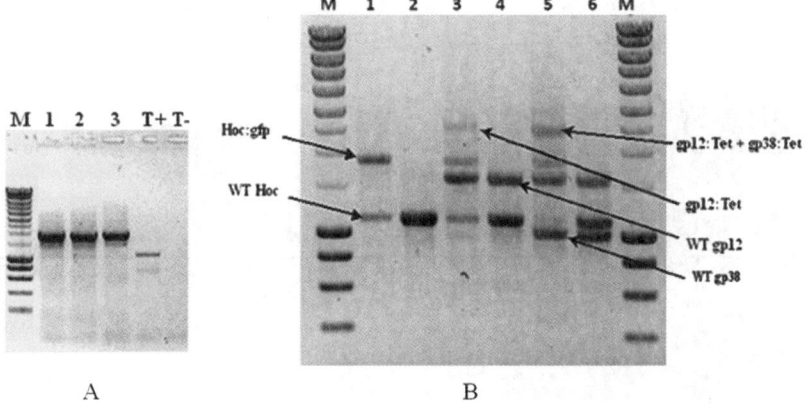

Figure 8.7　Inhibition of phage replication mediated by Rho* over-expression allows high-efficiency, stable homologous recombination of a phage's genes. A *hoc–GFP* construct was electroporated into Rho* over-expressing DK8 cells harboring the cloned Red–recombinase system together with a 'deactivated' wild-type T4 genome. Following the induction of Red-mediated homologous recombination, part of the cells was used for PCR analysis directed towards the *hoc–GFP* construct (A, lane 1). The remainder of the cells were left to grow for 12 h and the T4 lytic cycle restored (see text). An aliquot of the ensuing phage progeny was used for PCR analysis of the *hoc* genetic locus (A, lane 2) and the remainder used to infect standard DK8 host cells. The second-generation progeny was analyzed by PCR as above (A, lane 3). PCR analysis of the *hoc* locus in the first-generation progeny (lane 2) revealed a remarkably high apparent recombination rate. The strong PCR band corresponding to the *hoc–GFP* construct with the concurrent presence of a weak wild-type *hoc* signal (*cf.* lane T+) indicates that a very significant proportion of the first-generation progeny appeared to carry the recombinant construct instead of the wild-type gene. This was confirmed by the analytical results obtained from the second-generation progeny (lane 3) in which the recombinant construct was overwhelmingly present and the wild-type *hoc* sequence hardly detectable. T+, wild-type PCR *hoc* signal from DK8 hosts harboring a 'deactivated' T4 genome; T–, non-infected control DK8 cells. Simultaneous recombinations at multiple loci within the phage's genome were efficiently achieved (B, lanes 1, 3 and 5) following concurrent electroporation of *hoc–GFP* and constructs consisting of the *gp12* and *gp38* genes fused to a tetracycline resistance encoding insert (gp12:Tet and gp38:Tet, respectively), as demonstrated by post-recombination PCR analyses of the ensuing T4 progeny shown in (B). M, molecular weight marker; lane 1, Hoc:GFP insert into the T4 genome; Lane 2, wild-type (WT) *hoc* control; lane 3, Hoc:GFP + gp12:Tet inserts into the T4 genome; lane 4, WT *hoc* + WT *gp12* control; lane 5, Hoc:GFP + gp12:Tet + gp38:Tet inserts into the T4 genome; lane 6, WT *hoc* + WT *gp12* + WT *gp38* control.

co-evolve, indicate that for phage populations with minimal genetic loads, stochastically engineered moderate increases in mutation rate beyond the optimal value of mutation–selection balance (10 deleterious mutation per genome per generation) may provide even greater protection against emergent

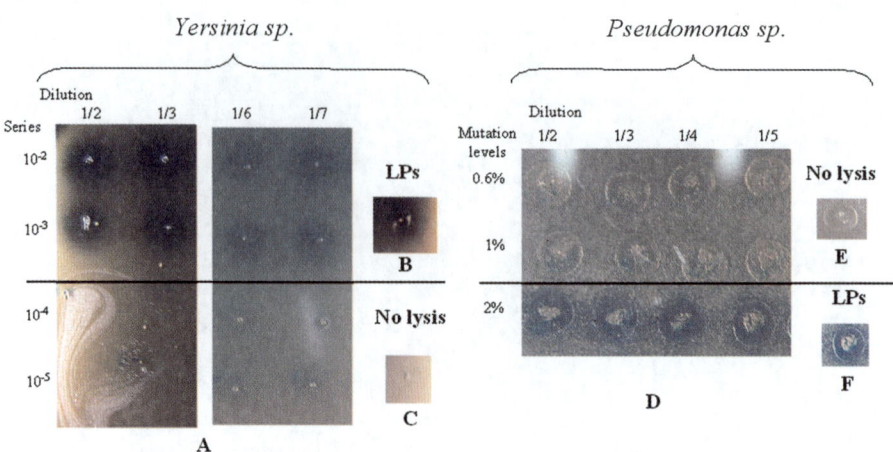

Figure 8.8 Recombinant T4 banks engineered for extended host range contain particles capable of infecting and killing bacteria very different from the
original DK8 host. In the T4 phage bank used for these assays, generated
from a T4 strain (ATCC 110303-B4) known to be restricted to *E. coli*
hosts, both gp38 and gp37 were modified using TAPE. For the assays
using *Yersinia ruckeri* as target cells (A, B, C), the bank (100 ml in TE) was
lyophilized to 1 ml that was then used to produce 10-fold dilution series
(100 μl into 1 ml final volume, labeled 10^{-2}, 10^{-3}, *etc.*, in the figure), each
of which was in turn diluted in 10 successive twofold steps (100 μl into
200 μl final volume, *etc.*, labeled 1/2, 1/3, *etc.*, in the figure). For the assays
using *Pseudomonas aeruginosa* as target cells (D, E, F), banks constructed
using different mutations levels (0.6–2%) introduced in both gp38 and
gp37 were diluted as above. For each dilution, 100 μl were aliquoted on to
absorbent discs that were arrayed on agar plates The discs were then
overlaid with soft agar containing, as indicator cells, bacteria belonging to
species far removed from the original *E. coli* K12 host strain (DK8). With
both *Yersinia* and *Pseudomonas* indicator cells, multiple lysis plaques
(LPs), in the form of dark halos over a clearer background [parts of (A)
and (D) above and below the black horizontal bar, respectively, and (B)
and (F) showing isolated individual LPs], were observed, indicating the
presence, within the bank, of numerous phage particles capable of initi-
ating and maintaining their lytic replication cycles in the new hosts species.
Areas of unaffected indicator cells (no lysis), appearing as an absence of
distinction with the general background [lower half of (A), upper part of
(D) (absorbent disc visibly overgrown by indicator bacteria) and (C) and
(E)], were associated with higher dilution levels (A) or low mutation rates
(D), thus providing evidence that the approach described here allows the
production, within a single bank, of a spectrum of phage particles capable
of controlling bacterial cells belonging to multiples and very different
species. For the assays using *Yersinia ruckeri* as target cells, the second
(dilution to 10^{-2}) to fifth (dilution to 10^{-5}) 10-fold dilution series are
shown. Lysis of *Yersinia ruckeri* cells remained observable up to an
effective 7×10^3-fold dilution of the bank. For the assays using *Pseudomonas aeruginosa* as target cells, only the higher mutation rates for both
gp37 and gp38 produced banks containing T4 variants capable of infecting
and lysing the target cells. Lysis remained observable up to an 80-fold
effective dilution of these banks (100 μl into 8 ml final volume).

bacterial types, albeit only with very weak selective coefficients for *de novo* genome-wide deleterious mutations (below a b/d ratio of 0.01 per genome).[38] However, in such studies, it is the entire viral genome that is being considered and not hyper-variable domains within genes encoding tail fibers proteins exclusively. Here, different tailed phage that are structurally very similar and can infect the same hosts, such as T4, RB69, K3 and T5, carry tail fiber genes that have very little sequence similarities to each other.[39] Indeed, between K3 and T4, two closely homologous phage, the global long tail fiber proteins (gp37) sequence similarities of 76% falls to 21.7% over the last 231 C-terminal amino acids. The same can be said of a comparison between adhesin protein sequences in T4 (gp37) and T2 (gp38), yielding 50% homology over the last 98 amino acids (blast analyses carried out for the design of constant domains within the T4 gp37 C-terminus; data not shown). Furthermore, previous studies have demonstrated that rearrangements, fusions and insertions can be made to gp37 without disrupting the functional integrity of the mature protein structure.[40,41] Hence there is considerable scope for mutagenesis beyond the 0.6–2% levels conservatively used in this study, let alone beyond the levels advocated by mutation–selection balance studies. All the more so since, in the present study, desirable variants were selected by direct exposure to various bacterial strains, thereby *de facto* bypassing all non-functional variants in the phage bank, irrespective of their actual frequencies. Thus, once isolated from the phage bank, the desirable novel variants form a population with a genetic load practically identical with that of any wild-type T4 population, since, apart from the tail fibers' coding sequences, their genome is that of the T4 of origin, all mutations with deleterious effects upon the tail fiber proteins having been largely weeded out by the selection process.

Although active phage therapies could be very effective at suppressing susceptible bacteria, phage-resistant cells are readily found both *in vitro* and *in vivo* and are a particular problem when the bacterial population is large or growing. Here, single-strain active phage therapies are unlikely to avoid problems of resistance. Single-step mutants can be highly or totally resistant to infection by the phage and combinations of multiple phage for which cross-resistance does not arise will probably be the only option for closing the mutant selection window.[14,42] Combinations of phage that are used in an active mode (the concentration of phage exceeds the rate at which bacterial growth is equaled by the rate of phage infection) must be closely matched for their lifecycle properties in order to avoid the 'faster' phage dominating the treatment and making the combination little more effective than this faster phage would have been alone, thereby favoring the appearance of resistant bacterial mutants. Furthermore, as the differences between phage strains increase, there is a rapid transition from a relatively low probability of multistrain-resistant cells emerging to a very high probability.[14,42] Our approach provides an elegant solution to this problem.

The technological approaches (named TAPE, RIPh and AB-Accus) that we have developed allow the efficient production of very large, stochastically engineered, virulent phage banks containing a very wide spectrum of variants that differ for their host-targeting specificities but are all identical in their lifecycle properties and from which multiple particles capable of controlling a pathogenic, uncharacterized Gram-negative bacterial population can be easily and rapidly isolated, as demonstrated by our results. The large lysis plaques that were obtained with multiple dilution series of the phage bank on both *Yersinia* and *Pseudomonas* cells strongly suggest the concurrent presence, within the bank, of multiple variants capable of targeting each bacterial species. Indeed, within such a bank, the various constituents of the multitude of variants produced are highly unlikely to be represented with quantitative homogeneity and any given variant cannot be expected to dominate quantitatively all others. Hence, for lysis to remain highly effective across several dilutions levels of the bank, multiple variants represented with quantitative heterogeneity must be present. This amounts to producing complex therapeutic cocktails consisting of phage perfectly matched in terms of lifecycle properties. From a technological standpoint, our novel methodologies appear to be robust, trustworthy and easy to implement and manipulate and they could be directly applicable to a wide spectrum of obligate lytic phage.

Nevertheless, any given form of phage-based treatment cannot be expected to be exempt from all types of resistance problems, and this must be taken into account if the full potential of phage therapy is to be realized.

Activation of existing restriction-modification or abortive infection mechanisms, such as the ToxN and Abi systems,[43,44] and stationary or other phases of growth that confer temporary resistance to phage,[43] figure prominently among the arsenal of very effective escape strategies developed by bacteria. The approach we described here most certainly cannot allow, in its present form, these effective defense mechanisms to be bypassed or counteracted. However, some of the means whereby phage can counteract the induction of Abi-mediated 'altruistic suicide' of infected cells, active across multiple genera of Gram-negative bacteria and against a spectrum of phage, have been described,[45–47] and can be engineered relatively easily into the genome of a lytic phage, using some of the techniques described here, prior to modifying its host-targeting mechanism.

From a fundamental standpoint, this work constitutes the first direct experimental demonstration explaining the mechanisms governing the expression of T4 very early proteins in *E. coli*, their entire dependence upon the immediate availability of functional, host-encoded transcription termination factors, such as the Rho homo-hexameric complex, and the effects of their functional absence upon the T4 replication program.

8.8 Conclusion and Perspectives

The methods that we have developed allow the efficient production of very large, stochastically engineered, virulent phage banks containing a very wide spectrum of variants for any chosen phage-associated function. The banks that can be produced using these approaches can naturally address a wide variety of virulent phage, the genomes of which have been sequenced and annotated. A collection of such banks, comprising Gram-negative and Gram-positive phage, could conceivably provide the means largely to precede and counteract the escape strategy that may be implemented by any given bacterial population, without recourse to natural phage pools and the lengthy, labor-intensive isolation processes necessarily attached. In addition, these approaches also open the way to an entirely novel utilization of virulent phage.

Within such a bank, a significant number of phage will be capable of adsorbing on a variety of molecules never encountered on bacterial surfaces. Since the capsids of these phage may be made to contain a variety of molecules other than the usual genomic DNA,[44] these particles, in effect, constitute targetable nano-devices that could be used to deliver cargoes of small molecules on to materials bearing the surfactants appropriate for efficient phage adsorption.

Thus, in addition to providing a broadly applicable solution for the long-term maintenance of effective phage therapies (T4 being merely one such example) against emergent, multi-resistant bacterial pathogens, these engineered phage banks also provide a source of widely diversified targetable delivery nano-devices. To our knowledge, this is the first time that such an approach to broadly improved biosecurity has been successfully developed and implemented.

8.9   Methods

All standard laboratory procedures used in this work have been extensively described by Sambrook and Russell.[48]

8.9.1   High-fidelity PCR

T4 *gp37* (NC_000866.4, GeneID: 1258629) and *gp38* (NC_000866.4, GeneID: 1258706) genes DNA served as a template. The primers used in this study are listed in Table 8.2. PCRs were performed with 1 unit of Taq polymerase (Roche) in the supplier's buffer. Reaction mixes contained 10 µM each primer and 1 mM dNTPs. The PCR program involved one step at 94 °C for 2 min, followed by 30 cycles of amplification of three steps at (i) 94 °C for 30 s, (ii) 55 °C for 30 s and (iii) 72 °C for 1–3 min, depending on the fragment length. PCR products were maintained at 72 °C for 5 min.

8.9.2   Error-prone PCR

The JBS Error-Prone Kit (Jena Bioscience) was used to introduce stochastic mutations by decreasing the fidelity of Taq polymerase (method 1). These reactions were carried out using the supplier's buffer (1×) mixed with 1 µl Taq polymerase (1 U), 50 ng DNA template, 2 µl dNTP Error-prone Mix, 1× Error-Prone Solution and 10 mM of a primer pair amplifying an internal portion of

Table 8.2 Primers used for PCR amplification.

Primer pairs	Target sequence	Variable domain size (bp)	Nucleotide sequences
37 P1-F	P1	225	5'-GATTCAGGAAATATCATCGATCTAGG-3'
37 P1-R			5'-GGGCGAGCATATATAACGCCACGTTC-3'
37 P2-F	P2	255	5'-GAACGTGGCGTTATATATGCTCGCCC-3'
37 P2-R			5'-GCGAACTTTACGAAGATAGTTTACA-3'
37 P3-F	P3	57	5'-GGTGTAAACTATCTTCGTAAAGTTCGC-3'
37 P3-R			5'-TACACCAGACCACCAAGAAACTTCATC-3'
37 P4-F	P4	54	5'-GATGAAGTTTCTTGGTGGTCTGGTGAT-3'
37 P4-R			5'-TGTACCTAATGCAAGGCTATTACG-3'
37 P5-F	P5	267	5'-CGTAATAGCCTTGCATTAGGTACA-3'
37 P5-R			5'-ATATGGGGTTTGTCCGTCTCCAGCATTGTTATTATC-3'
37 P6-F	P6	624	5'-GCTGGAGACGGACAAACCCATATCGGGTAC-3'
37 P6-R			5'-TTCCGAACGACGGAAAATAGCACCATATTCAGCGT-3'
37 P7-F	P7	246	5'-AACGCTGAATATGGTGCTATTTTCCGTCGTTCGGA-3'
37 P7-R			5'-GCAAATGAACCTGCACCCG-3'
37 P8-F	P8	75	5'-CCCGGCGGGTGCAGGTTCAT-3'
37 P8-R			5'-ATTGCCTTGAACATAACGTTGTTTCAAAATAGG-3'

37 P9-F	P9	180	5'-CCTATTTTGAAACAACGTTATGTTCAAGGCAA-3'
37 P9-R			5'-TTTAAGTTAGCAAAATTACCAGAACCACCAGTGATATTAC-3'
37 P10-F	P10	135	5'-GGTAATTTTGCTAACTAAACAGTACAATTGAATCACTAAAACTGA-TATC-3'
37 P10-R			5'-GGTTTACCCTGATAGTTTGCCCGCGCATATCTGGAAT-3'
37 P11-F	P11	51	5'-GGGCAAACTATCAAGGGTAAACCAAGTGGTCGTGCTGTTTTGAGC-3'
37 P11-R			5'-GTCAAAGCTTGATGTGGTTTTAGTACCTAAGTCAGT-3'
37 P12-F	P12	260	5'-GGTACTAAAACCACATCAAGCTTTGACTATGGTACGAAG-3'
37 P12-R			5'-CCTGTACTATTTACAGTGATAGTAGTATGACCATGTGATCCAATT-3'
38 P1-F	P1	60	5'-CGGCCCTTCTAAATATGAAAATATATCAT-3'
38 P1-R			5'-TGTTGAATGAGCAGGAAGACCG-3'
38 P2-F	P2	75	5'-CGGTCTTCCTGCTCATTCAACA-3'
38 P2-R			5'-TTTTCCGCGATGGTCTTCT-3'
38 P3-F	P3	27	5'-GAAGACCATCGCGGAAAACG-3'
38 P3-R			5'-TTCCAATGTCACTTATAAAAATT-3'
38 P4-F	P4	258	5'-TTTTTATAAGTGACATTGGA-3'
38 P4-R			5'-CTTCTGTTCTTTTCAGGCCA-3'
38 P5-F	P5	33	5'-TGGCCTGAAAAGAACACAGAAG-3'
38 P5-R			5'-CCCTCCTTTTAAGATATTT

gp37 or *gp38*. Each amplification reaction involved one step at 94 °C for 2 min, followed by 30 cycles of amplification of three steps at (i) 94 °C for 30 s, (ii) 55 °C for 30 s and (iii) 72 °C for 1–3 min, depending on the fragment length. PCR products were maintained at 72 °C for 5 min. The Error-Prone Solution enhances the mutational rate by modifying parameters of the PCR reaction such as a higher Mg^{2+} concentration of up to 7 mM, partial substitution of Mg^{2+} by Mn^{2+} and unbalanced rates of dNTPs, resulting in a rate of mutagenesis of 0.6–2%.

Stochastic insertion of nucleotide analogs (method 2) was performed using the JBS dNTP-Mutagenesis Kit (Jena Bioscience). These reactions were carried out using the supplier's buffer (1×) mixed with 1 µl Taq polymerase (1 U), 50 ng DNA template, 1 mM dNTPs, 0.5 mM 8-oxo-dGTP, a range of 0.1–0.5 mM dPTP and 10 mM of a primer pairs amplifying an internal portion of *gp37* or *gp38*. Each amplification reaction involved one step at 94 °C for 2 min, followed by 5–30 amplification cycles of three steps at (i) 94 °C for 30 s, (ii) 55 °C for 30 s and (iii) 72 °C for 1–3 min, depending on the fragment length. PCR products were maintained at 72 °C for 5 min. 8-oxo-dGTP is incorporated opposite template adenine, yielding two transition mutations: A → C and T → G. dPTP has been reported[49,50] to be approximately 10-fold more mutagenic than 8-oxo-dGTP and yields four type of transition mutations: A → G, T → C, G → A and C → T. The resulting rate of mutagenesis is between 3 and 17%, depending on the dPTP concentration and number of PCR cycles. For elimination of the mutagenic dNTPs and to conserve the mutations introduced, an aliquot of 1 µl of this first PCR reaction was used in a second high fidelity PCR amplification.

8.9.3   Selective High-fidelity Amplification of Desired Fragments

In order to select fragments with priming sites preserved, fragments produced by the error-prone reactions were re-amplified using a 1:1 mixture of Taq (Roche) and Isis (MP Biomedicals, proofreading) polymerases and the appropriate primer pairs (Table 8.2). The reaction mixtures and the cycling profiles were as described above (high-fidelity PCR).

8.9.4   Reconstruction of Sequence Through PCR

Following error-prone and high-fidelity PCR amplifications, selected fragments in partial overlap were used as template in equal proportions and PCR amplifications were performed with 1 unit of Taq polymerase (Roche) in the supplier's buffer and 1 mM dNTPs. After 15 amplification cycles, 1 µM each of the primers corresponding to the termini of the reconstructed fragments were added and the reaction was resumed for a further 15 cycles. The cycling protocol was (i) 94 °C for 30 s, (ii) 55 °C for 30 s and (iii) 72 °C for 1–3 min, depending on the fragment length.

8.9.5   DNA Sequencing and Analysis

Reconstructed fragments were batch-sequenced by BIOFIDAL (http://www.biofidal.com/) and the ApE program was used to analyze the sequencing chromatograms (http://www.biology.utah.edu/jorgensen/wayned/ape/).

8.9.6   Production and Expression of Non-functional *E. coli* Rho Genes

The *E. coli* genome of DK8 cells [F-Δ(srl-recA)306lacZya536(lacZam)rpsL^{51}] was used to PCR amplify and isolate the rho coding sequence (NC_000913.2, GeneID: 948297). The primers used were 5'-TACTTAGAATGGCTTAATTTCT-TATGC (ERf) and 5'-TTATGAGCGTTTCATATTTCGA (ERr). The PCR product was then subjected to error-prone mutagenesis as above. Following sequencing, a mutant bearing an in-frame stop codon at position 634 (Rho*) was cloned into the IPTG-inducible expression vector pHSG299 (Takara) and electroporated[26] into DK8 cells that were then plated on to LB X-gal/IPTG plates containing 30 µg ml^{-1} kanamycin. Following isolation of a white colony and PCR verification for the presence of Rho*, expression of the truncated gene was obtained by exposure of the cells to a final concentration of 0.1 mM IPTG. To inhibit Rho* expression, the cells (15 ml) were centrifuged (5000*g*) and thoroughly resuspended three consecutive times in 40 ml of minimal medium[26] with a final resuspension in 20 ml of LB.

8.9.7   Construction and Expression of the Heat-inducible Red–Recombinase System

Genomic DNA from the phage λcI857 was used as template in a PCR reaction amplifying the attP-cro region that contains the *exo*, *bet* and *gam* under the control of the heat-sensitive cI repressor. The PCR primers utilized were 5'-GTATGCATGCTGGGTGTGG (MλRf) and 5'-CGCACTCTCGATTCG-TAGAGCCTCG (MλRr). The PCR product was then cloned into the low copy number, constitutive expression vector pFN476 (ATCC 86962) and electroporated as above into DK8 cells. The cells were then grown in LB at 26 °C where the cI repressor is expressed and actively prevents *exo*, *bet* and *gam* induction. Upon warming to 34 °C, the cI repressor becomes inactivated, allowing implementation of the *exo*, *bet*, *gam*-mediated recombinase system.

Acknowledgements

We thank Thierry Reynaud and Jérôme Gabbard for their valued support throughout this work.

References

1. W. C. Summers, *Annu. Rev. Microbiol.*, 2001, **55**, 437.
2. M. Uzan, *Prog. Mol. Biol. Transl. Sci.*, 2009, **85**, 43.
3. C. Desplats and H. M. Krisch, *Res. Microbiol.*, 2003, **154**(4), 259.
4. I. Riede, M. Degen and U. Henning, *EMBO J.*, 1985, **4**(9), 2343.
5. K. Drexler, J. Dannull and I. Hindennach, *et al.*, *J. Mol. Biol.*, 1991, **219**(4), 655.
6. M. Morita, C. R. Fischer and K. Mizoguchi *et al.*, *FEMS Microbiol. Lett.*, 2002, **216**(2), 243.
7. M. Snyder and W. B. Wood, *Genetics*, 1989, **122**(3), 471.
8. F. Tetart, C. Desplats and H. M. Krisch, *J. Mol. Biol.*, 1998, **282**(3), 543.
9. F. Tetart, F. Repoila and C. Monod *et al.*, *J. Mol. Biol.*, 1996, **258**(5), 726.
10. H. Brussow and R. W. Hendrix, *Cell*, 2002, **108**(1), 13.
11. H. Brussow, C. Canchaya and W. D. Hardt, *Microbiol. Mol. Biol. Rev.*, 2004, **68**(3), 560.
12. H. W. Ackermann, *Res. Microbiol.*, 2003, **154**(4), 245.
13. L. G. Pell, V. Kanelis and L. W. Donaldson *et al.*, *Proc. Natl. Acad. Sci. USA*, 2009, **106**(11), 4160.
14. B. J. Cairns and R. J. Payne, *Antimicrob. Agents Chemother.*, 2008, **52**(12), 4344.
15. Y. Tanji, K. Hattori and K. Suzuki *et al.*, *Appl. Environ. Microbiol.*, 2008, **74**(14), 4256.
16. S. Matsuzaki, M. Rashel and J. Uchiyama *et al.*, *J. Infect. Chemother.*, 2005, **11**(5), 211.
17. M. L. Pedulla, M. E. Ford and J. M. Houtz *et al.*, *Cell*, 2003, **113**(2), 171.
18. R. M. Donlan, *Trends Microbiol.*, 2009, **17**(2), 66.
19. T. K. Lu and J. J. Collins, *Proc. Natl. Acad. Sci. USA*, 2009, **106**(12), 4629.
20. T. K. Lu and J. J. Collins, *Proc. Natl. Acad. Sci. USA*, 2007, **104**(27), 11197.
21. S. J. Labrie and S. Moineau, *J. Bacteriol.*, 2007, **189**(4), 1482.
22. R. Gallet, Y. Shao and I. N. Wang, *BMC Evol. Biol.*, 2009, **9**, 241.
23. B. J. Cairns, A. R. Timms and V. A. Jansen *et al.*, *PLoS Pathog.*, 2009, **5**(1), e1000253.
24. B. Lesic and L. G. Rahme, *BMC Mol. Biol.*, 2008, **9**, 20.
25. K. A. Datsenko and B. L. Wanner, *Proc. Natl. Acad. Sci. USA*, 2000, **97**(12), 6640.
26. U. K. Sharma and D. Chatterji, *J. Bacteriol.*, 2008, **190**(10), 3434.
27. D. M. Hinton, S. Pande and N. Wais, *et al.*, *Microbiology*, 2005, **151**(Pt 6), 1729.
28. B. Tiemann, R. Depping and E. Gineikiene, *et al.*, *J. Bacteriol.*, 2004, **186**(21), 7262.
29. K. Wilkens, B. Tiemann and F. Bazan, *et al.*, *Adv. Exp. Med. Biol.*, 1997, **419**, 71.
30. J. Chalissery, S. Banerjee and I. Bandey, *et al.*, *J. Mol. Biol.*, 2007, **371**(4), 855.

31. D. L. Kaplan and M. O'Donnell, *Curr. Biol.*, 2003, **13**(18), R714.
32. E. Skordalakes and J. M. Berger, *Cell*, 2003, **114**(1), 135.
33. J. P. Richardson, *Cell*, 2003, **114**(2), 157.
34. S. Banerjee, J. Chalissery and I. Bandey, *et al.*, *J. Microbiol.*, 2006, **44**(1), 11.
35. N. C. Kalarickal, A. Ranjan, B. S. Kalyani, *et al.*, *J. Mol. Biol.*, 2010, **395**(5), 966.
36. S. Yamamoto, H. Izumiya and M. Morita, *et al.*, *Gene*, 2009, **438**(1–2), 57.
37. M. de la Pena, S. F. Elena and A. Moya, *Evolution*, 2000, **54**(2), 686.
38. D. T. Kysela and P. E. Turner, *J. Theor. Biol.*, 2007, **249**(3), 411.
39. S. R. Casjens, *Res. Microbiol.*, 2008, **159**(5), 340.
40. P. Hyman, R. Valluzzi and E. Goldberg, *Proc. Natl. Acad. Sci. USA*, 2002, **99**(13), 8488.
41. S. T. Abedon, *Foodborne Pathog. Dis.*, 2009, **6**(7), 807.
42. T. R. Blower, P. C. Fineran and M. J. Johnson, *et al.*, *J. Bacteriol.*, 2009, **191**(19), 6029.
43. E. Bidnenko, A. Chopin and S. D. Ehrlich, *et al.*, *BMC Mol. Biol.*, 2009, **10**, 4.
44. T. E. Keller, I. J. Molineux and J. J. Bull, *Mol. Biol. Evol.*, 2009, **26**(9), 2041.
45. A. Sanguino, G. Lopez-Berestein and A. K. Sood, *Mini Rev. Med. Chem.*, 2008, **8**(3), 248.
46. P. C. Fineran, T. R. Blower and I. J. Foulds, *et al.*, *Proc. Natl. Acad. Sci. USA*, 2009, **106**(3), 894.
47. L. Van Melderen and M. Saavedra De Bast, *PLoS Genet.*, 2009, **5**(3), e1000437.
48. J. Sambrook and D. W. Russell, *Molecular Cloning: a Laboratory Manual*, Cold Spring Harbor Laboratory Press, Cold Spring Harbor, NY, 2001.
49. M. Zaccolo, D. M. Williams and D. M. Brown, *et al.*, *J. Mol. Biol.*, 1996, **255**(4), 589.
50. L. Pritchard, D. Corne and D. Kell, *et al.*, *J. Theor. Biol.*, 2005, **234**(4), 497.
51. D. M. Kurnit, *Gene*, 1989, **82**(2), 313.

CHAPTER 9

Site-directed Chemical Modification of Phage Particles

LANA SALEH AND CHRISTOPHER J. NOREN

New England Biolabs, 240 County Road, Ipswich, MA 01938, USA

9.1 Introduction

Combinatorial peptide libraries offer a powerful tool for studying molecular recognition by a diverse set of macromolecules, including antibodies, enzymes and cell-surface receptors. As an alternative to chemical synthesis, biologically expressed peptide libraries can be prepared by fusing randomized amino acid sequences to coat proteins of filamentous bacteriophages such as fd and M13.[1] Such libraries are constructed by engineering randomized oligonucleotides into the gene of either the minor coat protein pIII or major coat protein pVIII. Fusion to pIII results in five copies of displayed peptide/virion whereas fusion to pVIII results in ~2700 copies of displayed peptide/virion. Since a physical linkage exists between the displayed peptide and the encoding phage gene, the amino acid sequence of the selected peptide can be readily identified by sequencing the corresponding coding region of the viral DNA.[2,3]

Selection for phage-displayed peptide ligands, with specific binding properties to the antibody or protein of interest (target), usually involves several rounds of panning, elution and amplification. Panning is an *in vitro* selection process in which ligand-containing phage are affinity purified out of a library.[4] The unbound phage are washed away, followed by elution of the bound phage. The enriched pool of bound phage is then amplified through

RSC Nanoscience & Nanotechnology No. 17
Phage Nanobiotechnology
Edited by Valery A. Petrenko and George P. Smith
© Royal Society of Chemistry 2011
Published by the Royal Society of Chemistry, www.rsc.org

infection of bacterial culture. Several rounds of binding and amplification result in iterative enrichment of the pool for phage displaying target-binding peptides.

One crucial condition for the success of a panning experiment lies in the greater diversity of clones within a given library, *i.e.*, the complexity of the library. Methods for construction of high-complexity phage-displayed peptide libraries ($\geq 10^9$ independent clones) have been described in the literature.[5,6] Such high complexity offers phage display an advantage over synthetic combinatorial peptide libraries, which can incorporate extremely diverse functionalities but are size limited (complexity $< 10^6$) by the required spatial addressing of molecules within the library (*e.g.*, position within a grid). Nonetheless, an inherent limitation of phage-displayed libraries resides in the restricted diversity of the displayed peptides to combinations of the 20 canonical amino acids. To increase the functional diversity of phage libraries, small molecules can be introduced *via* covalent linkers to reactive moieties (such as amines, thiols and hydroxyl groups)[7] on coat proteins. Despite the wide range of functionalities that can be introduced using such methods, a major disadvantage is non-specific targeting of the molecules to reactive residues throughout the native coat proteins. In this chapter, we describe a site-specific labeling methodology which utilizes the uniquely reactive 21st amino acid residue,[8] selenocysteine, encoded into the pIII peptides of M13 phage. The specific tethering of molecules to Sec prior to each round of panning creates additional functional diversity while maintaining the benefits of phage-displayed libraries.

9.2 Unique Chemical Properties of Selenocysteine Compared with Cysteine

Selenocysteine (Sec) is an analog of cysteine (Cys), with a single chalcogen atom difference, selenium in place of sulfur. These two elements share several chemical properties [*e.g.*, chemical group, electronegativity (Se, 2.4; S, 2.5), oxidation states] and vary in many others [*e.g.*, atomic radius (Se, 1.17 Å; S, 1.04 Å), electron configuration, bond length, polarity and pK_a (3.73 for H_2Se; 6.96 for H_2S)].[9,10] The differences result in Sec displaying unique nucleophilic and redox properties as compared with Cys. At physiological pH, the selenol group of free Sec ($pK_a = 5.2$) is 99% ionized, whereas the thiol of free Cys ($pK_a = 8.3$) is mostly protonated. This suggests higher chemical reactivity for Sec compared with Cys, at least in aqueous environments. The situation may vary in protein microenvironments, where pK_a values may be altered by neighboring amino acid residues. Therefore, as has been stressed in recent reviews on the subject,[11,12] the increased reactivity of Sec compared with Cys cannot be attributed exclusively to a lower pK_a value but probably to a combination of chemical and environmental factors.

Differences in chemical properties also result in diselenenyl bonds possessing a lower redox potential ($E_0 = -488$ mV; Sec–Sec *versus* Sec) compared with disulfide bonds ($E_0 = -233$ mV; Cys–Cys *versus* Cys), as measured at neutral pH.[9] Nature does not seem to exploit diselenide bonds as extensively as disulfide bonds for protein structure, stability or redox chemistry. It has been hypothesized that the rare occurrence of such proteins is due to the very low redox potential of diselenide bonds, which may require a very strong reductant. To date, only one family of selenoproteins, SelL, has been identified to harbor a diselenide bond, which is formed between two Sec residues in a UXXU motif (U being Sec).[13] SelL is present in diverse aquatic organisms, including fish, invertebrates and marine bacteria. The fold bearing the UXXU motif of SelL mimics that harboring the catalytic CXXC of thioredoxin. It is believed that, similarly to thioredoxin, SelL utilizes its diselenide bond in a catalytic redox function that is yet to be identified.

Proteins containing Sec are expressed in all kingdoms of life. Genomes of certain organisms, such as yeasts and higher plants, do not encode selenoproteins.[9] Most selenoproteins are involved in redox reactions.[9] Examples of identified selenoproteins include glutathione peroxidase (GPx), which is the first selenoprotein identified in mammals[14] and which catalyzes the reduction of hydroperoxides using an active-site Sec and a glutathione substrate molecule;[15–17] mammalian thioredoxin reductases (TrxR isoenzymes), which are responsible for NADPH-dependent reduction of a wide variety of substrates, including thioredoxin (Trx) and lipid hydroperoxides *via* a catalytically essential Sec residue located at the penultimate position at the C-terminus of the protein to perform their function;[18–23] and *Escherichia coli* formate dehydrogenase H (fdhH), which catalyzes the conversion of formate to carbon dioxide by utilizing an iron–sulfur cluster (Fe_4S_4), a Sec-coordinated Mo atom and two molybdopterin–guanine dinucleotide cofactors.[24–28]

Interestingly, the majority of selenoproteins identified to date in certain organisms have homologs that contain Cys in place of Sec in other organisms (*e.g.*, lower eukaryotes often contain Cys homologs of vertebrate selenoproteins).[16] The replacement of Sec with Cys in variants of studied selenoenzymes has almost always resulted in a decrease in the catalytic activity of the corresponding enzyme.[29] Also, comparison of the determined kinetic parameters for several selenoenzymes with their natural Cys homologs shows higher k_{cat} values for selenoproteins.[29,30] As a result, it has been suggested that selenoproteins evolved from their Cys counterparts as a result of an increased reactivity requirement.[9,31] However, this view is open to debate since increased reactivity could be compensated for by a higher number of enzyme molecules without employing the metabolically expensive Sec incorporation system. Regardless of the questions that remain unanswered concerning the evolution of selenoproteins and their representative presence in certain domains of life, selenoproteins are unique compared with their Cys homologs due to their inherent reactivity over a broad pH range and towards a wide range of substrates,[32] a characteristic that makes them very appealing for biotechnological use.

9.3 *In vivo* Incorporation of Sec by *E. coli*

Sec is incorporated in response to a UGA (opal) codon, which is translationally recoded from a stop codon to a Sec codon by sequence context in mRNA. In prokaryotes, the co-translational incorporation of Sec at the UGA codon[33] requires the products of four genes, *selA*, *selB*, *selC* and *selD*,[29,34] and also a downstream *cis*-acting stem–loop structure [selenocysteine insertion sequence (SECIS)] in mRNA[35,36] (Figure 9.1A). The *selC* gene product, a Sec-specific tRNA (tRNA^{Sec}) containing a UCA anticodon, is first charged with a Ser residue by seryl-tRNA synthetase.[37,38] The loaded serine is in turn converted into a selenocysteinyl moiety by selenocysteine synthase (SelA), which uses selenophosphate, provided by selenophosphate synthetase (SelD), as the selenium donor.[37,39,40] The Sec-specific elongation factor, SelB, mediates Sec insertion at the UGA codon by binding GTP and Sec-tRNA^{Sec},[41,42] and also the mRNA SECIS,[43] in place of EF-Tu. The formation of this quaternary complex, SelB–Sec-tRNA^{Sec}–GTP–[mRNA SECIS], is believed to promote the situation of Sec-tRNA^{Sec} in close proximity to the UGA codon at the ribosomal A site for efficient translation.[43–45]

Several groups have attempted to define the recognition elements within SECIS to elucidate further the mechanism of Sec insertion and to minimize sequence constraints. In *E. coli*, the UGA stop codon is recoded to Sec in genes expressing formate dehydrogenases H, N and O.[46–48] For both fdhH and formate dehydrogenase N (fdhN), which are encoded by *fdhF* and *fdnG*, respectively, the presence of a 38-nucleotide long SECIS immediately downstream of the UGA-stop codon has been confirmed to be essential for Sec incorporation.[36] *In vitro* studies, carried out by Böck and co-workers, on *fdhF* SECIS have shown that the mRNA motif recognized by the C-terminus of SelB can be reduced to a 17-nucleotide mini-helix, consisting of the upper part of the stem–loop RNA structure (a loop structure of six nucleotides) and an adjacent five base-paired helical stem with one bulged nucleotide (Figure 9.1B), without loss of binding affinity or specificity.[49] The same study demonstrated that the N-terminus of *E. coli* SelB is homologous to the elongation factor EF-Tu and is responsible for binding of Sec-tRNA^{Sec}. A later *in vivo* study, performed by Engelberg-Kulka and co-workers, confirmed this finding and provided further evidence that the loop structure has to be comprised of specific nucleotides, hence the name 'invariant loop', whereas the stem structure requires non-specific but paired nucleotides with a bulged U at position $+17$ (Figure 9.1.B).[50] They further demonstrated that the lower stem domain of *fdhF* SECIS, comprised of nucleotides $+4$ to $+14$ and $+32$ to $+41$ (Figure 9.1B), has no pairing or sequence requirements to ensure Sec incorporation. The only identified role of this domain is to assure a crucial distance of 11 nucleotides between the UGA codon and the upper stem of the SECIS.

Major differences in location and structural features of SECIS elements appear in eukaryotes[51,52] and archaea[53–55] compared with bacteria. The SECIS element is located in the 3′-untranslated (3′-UTR) region of the selenoprotein mRNA at a variable distance of 51–111 nucleotides from the Sec-encoding

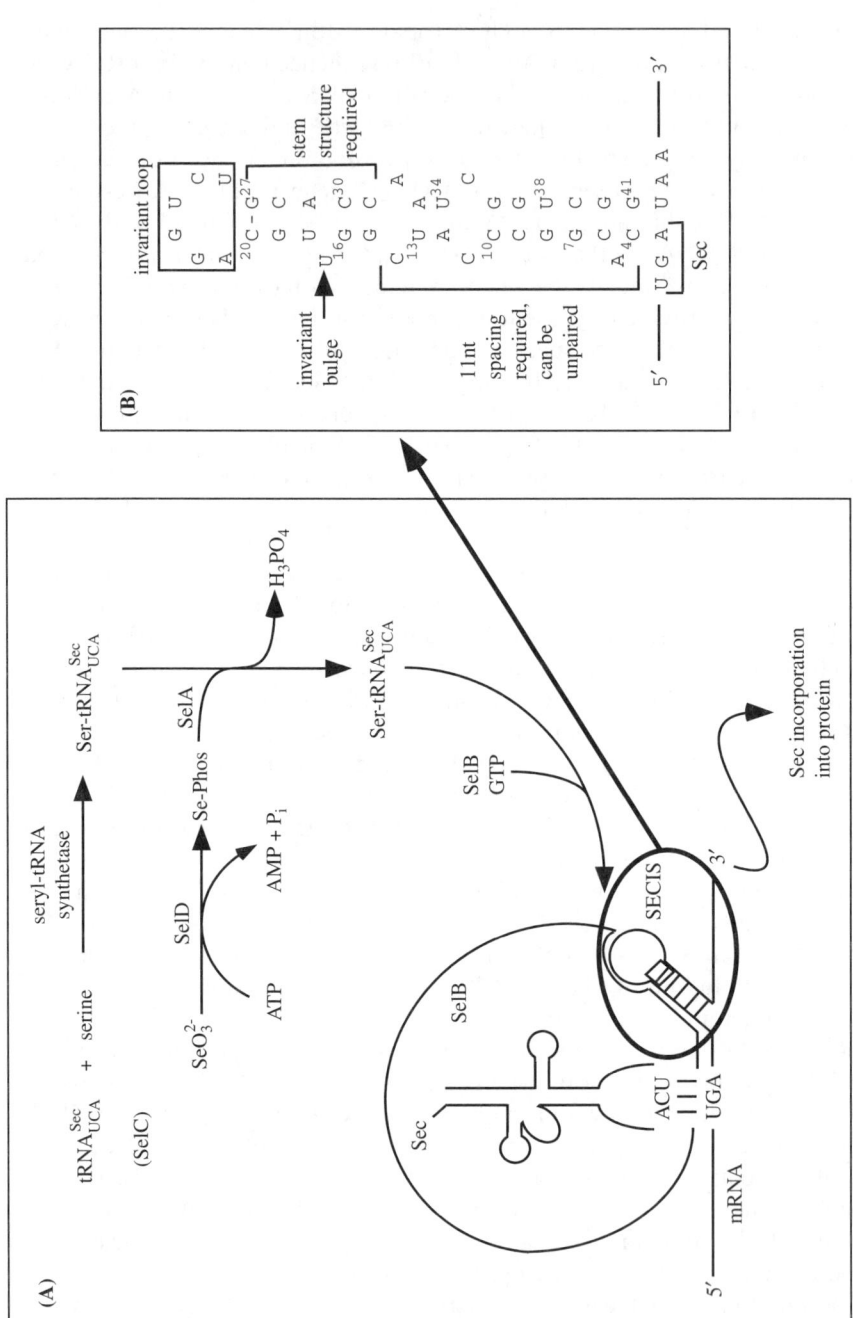

Figure 9.1 (A) Biosynthetic pathway for co-translational Sec incorporation in *E. coli* involves four essential proteins, SelA, SelB, SelC and SelD. (B) The secondary structure of the *E. coli fdhF* mRNA (SECIS), with allowable mutations (as reported in Heider *et al.*,[44] Liu *et al.*[50] and Klug *et al.*[65]).

UGA codon (compared with being immediately adjacent to the UGA codon in known selenoprotein bacterial genes).[16,35,56] More than one Sec–UGA codon can be present in eukaryotic and archaeal selenoprotein mRNAs, with a single SECIS element directing Sec insertion. Furthermore, two proteins, SBP2 and SelB, carry out the identified dual function of *E. coli* SelB, where SBP2 is the SECIS binding protein[57,58] and SelB is the translation elongation factor. These characteristics allow for the insertion of Sec at any UGA codon in the coding region, regardless of the adjacent sequence, and permit the utilization of one stem–loop structure for the insertion of multiple Sec residues.

On an evolutionary note, while the close proximity of the *cis*-acting factors in *E. coli* and other bacteria ensures Sec insertion with high efficiency, it imposes certain downstream amino acid constraints by requiring specific primary sequence and secondary structure of the SECIS element immediately downstream of the UGA codon. Such a compromise results in the restriction of selenoprotein diversity in bacteria. The lack of such evolutionary constraints in archaea and eukaryotes may explain the wider variety of selenoproteins observed in these organisms.

The divergence of SECIS elements in the three domains of life has made recombinant selenoprotein expression in *E. coli* challenging. *E. coli* SelB is highly specific for the SECIS elements of the *E. coli* formate dehydrogenase isoenzymes,[44,50] which in turn have little sequence conservation with SECIS elements from other organisms, even those from other bacteria. Furthermore, 'rules' for a SECIS structure are not easily defined. Despite the growing number of alternative approaches for obtaining selenoproteins, such as intein-mediated protein expression technology or expression in cultured *Chlamydomonas reinhardtii* plant cells, expression in *E. coli* remains a viable and attractive alternative. In Section 9.5, we discuss the implementation of a phage display method to examine the strictness of the *E. coli* SECIS requirements in an attempt to establish the minimum conditions for Sec insertion into translated gene products.

9.4 Construction of Selenopeptide-displayed Phage Libraries

Introduction of Sec into randomized peptides displayed on the surface of phage vastly expands the potential molecular and functional diversity of the resulting libraries. This is due to the ability to modify Sec selectively with molecules of diverse chemistry prior to panning, by virtue of its enhanced nucleophilicity at neutral pH compared with Cys. Our approach for constructing an M13 phage-displayed selenopeptide library in *E. coli* utilizes two important features of M13: (1) M13-phage infectivity is dependent on the expression of the infectivity protein pIII and (2) M13 proteins do not naturally contain Sec. The fusion of selenopeptides to the N-terminus of pIII should, therefore, couple UGA-opal suppression to phage production, resulting in quantitative insertion of a uniquely reactive group at a defined position in the phage coat. Based on these

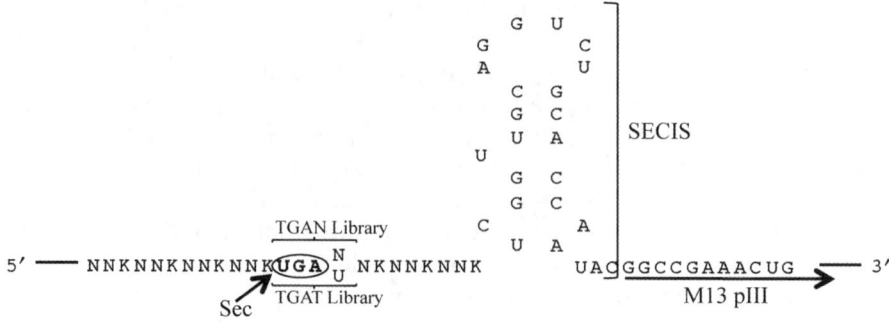

Figure 9.2 Description of the randomized SECIS (TGAT) and (TGAN) library inserts expressed as fusions to pIII.[60] N = A, G, C or U; K = G or U.

requirements, we constructed a library, which consists of the UGA codon with four upstream and three downstream randomized codons and a minimal mRNA SECIS (as defined by the work of Böck and co-workers[49] and Engelberg-Kulka and co-workers[50]) into the commonly used phage display vector M13KE[59,60] (Figure 9.2). Two other criteria had to be considered upon planning this library: (1) high levels of protein synthesis, required for plaque formation, would necessitate a selenium source for selenopeptide-phage display; and (2) enhancement of the endogenous tryptophan-inserting opal-suppression pathway by downstream purines and CUG codons would demand avoiding certain downstream nucleotides upon construction of the library. The first factor was addressed by supplementing the medium with sodium selenite,[59,60] and the second was tackled by preparing the library with the immediate-downstream nucleotide fixed as U (TGAT library) and the eight following nucleotides randomized[59,60] (Figure 9.2). The diversity at the amino acid position downstream of Sec in the TGAT library is thus restricted to F, L, S, Y, C and W.

The library insert was ligated into M13KE vector, which is derived from M13mp19 and is designed with *Acc65I* and *EagI* restriction sites for N-terminal pIII display.[61] This vector also carries the *lacZα* fragment, which results in characteristic blue plaques when plated with an α-complementing strain on X-gal medium. As controls, we used empty M13KE phage and clones displaying either a single Sec (Sec-1, displayed peptide SARV-Sec-HGP, corresponding to the native *E. coli* fdhH SECIS with no randomized codons and a full-length stem) or a single Cys (Cys-1, displayed peptide SARVL-Cys-NH, which carries a G to T point mutation within the opal codon).

To determine whether Sec was being inserted into the N-terminal pIII fusions of the identified phage clones, each clone was assayed with EZ-link-iodoacetyl-LC-biotin (I-Bt) (Thermo Scientific) at pH 2.5.[59,60] I-Bt is an electrophilic reagent that should react selectively with Sec and not Cys at this pH. Immunoblotting with an anti-biotin antibody was carried out to detect biotin modification on Sec residues on pIII. Parallel blots with anti-pIII monoclonal antibody (New England Biolabs), which measures pIII expression independent of its modification, was carried out to control for differences in phage loading. Our results showed specific biotinylation of clones with detected Sec in the sequence. No biotinylation signals were observed from clones with no Sec or from the negative control Cys-1.

All of the clones assumed to display selenopeptides formed plaques only in the presence of supplemental selenium. Plaque count and size were dependent on the concentration of selenite added to the plating medium. Furthermore, and as explained in more detail in Section 9.5.1.1, a high occurrence of clones encoding Cys proximal to the TGA codon was detected. This is further evidence for the incorporation of Sec since selenide–sulfide bridging would stabilize the otherwise unpaired Sec residue.

9.5 Applications Using Selenopeptide Phage Display

The Sec handle on the displayed peptides of these phage libraries is a powerful tool for various applications. Examples of such applications include, but are not limited to, screening for enzyme inhibitors by appending substrate- or transition-state analogs to the displayed Sec, then panning the modified-peptide libraries against the enzyme of interest to select flanking amino acids which increase the affinity and specificity of these analogs; discovery of new peptide–drug complexes by linking cytotoxic agents to the Sec followed by specific delivery to the target *in vivo*; screening for Sec insertion requirements by a phage-based *in vivo* screen; directed evolution of enzymes with novel or enhanced functionalities; and single-molecule biophysics.

Our laboratory is involved in the development and testing of several of these applications, some of which are discussed in more detail in the following sections.

9.5.1 Screening for Sec Insertion *in vivo:* Investigating the Stringency of *E. coli* SECIS Requirements Using Phage Display

9.5.1.1 Effect of the Immediate Downstream Nucleotide to the Opal Codon on Sec Insertion

For this study, we prepared a library with the position downstream of the UGA-opal codon fully randomized (TGAN library)[60] (Figure 9.2). This allowed us to identify any bias in nucleotide selection at this position and to understand the effect exerted by different nucleotide combinations on the opal suppression pathway. Both libraries, the TGAT library and TGAN library, were tested for Sec insertion into pIII, and also the effect of sodium selenite addition on phage production.

DNA sequencing of clones selected from the TGAN library revealed that 74% of the clones possessed a downstream purine and that the frequency of

single Cys insertion into the displayed peptides was <0.5% (the expected frequency is 3.1% based on random codon usage).[60] It was also discovered that
the presence of sodium selenite in the medium did not affect phage production
or plaque size for this library, with the exception of clones containing a single
Cys codon, but did increase the level of Sec insertion.[60] This was particularly
true for a library that combined the immediate downstream CTG codon and
the mRNA SECIS. Clones with an opal codon immediately followed by a
pyrimidine grew in an Se-dependent fashion. On the other hand, plaque count
and size of clones obtained from the TGAT library were strictly dependent on
selenite, an indication of a dominating Sec-insertion opal suppression pathway.[60] It should be noted, however, that occasional opal-codon mutations
(<10% frequency) were detected during clone growth and amplification even
with supplemental selenite. Such spontaneous mutations are probably due to
the inefficient process of Sec insertion observed in *E. coli*. The observed Cys-
insertion frequency in the displayed peptides of the TGAT library was higher
than the normal expected frequency of 3.1% of all random amino acids. This
effect is presumably a result of formation of stable selenosulfide linkages, which
would stabilize both the Cys and Sec in the M13 display system.

Comparison of results from the TGAN and TGAT libraries led us to the
conclusion that a U or C (but not CUG) downstream from the UGA-opal
codon, followed by a SECIS, exclusively directs Sec insertion. In the case of
UGA U, the Sec insertion pathway is strongly favored, resulting in more
homogeneous-displayed peptides. The disadvantage of this homogeneity is the
increased probability of the selection of adventitious mutations. Alternatively,
in the case of a downstream purine (A or G), phage production is maximized
due to maximum employment of the endogenous Trp-opal suppression pathway accompanied with co-translational Sec insertion. It is important to note,
however, that in the case of a UGA codon, which is followed by an immediate
G and preceded by codons encoding for Arg–Val (as in the native *E. coli fdhf*
mRNA sequence), the tendency of the downstream G to direct Trp-inserting
opal suppression in the absence of selenium is counteracted by the effect of the
Arg and Val codons to prevent UGA readthrough.[62,63]

9.5.1.2   Essential Elements of the E. coli SECIS

Nucleotides in positions +4 to +12 (numbering from the first nucleotide of
the UGA codon) were randomized as NNK for each codon in the TGAN
library (Figure 9.2), where N is any nucleotide and K is either a G or T. The
SECIS variants obtained showed no base pairing with the sequence downstream from the stem–loop structure and with the Sec-insertion machinery still
operable.[59,60] In other words, there is no selection bias in favor of a paired
lower stem. We further prepared libraries in which we randomized the lower
stem nucleotides +13, +14, +32 and +33 to obtain SECIS variants with all
possible nucleotide combinations at these positions.[64] Again, no combination
blocked Sec insertion at the opal codon, confirming the previously suggested

notion that the lower stem nucleotides in positions $+4$ to $+12$ could be randomized in the context of the wild-type (wt) upper stem and loop structures of SECIS and still allow efficient Sec insertion.[50]

Finally, we prepared a SECIS library with no variations in the nucleotides constituting the six-nucleotide loop and the C_{20}–G_{27} pair (no mutations were introduced into this sequence of eight nucleotides in any of the characterized clones from previously studied libraries) and with a 70:10:10:10 randomization of the stem in positions $+13$ to $+19$ and $+27$ to $+33$ (bold nucleotides in Figure 9.3), where positions randomized had 70% of the native nucleotide at each position and 10% of the other three nucleotides.[64] Our findings revealed that the upper stem was unexpectedly more tolerant to mutations than previously predicted. The flexibility for these variations increased with more spacing from the loop structure [compare the number of functional variants for positions ($+19$ and $+28$) and ($+18$ and $+29$) with those of ($+15$ and $+31$) and ($+16$ and $+30$) in Figure 9.3]. Interestingly, clones prepared with a non-native upper stem and an unpaired lower stem showed permissible selenium-dependent UGA readthrough in a lacZ fusion reporter assay, but at less than half the levels as their parent clones (non-native upper stem with a paired lower stem). Similar observations had been reported previously: a SECIS with G at position $+15$ mutated to C and an unpaired lower stem had no UGA readthrough;[50] also, a clone selected from an RNA aptamer library for its SelB-binding ability and prepared with variant upper and lower stems had no UGA readthrough. On the other hand, 17% suppression of the opal codon was observed with the same variant upper loop and wt lower loop.[65] It is therefore safe to conclude that a minimal level of pairing is required for the lower stem to allow for the SECIS element to be recognized by SelB during translation.

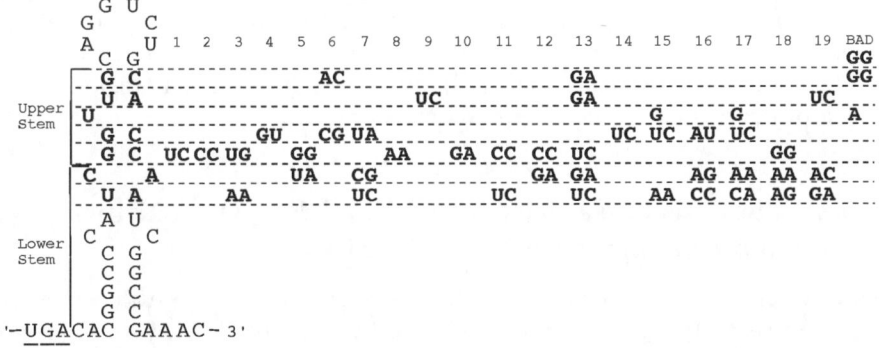

Figure 9.3 Summary of the wt *fdhF* SECIS randomized stem positions (indicated in brackets) and sequences of the SECIS functional variant clones (1–19) as tested in Sandman *et al.*[64] Sequence variations are shown in the position of the corresponding nucleotide in the wt fdh SECIS, although the pairings may differ from the native structure. BAD corresponds to a non-functional SECIS variant clone studied in the same work.

It has been reported previously that a 'bulged' U17 is essential for SelB recognition of the SECIS element.[50,66] From our 70:10:10:10 SECIS library, we identified two functional variants, clone 15 and clone 17 (Figure 9.3), with U17 mutated into G.[64] Interestingly, G in position + 16 has been replaced with U in both of these variants. In addition, U in position + 18 did not undergo any change in both of these variants. Hence despite the absence of U at position + 17, the presence of U in either position + 16 or + 18 might serve as a recognition element for SECIS binding by SelB.

9.5.1.3   *Enhancement of Selenopeptide Expression in the Presence of pSelABC*

In our earlier work, we screened for functional SECIS variants from our combinatorial libraries using *E. coli* ER2738 cells [F′ *proA*$^+$*B*$^+$ *lacI*^{q} *Δ(lacZ)M15 zzf::Tn10 (Tet*^{R}*)/fhuA2 glnV thi-1 Δ (lac-proAB) Δ (hsdS-mcrB)5*].[59,60] The majority of phage clones amplified in this strain carried mutations in the UGA codon, suggesting instability of the Sec-inserting clones. We therefore resorted to propagating phage in *E. coli* ER2738 cells carrying the pSelABC plasmid, which overexpresses the *selA*, *selB* and *selC* genes involved in Sec incorporation (SelD being naturally present and sufficiently expressed in *E. coli*). The constructed strain significantly enhances the genetic stability of phage containing the TGA codon, leading to increased numbers of Sec-containing phage clones upon amplification.[64] SECIS variant clones, with incorporated Sec, obtained *via* this amplification procedure were functional in the normal *E. coli* background, as shown by β-galactosidase reporter assays performed in ER2738 cells without pSelABC.

In summary, our studies on combinatorial phage libraries of SECIS variants have established minimal conditions for overexpression of heterologous selenoproteins in *E. coli*. These conditions simply entail (1) overexpressing the *selABC* genes and (2) designing a minimum SECIS sequence that allows expression of most of the native downstream amino acid sequence of the protein, but which preserves the base-pairing, length of the stem region and as much of the sequence of the loop region as possible.

9.5.2   Catalysis-based Selection of Novel Enzyme Activities from Substrate-appended Phage Libraries

Directed enzyme evolution studies often use *in vivo* procedures such as random mutagenesis and/or gene shuffling followed by screening for new or improved properties of the gene products using cell-based selection techniques. Phage display is normally used for affinity-based selections (panning), but it can also be adapted for catalysis-based selection by basing the panning step on affinity selection for the desired reaction product. This requires simultaneous display of the enzyme library and the substrate on the phage, with the substrate spatially accessible to the enzyme. An additional requirement is an affinity-selection method for the desired product. This technique can be used to screen expressed cDNA libraries for a desired activity, or for evolving enzymes to have different substrate specificity or reaction chemistry (*i.e.*, generate a different product). Schultz and co-workers pioneered this technique, attaching the substrate to the phage *via* a non-covalent coiled-coil interaction in which one helix is displayed on the phage and the complementary helix is chemically synthesized and appended to the substrate molecule.[67] This technique was subsequently used to evolve DNA polymerases to accept ribonucleotides, with the affinity selection based on incorporation of a biotinylated ribouridine.[68] The drawbacks of this approach are the non-covalent attachment of the substrate to the phage, which precludes panning under chemically stringent conditions, and the size of the substrate–α-helix chimera, which precludes high concentrations necessary to modify all of the phage in the pool.

An alternative methodology, developed by Kossiakoff and co-workers, couples synthesized peptide fragments to protein fragments bearing randomized sequences expressed either monovalently or multivalently on filamentous phage *via* intein-mediated protein ligation (also known as native chemical ligation).[69] The advantage of such a method is the ability to introduce unnatural amino acids with modifications linked covalently. However, one major drawback is the long distance between the prosthetic group and the randomized library.

We envisioned that incorporation of the uniquely reactive functional group, Sec, on the surface of the phage would present solutions for the above problems since this amino acid permits the use of small electrophilic compounds for regiospecific covalent phage modification. The *modus operandi* would include immobilizing a substrate molecule *via* Sec attachment proximal to a phage-bound library of enzyme variants, thereby allowing affinity capture of the desired product. Use of an M13 phagemid/helper phage system yields two different types of pIII fusions displayed on the surface of the phage: the pIII expressed from the phagemid would have the enzyme library fused to its N-terminus, while the adjacent pIII peptides expressed from the helper phage would have the substrate tethered to the displayed Sec (Figure 9.4). The covalent linkage of substrates to phage-displayed peptides is expected to be more stable to rigorous reactions and screening conditions that have been shown to disrupt non-covalent linkages in other tested phage-mediated enzyme evolution procedures. The simultaneous display of enzyme and Sec-linked substrate on the same phage surface would solve the spatial accessibility issue between the two.

This method was subsequently used, in collaboration with Suzanne Walker and co-workers, in setting up a system for glycosyltransferase evolution[70] using *E. coli* glycosyltransferase (Gtf) MurG, an enzyme essential in bacterial cell wall biosynthesis, as the prototype. MurG condenses cell wall precursor lipid I with UDP–*N*-acetylglucosamine (UDP-GlcNAc) to form lipid II with release of UDP. The choice of MurG was based on its being a member of the GT-B superfamily of Gtfs, which share related three-dimensional structures but different-substrate selectivity (Gtfs are classified into two structural superfamilies, GT-A and GT-B, each containing various families, which do not necessarily

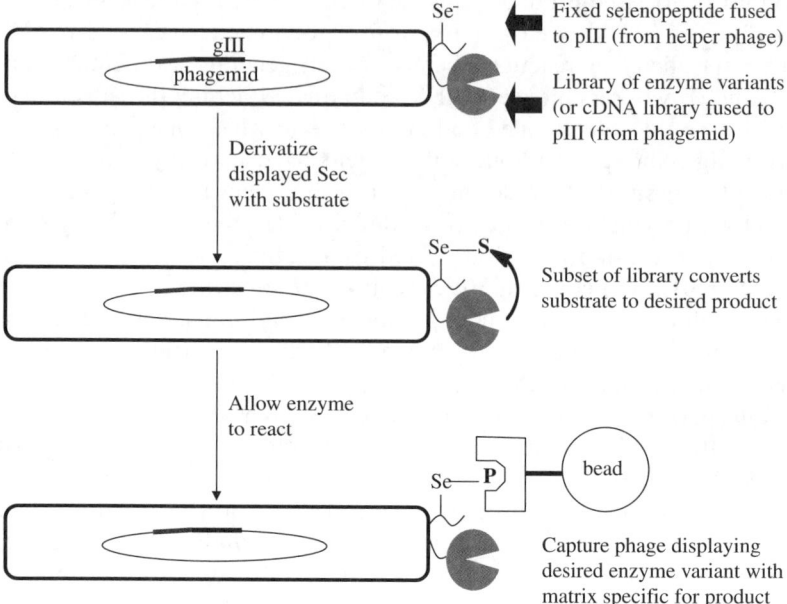

Figure 9.4 Method summarizing the concomitant display of enzyme and substrate on the surface of M13 phage and the capture of enzyme-bound product by affinity-based procedures.

share significant sequence similarities), indicating a modular structure amenable to evolution.

pMurG–pIII phagemid was constructed by inserting a PCR product bearing *MurG* to the 3′-end of a *pelB* leader sequence and the 5′-end of a truncated *gIII* into a similarly digested-phagemid vector (pFAB5cHis.TT.HUI). The M13–Sec helper phage vector was the ligation product of an M13KO7 helper-phage fragment containing the attenuated origin of replication and an M13KE fragment containing the SECIS sequence fused to *gIII*. Production of phage bearing both MurG and the Sec handle on the same end of the phage particle was demonstrated by infecting cultures containing the pMurG–pIII phagemid and M13–Sec helper phage in the presence of sodium selenite (as a negative control, phage displaying wt pIII was propagated using media infected with pMurG–pIII phagemid and M13KO7 helper phage). Incubation of this phage with [^{14}C]UDP–GlcNAc and biotinylated lipid I resulted in the formation of radioactive lipid II, which was captured on a streptavidin-coated membrane. Analysis of the time course of formation of radioactive lipid II showed that the enzyme was active and performed between 10^3 and 10^4 turnovers. The presence of Sec in the pIII fusions was demonstrated using I-Bt as substrate at pH 2.5, as described earlier in this chapter (immunoblots of biotinylated Sec- and M13KO7-derived phage show a positive chemiluminescent signal for the former and no signal for the latter with anti-biotin antibody).

In another experiment, the biotin-modified Sec-derived phage was trapped on avidin resin. Recovery of the modified phage was $\sim$ 100-fold greater than that of control phage (untreated Sec phage or M13KO7 control phage). This result is very encouraging since the ability to capture the phage after Sec modification is a prerequisite in the selection process to allow for enrichment of the pool with the desired modified phage. The recovered phage was tested for infectivity in its bead-bound form and free form. As expected, the infectivity of the bead-bound form was compromised compared with free phage.

The above study demonstrated the ability to display an active form of an enzyme on the surface of phage and that a substrate could be appended *via* the displayed Sec in a stable manner, thus establishing a solid basis for future directed-enzyme evolution work. Since our publication, a study utilizing a similar phagemid/helper phage approach successfully demonstrated the simultaneous display of an Sfp phosphopantetheinyl transferase library and its peptide substrate, ybbR, on the surface of M13 phage. Both the Sfp enzyme and ybbR peptide were expressed as N-terminal fusions to pIII of both pha-gemid and helper phage, respectively.[71] Phage selection for Sfp mutants, with activity towards biotin-conjugated coenzyme A analogs, was demon-strated by covalent transfer of biotin from the CoA analog to the displayed ybbR peptide. Biotin-labeled phage particles carrying mutant enzyme with desired activity were then pulled down on streptavidin beads and used for further rounds of pool enrichment with phage displaying Sfp mutants with activity toward the substrate analog. This system could be extended towards evolving any enzyme of interest by displaying a library of the enzyme on phage and using Sfp in solution to transfer the desired substrate from a CoA con-jugate to phage-borne ybbR peptide, then selecting for phage-bound product as described above. The disadvantage of this system compared with the Sec-based system described here is that it requires separate steps of modifying CoA with the desired substrate and then enzymatically transferring the substrate to the phage, whereas the Sec method allows modification of the phage with the substrate in a single step.

9.5.3 Mechanical Manipulation of M13 Phage

M13 has been widely used in the preparation of combinatorial libraries and display of various molecules. A recent and rapidly growing application of M13 includes the exploitation of this phage in the development of semiconducting/magnetic nanowires, lithium ion battery electrodes and other nanotechnology applications.[72-74] To facilitate advancement of such technologies and possible innovation of other applications, understanding the mechanical properties of M13 becomes essential.

For direct mechanical manipulation of single M13 virions, Belcher and co-workers utilized a hetero-bifunctional M13 phage bearing biotin molecules linked to a selenopeptide at its proximal end (selenopeptide phage was prepared in our laboratory) and (His)$_6$ epitopes fused to pIX at its remote end and suspended it between an anti-(His)$_6$ antibody-coated surface and a streptavidin-coated microsphere.[75] The M13 molecules were then stretched using a focused laser in an optical trap with nanometer-scale resolution. Physical parameters such as applied force and extension components were computed using the appropriate mathematical equations and models.

9.6 Conclusion

As phage display is uniquely suited for screening vast combinatorial libraries, tethering small molecules to the surface of phage in a stable and selective fashion permits both selection of chimeric semisynthetic ligands and catalysis-based selection of novel enzyme activities. The insertion of a DNA sequence containing the UGA-opal codon and a SECIS element at the 5'-end of *gIII* of M13 phage results in the expression of an N-terminal selenopeptide fused to the coat protein pIII. The superior and orthogonal chemical reactivity of Sec, as compared with other residues, such as Cys and Lys, permits its selective modification with a variety of small molecules, such as pharmacophores, prior to each panning step. Such covalent modification vastly increases the molecular diversity of the displayed peptides in the library and is stable to rigorous manipulations during *in vitro* selection. For these reasons, selenopeptide-displayed phage libraries should be considered for applications involving enzyme evolution, phage nanofabrication, selection of novel semisynthetic ligands for a target of interest, and medical imaging.

References

1. G. P. Smith and V. A. Petrenko, *Chem. Rev.*, 1997, **97**, 391–410.
2. D. J. Rodi and L. Makowski, *Curr. Opin. Biotechnol.*, 1999, **10**, 87–93.
3. D. R. Wilson and B. B. Finlay, *Can. J. Microbiol.*, 1998, **44**, 313–329.
4. S. F. Parmley and G. P. Smith, *Gene*, 1988, **73**, 305–318.
5. K. A. Noren and C. J. Noren, *Methods*, 2001, **23**, 169–178.
6. S. S. Sidhu, H. B. Lowman, B. C. Cunningham and J. A. Wells, *Methods Enzymol.*, 2000, **328**, 333–363.
7. L. Dente, C. Vetriani, A. Zucconi, G. Pelicci, L. Lanfrancone, P. G. Pelicci and G. Cesareni, *J. Mol. Biol.*, 1997, **269**, 694–703.
8. A. Böck, K. Forchhammer, J. Heider and C. Baron, *Trends Biochem. Sci.*, 1991, **16**, 463–467.
9. L. Johansson, G. Gafvelin and E. S. Arnér, *Biochim. Biophys. Acta*, 2005, **1726**, 1–13.
10. L. A. Wessjohann, A. Schneider, M. Abbas and W. Brandt, *Biol. Chem.*, 2007, **388**, 997–1006.
11. E. S. J. Arnér, *Exp. Cell Res.*, 2010, **316**, 1296–1303.
12. R. J. Hondal and E. L. Ruggles, *Amino Acids*, 2010, DOI:10.1007/s00726-010-0494-6.

13. V. A. Shchedrina, S. V. Novoselov, M. Y. Malinouski and V. N. Glady-shev, *Proc. Natl. Acad. Sci. USA*, 2007, **104**, 13919–13924.
14. L. Flohe, W. A. Gunzler and H. H. Schock, *FEBS Lett.*, 1973, **32**, 132–134.
15. J. R. Arthur, *Cell Mol. Life Sci.*, 2000, **57**, 1825–1835.
16. G. V. Kryukov, S. Castellano, S. V. Novoselov, A. V. Lobanov, O. Zehtab, R. Guigo and V. N. Gladyshev, *Science*, 2003, **300**, 1439–1443.
17. F. Ursini, M. Maiorino, R. Brigelius-Flohe, K. D. Aumann, A. Roveri, D. Schomburg and L. Flohe, *Methods Enzymol.*, 1995, **252**, 38–53.
18. L. D. Arscott, S. Gromer, R. H. Schirmer, K. Becker and C. H. Williams Jr, *Proc. Natl. Acad. Sci. USA*, 1997, **94**, 3621–3626.
19. S. Gromer, L. D. Arscott, C. H. Williams Jr, R. H. Schirmer and K. Becker, *J. Biol. Chem.*, 1998, **273**, 20096–20101.
20. T. Sandalova, L. Zhong, Y. Lindqvist, A. Holmgren and G. Schneider, *Proc. Natl. Acad. Sci. USA*, 2001, **98**, 9533–9538.
21. L. Zhong, E. S. Arnér and A. Holmgren, *Proc. Natl. Acad. Sci. USA*, 2000, **97**, 5854–5859.
22. L. Zhong, E. S. Arnér, J. Ljung, F. Åslund and A. Holmgren, *J. Biol. Chem.*, 1998, **273**, 8581–8591.
23. L. Zhong and A. Holmgren, *J. Biol. Chem.*, 2000, **275**, 18121–18128.
24. M. J. Axley, A. Böck and T. C. Stadtman, *Proc. Natl. Acad. Sci. USA*, 1991, **88**, 8450–8454.
25. S. Bar-Noy and J. Moskovitz, *Biochem. Biophys. Res. Commun.*, 2002, **297**, 956–961.
26. J. C. Boyington, V. N. Gladyshev, S. V. Khangulov, T. C. Stadtman and P. D. Sun, *Science*, 1997, **275**, 1305–1308.
27. V. N. Gladyshev, S. V. Khangulov, M. J. Axley and T. C. Stadtman, *Proc. Natl. Acad. Sci. USA*, 1994, **91**, 7708–7711.
28. Y. Zhang and V. N. Gladyshev, *Bioinformatics*, 2005, **21**, 2580–2589.
29. T. C. Stadtman, *Annu. Rev. Biochem.*, 1996, **65**, 83–100.
30. D. L. Hatfield, V. N. Gladyshev, J. Park, S. I. Park, H. S. Chittum, H. J. Baek, B. A. Carlson, E. S. Yang, M. E. Moustafa and B. J. Lee, *Comp. Nat. Prod. Chem.*, 1999, **4**, 353–380.
31. V. N. Gladyshev, in *Selenium: Its Molecular Biology and Role in Human Health*, ed. D. L. Hatfield, Kluwer, Dordrecht, 2001, pp. 99–113.
32. S. Gromer, L. Johansson, H. Bauer, L. D. Arscott, S. Rauch, D. P. Ballou, C. H. Williams, Jr., R. H. Schirmer and E. S. Arnér, *Proc. Natl. Acad. Sci. USA*, 2003, **100**, 12618–12623.
33. F. Zinoni, A. Birkmann, W. Leinfelder and A. Böck, *Proc. Natl. Acad. Sci. USA*, 1987, **84**, 3156–3160.
34. A. Böck, K. Forchhammer, J. Heider, W. Leinfelder, G. Sawers, B. Veprek and F. Zinoni, *Mol. Microbiol.*, 1991, **5**, 515–520.
35. M. J. Berry, L. Banu, Y. Y. Chen, S. J. Mandel, J. D. Kieffer, J. W. Harney and P. R. Larsen, *Nature*, 1991, **353**, 273–276.
36. F. Zinoni, J. Heider and A. Böck, *Proc. Natl. Acad. Sci. USA*, 1990, **87**, 4660–4664.

37. W. Leinfelder, E. Zehelein, M. A. Mandrand-Berthelot and A. Böck, *Nature*, 1988, **331**, 723–725.
38. A. Schön, A. Böck, G. Ott, M. Sprinzl and D. Soll, *Nucleic Acids Res.*, 1989, **17**, 7159–7165.
39. W. Leinfelder, K. Forchhammer, B. Veprek, E. Zehelein and A. Böck, *Proc. Natl. Acad. Sci. USA*, 1990, **87**, 543–547.
40. W. Leinfelder, T. C. Stadtman and A. Böck, *J. Biol. Chem.*, 1989, **264**, 9720–9723.
41. K. Forchhammer, W. Leinfelder and A. Böck, *Nature*, 1989, **342**, 453–456.
42. K. Forchhammer, K. P. Rucknagel and A. Böck, *J. Biol. Chem.*, 1990, **265**, 9346–9350.
43. C. Baron, J. Heider and A. Böck, *Proc. Natl. Acad. Sci. USA*, 1993, **90**, 4181–4185.
44. J. Heider, C. Baron and A. Böck, *EMBO J.*, 1992, **11**, 3759–3766.
45. S. Ringquist, D. Schneider, T. Gibson, C. Baron, A. Böck and L. Gold, *Genes Dev.*, 1994, **8**, 376–385.
46. B. L. Berg, C. Baron and V. Stewart, *J. Biol. Chem.*, 1991, **266**, 22386–22391.
47. G. Sawers, J. Heider, E. Zehelein and A. Böck, *J. Bacteriol.*, 1991, **173**, 4983–4993.
48. F. Zinoni, A. Birkmann, T. C. Stadtman and A. Böck, *Proc. Natl. Acad. Sci. USA*, 1986, **83**, 4650–4654.
49. M. Kromayer, R. Wilting, P. Tormay and A. Böck, *J. Mol. Biol.*, 1996, **262**, 413–420.
50. Z. Liu, M. Reches, I. Groisman and H. Engelberg-Kulka, *Nucleic Acids Res.*, 1998, **26**, 896–902.
51. P. R. Copeland, *Gene*, 2003, **312**, 17–25.
52. S. C. Low and M. J. Berry, *Trends Biochem. Sci.*, 1996, **21**, 203–208.
53. A. Böck, in *Selenium: Its Molecular Biology and Role in Human Health.*, ed. D. L. Hatfield, Kluwer, Dordrecht, 2001, pp. 7–22.
54. M. Leibundgut, C. Frick, M. Thanbichler, A. Böck and N. Ban, *EMBO J.*, 2005, **24**, 11–22.
55. M. Rother, A. Resch, R. Wilting and A. Böck, *Biofactors*, 2001, **14**, 75–83.
56. M. J. Berry and P. R. Larsen, *Biochem. Soc. Trans.*, 1993, **21**, 827–832.
57. P. R. Copeland and D. M. Driscoll, *J. Biol. Chem.*, 1999, **274**, 25447–25454.
58. P. R. Copeland, J. E. Fletcher, B. A. Carlson, D. L. Hatfield and D. M. Driscoll, *EMBO J.*, 2000, **19**, 306–314.
59. K. E. Sandman, J. S. Benner and C. J. Noren, *J. Am. Chem. Soc.*, 2000, **122**, 960–961.
60. K. E. Sandman and C. J. Noren, *Nucleic Acids Res.*, 2000, **28**, 755–761.
61. M. B. Zwick, L. L. Bonnycastle, K. A. Noren, S. Venturini, E. Leong, C. F. Barbas III, C. J. Noren and J. K. Scott, *Anal. Biochem.*, 1998, **264**, 87–97.
62. H. Engelberg-Kulka, Z. Liu, C. Li and M. Reches, *Biofactors*, 2001, **14**, 61–68.

63. Z. Liu, M. Reches and H. Engelberg-Kulka, *J. Mol. Biol.*, 1999, **294**, 1073–1086.
64. K. E. Sandman, D. F. Tardiff, L. A. Neely and C. J. Noren, *Nucleic Acids Res.*, 2003, **31**, 2234–2241.
65. S. J. Klug, A. Huttenhofer, M. Kromayer and M. Famulok, *Proc. Natl. Acad. Sci. USA*, 1997, **94**, 6676–6681.
66. C. Li, M. Reches and H. Engelberg-Kulka, *J. Bacteriol.*, 2000, **182**, 6302–6307.
67. H. Pedersen, S. Holder, D. P. Sutherlin, U. Schwitter, D. S. King and P. G. Schultz, *Proc. Natl. Acad. Sci. USA*, 1998, **95**, 10523–10528.
68. G. Xia, L. Chen, T. Sera, M. Fa, P. G. Schultz and F. E. Romesberg, *Proc. Natl. Acad. Sci. USA*, 2002, **99**, 6597–6602.
69. M. A. Dwyer, W. Lu, J. J. Dwyer and A. A. Kossiakoff, *Chem. Biol.*, 2000, **7**, 263–274.
70. K. R. Love, J. G. Swoboda, C. J. Noren and S. Walker, *ChemBioChem*, 2006, **7**, 753–756.
71. M. Sunbul, N. J. Marshall, Y. Zou, K. Zhang and J. Yin, *J. Mol. Biol.*, 2009, **387**, 883–898.
72. C. Mao, C. E. Flynn, A. Hayhurst, R. Sweeney, J. Qi, G. Georgiou, B. Iverson and A. M. Belcher, *Proc. Natl. Acad. Sci. USA*, 2003, **100**, 6946–6951.
73. C. Mao, D. J. Solis, B. D. Reiss, S. T. Kottmann, R. Y. Sweeney, A. Hayhurst, G. Georgiou, B. Iverson and A. M. Belcher, *Science*, 2004, **303**, 213–217.
74. K. T. Nam, D. W. Kim, P. J. Yoo, C. Y. Chiang, N. Meethong, P. T. Hammond, Y. M. Chiang and A. M. Belcher, *Science*, 2006, **312**, 885–888.
75. A. S. Khalil, J. M. Ferrer, R. R. Brau, S. T. Kottmann, C. J. Noren, M. J. Lang and A. M. Belcher, *Proc. Natl. Acad. Sci. USA*, 2007, **104**, 4892–4897.

CHAPTER 10

Filamentous Phage-templated Synthesis and Assembly of Inorganic Nanomaterials

BINRUI CAO AND CHUANBIN MAO

Department of Chemistry and Biochemistry, University of Oklahoma, Norman, OK 73019, USA

10.1 Introduction

Structured inorganic materials built at the nanometer scale have revealed unique properties that hold great promise for future technological development.[1-4] A recurring challenge has been assembling these inorganic materials into well-defined structures and integrating them into large-scale devices. The bottom-up approach to this challenge seeks to utilize nano-scaled building blocks for developing functional nanosystems.[5-7] However, it has proven difficult to achieve precise placement of nanomaterials and control specific recognition among the individual building blocks. Accordingly, significant efforts have been made towards the controlled assembly of nanocomponents.

One promising strategy exploits self-ordered templates for arranging nanomaterials into designed patterns. A wide variety of ordered nano-scale templates have been successfully employed for this purpose.[8-11] One branch of well-demonstrated templates includes refined biomolecules directly obtained from Nature such as nucleic acids, proteins and viruses, all of which have sophisticated hierarchical nanostructures with precise molecular recognition capabilities. The combination of functional inorganic nanomaterials with these

RSC Nanoscience & Nanotechnology No. 17
Phage Nanobiotechnology
Edited by Valery A. Petrenko and George P. Smith
© Royal Society of Chemistry 2011
Published by the Royal Society of Chemistry, www.rsc.org

biological templates has led to many two-dimensional (2D) and three-dimensional (3D) nanostructure organizations with controlled size, shape, alignment and orientation.[12–17]

Filamentous bacteriophage (also called phage) are viruses that specifically infect bacteria. The phage particles (virions) have precise, genetically programmable, physically robust nanostructures – all attractive properties to materials scientists. Although the natural functions of virions are storage of genetic material and efficient transfer of that material to new host cells, they have recently been employed to serve as biotemplates for the synthesis and assembly of nanomaterials.[13,14,18–33] Phage-based nanosystems can be integrated into many practical applications such as lithium ion batteries,[34] biosensors[33,35] and bone mimics (our unpublished data), which has greatly encouraged scientists to explore further in this promising field.

Of the various biotemplates studied for the synthesis and assembly of nanomaterials, filamentous phage has proven to be particularly efficient and powerful, for several reasons. First, a well-established technique, phage display,[36] can be used to identify and generate phage that are capable of specifically binding to a target nanomaterial. The process involves selecting a target-binding peptide from a large combinatorial library of peptides displayed along the surface of the tubular sheath that forms the virion's outer covering and then assembling nanomaterials along the peptide-bearing surface of the selected virion.[10,23] Numerous targeting peptides against various materials have been identified in this manner and are currently available for use (Table 10.1). Second, the virion is chemically stable in many acidic, basic and organic solvents and withstands temperatures up to 80 °C.[37] Third, error-free mass production of identical phage nanofibers from natural factories, bacteria, guarantees a level of monodispersity far higher than can be achieved with commercially available conventional polymers.[37] Fourth, up to three different peptides can be displayed on the phage particle's outer surface simultaneously, making it possible for an individual nanofiber to assemble more than one functional nanomaterial (Figure 10.1).[38,39] Fifth, individual phage nanofibers can self-assemble into ordered structures,[19,40–43] which can further direct the assembly of nanomaterials. In this chapter, we review recent accomplishments in filamentous phage templated synthesis and assembly of inorganic nanomaterials and their applications and future prospects.

10.2 Virion Structure and Phage Display

10.2.1 Biology

Phage that have been successfully used in templated nanosynthesis include most prominently the Ff class of filamentous phage, which encompasses wild-type strains M13, fd and f1. The wild-type M13 virion is a flexible nanofiber, 6.5 nm in diameter and 930 nm in length (Figure 10.2). A circular 6407- or 6408-base single-stranded DNA (ssDNA) genome is packed into a tubular sheath

Table 10.1 Summary of inorganic material-binding peptides, phage libraries used and corresponding references.

Library	Material	Sequence	Ref.
12-mer–pIII	GaAs	AQNPSDNNTHTH	18
12-mer–pIII	GaN	SVSVGMKPSPRP	64
12-mer–pIII	Ag	NPSSLFRYLPSD	65,66
C7C–pIII	Pt	PTSTGQA	67
C7C–pIII	Pd	SVTQNKY	68
		SPHPGPY	68
		HAPTPML	68
12-mer–pIII	Ti	RKLPDA	69
7-mer–pIII		TLHVSSY	70
8-mer–pVIII	Au	VSGSSPDS	28
12-mer–pIII	SiO_2	MSPHPHPRHHHT	71
		RGRRRRLSCRLL	71
		KPSHHHHHTGAN	71
12-mer–pIII	Quartz (SiO_2)	RLNPPSQMDPPF	72
12-mer–pIII	$CaCO_3$	HTQNMRMYEPWF	73
15-mer–pVIII		DVFSSFNLKHMR	73
		AYGSSGFYSASFTPR	74
C7C–pIII	ZnS	NNPMHQN	19
		VISNHAESSRRL	14
C7C–pIII	CdS	TYSRLHL	22
12-mer–pIII		SLTPLTTSHLRS	14,22
8-mer–pVIII	Co^{2+}	EPGHDAVP	29
12-mer–pIII	Co	HYPTLPLGSSTY	66
12-mer–pIII	TiO_2	RKKRTKNPTHKL	75
12-mer–pIII	ZnO	EAHVMHKVAPRP	76
12-mer–pIII	FePt	HNKHLPSTQPLA	13,77
12-mer–pIII	$BaTiO_3$	HQPANDPSWYTG	78
		NTISGLRYAPHM	78
12-mer–pIII	$CaMoO_4$	YESIRIGVAPSQ	79
		DSYSLKSQLPRQ	79
12-mer–pIII	CoPt	KTHEIHSPLLHK	13,80
12-mer–pIII	Hydroxyapatite	SVSVGMKPSPRP	81
C7C–pIII		MLPHHGA	82
12-mer–pIII		APWHLSSQYSRT	83
12-mer–pIII		STLPIPHEFSRE	83
12-mer–pIII		VTKHLNQISQSY	83
8-mer–pVIII		DSSTPSSTD	*
12-mer–pIII	Ge	SLKMPHWPHLLP	84
		TGHQSPGAYAAH	84
12-mer–pIII	SWNT	HWSAWWIRSNQS	85
		HWKHPSGAWDTL	85
12-mer–pIII	Carbon nanohorn	DYFSSPYYEQLF	86
C7C–pIII	C_{60}	NMSTVGR	87

*Our unpublished result

composed of 2760 subunits of the major coat protein pVIII and five subunits of the minor coat proteins pIII, pVI, pVII and pIX (Figure 10.1). The pVIII subunits form the side wall of the sheath, while five copies of pIII and pVI are

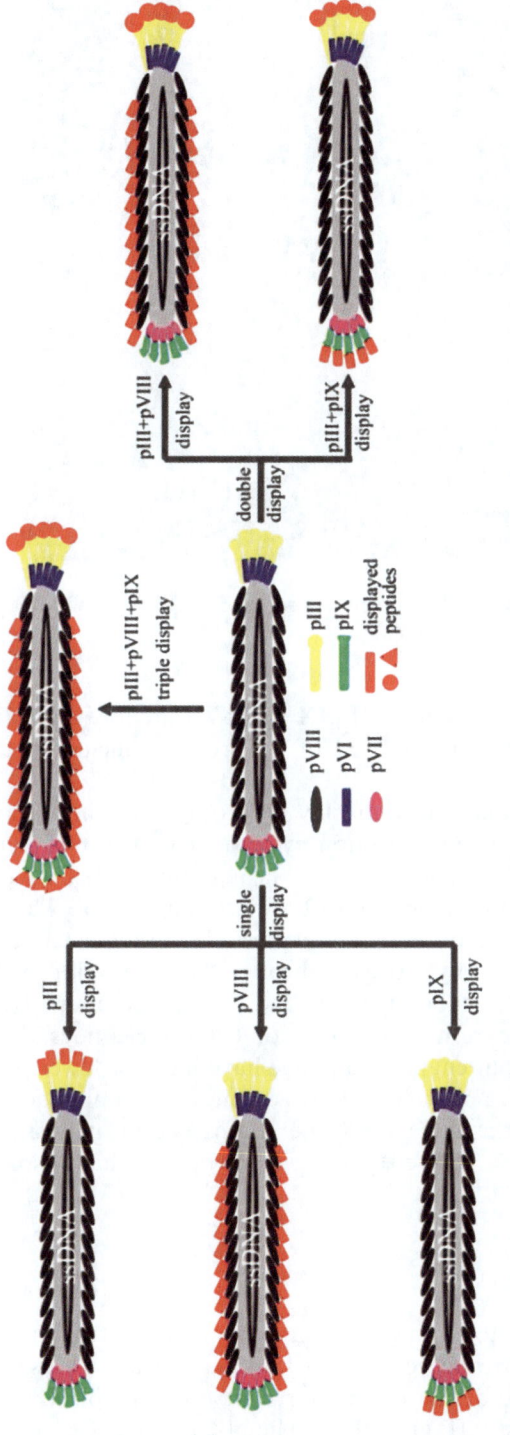

Figure 10.1 The structure of a filamentous phage (M13 or fd) and its genetically engineered fashion. An ssDNA is packed into a protein coat which is composed of thousands of copies of major coat proteins (pVIII) and five copies of each minor coat proteins (including pIII, pVI, pVII and pIX). Single, double or triple display of different peptides on the surface of a single phage fiber can be achieved, leading to site-specific modification of the phage surface. The foreign peptides fused to the coat proteins are highlighted in red.

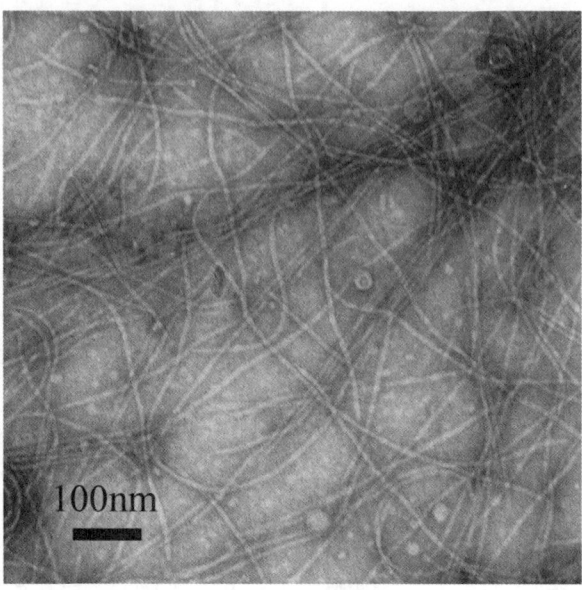

Figure 10.2   Transmission electron microscopy (TEM) image of M13 phage stained
with 1% uranyl acetate.

located at one tip and five copies of pIX and pVII are located at the other tip.
The phage genome encodes a total of 11 proteins, including the five coat
proteins.

In order to understand phage display technology, it is important to understand the filamentous phage infection cycle (reviewed in Chapter 1). Briefly,
phage coat protein pIII binds to bacteria through the F pilus, triggering a train
of events that injects the ssDNA into the bacterial cytosol. The phage DNA
replicates and the phage proteins are expressed, leading eventually to steady
release of progeny virions by extrusion through the cell envelope without killing
the cell. There are 0.431 subunits of the major coat protein pVIII per ssDNA
nucleotide; wild-type virions, with 6407 or 6408 nucleotides, therefore have
2760 pVIII subunits, but engineered phage have more or fewer pVIII subunits
in proportion to the size of their ssDNA genomes. The amino acid sequence of
pVIII is identical in the three wild-type strains except at position 12 of the
mature (signal peptidase cleaved) form, where f1 and fd have aspartic acid and
M13 has asparagine.[36,44]

10.2.2   Chemistry

The ability of filamentous phage to serve as an ideal template is primarily
determined by their surface chemistry. Since there are thousands of pVIII
subunits but only five of the other four coat proteins, pVIII dominates the
overall surface chemistry (Figure 10.1). The N-terminus of each pVIII subunit

is exposed on the surface, with the first five residues (AEGDD) extending away from the virion. There are thus three exposed side-chain carboxyls per subunit, which results in a negatively charged surface. The experimental isoelectric point of the M13 virion is 4.2, which confirms the negative phage surface. The surface carboxyl groups and the exposed ε-amino group at position 7 of the mature protein can be chemically modified under mild conditions without compromising the physical integrity of the particle.[36,44]

10.2.3 Site-specific Engineering of the Virion Surface

Designed foreign peptides can be surface-displayed at the tips of the virion by genetic fusion to the minor coat proteins (usually pIII but sometimes pVI, pVII or pIX) or along the entire length of the phage body by genetic fusion to the major coat protein pVIII. Double or triple display of different peptides on a single phage fiber can also be achieved (Figure 10.1). Therefore, using the phage display technique, the surface chemistry of a filamentous virion can be site-specifically controlled.[36,44]

10.2.4 Liquid Crystalline Behavior

In addition to the controllable surface chemistry of the virion, its unique physical properties are also very attractive. One in particular is the ability of filamentous virions at certain concentrations to form liquid crystal structures.[40] With increasing concentration, filamentous virions exhibit the following phase sequences: isotropic, nematic, cholesteric, smectic A and smectic C.[40] This ordered phage self-assembling system has been used to fabricate highly ordered functional nanoparticle composites.[19]

10.3 Exploiting Phage Display to Alter Surface Chemistry by Selection Rather Than Rational Design

10.3.1 Random Peptide Libraries

Single peptide display is achieved by splicing the peptide's coding sequence into the appropriate site in a coat protein gene; the resulting recombinant gene encodes a recombinant polypeptide in which the peptide is fused genetically to the coat protein on the virion surface. If instead a degenerate coding sequence is spliced into the same site, the result will be a library of up to billions of distinct peptide displayed by billions of distinct phage clones; such libraries are called random peptide libraries. Type 3 random peptide libraries, in which the foreign peptide is displayed on all five pIII subunits at the tip of the virion, are particularly popular and are commercially available from New England Biolabs. Type 8 libraries, in which the foreign peptide is displayed on all copies of the pVIII subunit, have been called 'landscape' libraries because the foreign peptide subtends a substantial fraction of the total surface molecular 'landscape' of the virion.[45] Type 88 libraries, in which the foreign peptide is displayed on a few percent of the pVIII subunits, are well suited to some applications.

10.3.2 Affinity Selection ('Biopanning')

Random peptide libraries can be surveyed for rare clones whose displayed peptides bind a chosen target, including inorganic materials. The survey is accomplished not *via* clone-by-clone screening, but rather by selection. Figure 10.3 illustrates affinity selection ('biopanning') from a landscape library.

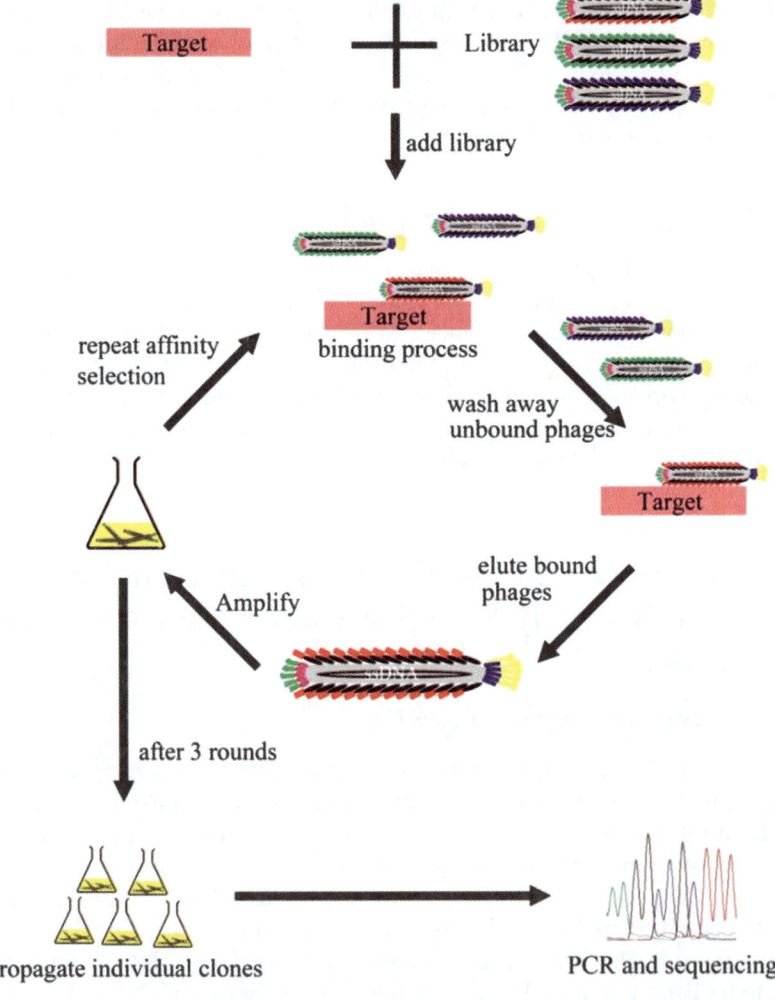

Figure 10.3 Procedure for selecting target-binding peptides from landscape phage library (biopanning).

The library phage are allowed to interact with the immobilized target, resulting in some phage binding to the target whereas others do not. Non-binding phage are washed away and binding phage are eluted from the target and amplified by infecting fresh bacteria, resulting in a sublibrary that is enriched for target-binding phage. In order to select for the best target-binding peptides that the library has to offer, additional rounds of affinity selection are carried out, in each case using as input the output of the previous round. Individual clones from the final output are sequenced to reveal the specific peptides that are responsible for enrichment.

Biopanning has successfully identified peptides that specifically bind inorganic materials such as GaAs, GaN, Ag, Pt, Au, Pd, Ge, Ti, SiO_2, quartz, $CaCO_3$, ZnS, CdS, Co, TiO_2, ZnO, CoPt, FePt, $BaTiO_3$, $CaMoO_4$, hydroxyapatite, C_{60} and carbon nanotubes. The binding peptides, source random peptide libraries and references are summarized in Table 10.1. The identified peptides have in turn been used to fashion inorganic materials. Virions displaying these peptides have also been used for the template synthesis of novel ordered, functional nanomaterials, as described in the next subsection.

10.4 Synthesis and Assembly of Inorganic Materials on Individual Virions

Nanowires made from inorganic materials have useful physical properties, including quantum conductance,[46] ballistic conduction,[46] low thermal conductivity,[47–49] localization,[46] negative magnetoresistance,[50] quantum size effects[51] and ferromagnetism.[52] Based on these unique properties, many applications can be expected in nanoelectronic circuitry, electrochemical energy storage and thermoelectric and photovoltaic devices.[53] Typical methods for the synthesis of inorganic nanowires such as templating, ligand control and oriented attachment have been reviewed by Cademartiri and Ozin.[54]

Recently, engineered filamentous virions have been used to fabricate inorganic nanowires.[13,14,28] The virions' filamentous morphology, genetically tunable target specificity and thermal and chemical stability commend them as a powerful new class of inorganic recognition templates in the synthesis of inorganic nanowires. The typical approach includes the following steps: first, a phage random peptide library, either pIII or pVIII, is selected against a target via a biopanning process to obtain a target-binding peptide (Figure 10.2); second, the identified peptide is fused to the N-terminus of the pVIII protein, resulting in an ordered arrangement of selected peptides down the length of the virion (Figure 10.1, pVIII display); third, the phage template is incubated with inorganic precursors under suitable conditions, leading to the final nucleated nanocrystals along the phage template. Avery *et al.* found that non-specific electrostatic interactions between the anionic carboxylate groups from the pVIII region of wild-type M13 virion and cationic aqueous metal complexes, followed by reduction, can also generate metallic nanowires without the need for genetic engineering.[55] This kind of nucleation, however, is without specificity, which may lower the efficiency or even cause problems for practical applications.[56]

Mao and co-workers found that the highly organized structure of the M13 virion can serve as a generic template for the direct synthesis of semiconductor and magnetic nanowires.[13,14] Specifically, they screened ZnS, CdS, FePt and CoPt substrates with pIII phage libraries (New England Biolabs) and identified the following substrate-binding peptides: CNNPMHQNC (ZnS), SLTPLTTSIILRS (CdS), HNKHLPSTQPLA (FePt) and CNAGDHANC (CoPt). These peptides were then fused to the N-terminus of pVIII proteins, resulting in an ordered arrangement of selected peptides down the length of the M13 virion (Figure 10.1, pVIII display). The virions displaying these substrate-binding peptides were incubated with corresponding metal salt precursors ($ZnCl_2$, $CdCl_2$, $FeCl_2 + H_2PtCl_6$, $CoCl_2 + H_2PtCl_6$) followed by adding other reactants (Na_2S, Na_2S, $NaBH_4$, $NaBH_4$) in suitable environments, finally leading to nucleated nanocrystals along the phage template.

Phage templates can be further removed by annealing at 350–400 °C, forming single-crystalline nanowires. Mineralization of ZnS along engineered virions displaying CNNPMHQNC on pVIII proteins leads to a preferred crystallographic orientation of nucleated nanocrystals (Figure 10.4a). The crystals grown on the surface of the virion were in a wurtzite structure and in close contact, showing preferential orientation with the (100)* direction parallel to the long axis of the virion (Figure 10.4a). The annealing process at temperatures higher than 350 °C and lower than 400 °C (melting point of ZnS) can slowly remove the organic phase of the virion, allowing the oriented attachment of the nanocrystals into single-crystalline nanowires (Figure 10.4b). An electron diffraction (ED) pattern (Figure 10.4b) and high-resolution TEM image (Figure 10.4c) confirmed the single-crystalline structure of an individual ZnS nanowire. The measured lattice spacing of 0.33 nm corresponds to the spacing between two (010) planes. CdS, FePt and CoPt nanowires could also be synthesized using similar methods (Figure 10.4c–e). This work demonstrated a new virus-based tool for the general synthesis of single-crystalline nanowires.

In addition to direct nucleation along the virion, phage can also first assemble inorganic nanoparticles into a necklace-like morphology. The substrate material will then be electroless deposited on the existing nanoparticle arrays, forming conductive nanowires. These nanowires may have potential applications in nanoelectronic circuitry.[28] In this work, a type 8 random peptide library was screened against a gold thin film, resulting in a gold-binding peptide (VSGSSPDS). After several minutes of incubation with pre-synthesized 5 nm gold nanoparticles (AuNP) with pVIII engineered M13 phage displaying VSGSSPDS, AuNPs were aligned into an ordered 1D array on pVIII proteins along the virion axis (Figure 10.5a). The AuNP arrays can grow into continuous gold nanowires through electroless deposition of gold on existing AuNP arrays. This process was further confirmed by TEM images of gold nanowires after 3 and 5 min deposition (Figure 10.5b and c). For the current–voltage (I–V) measurement, electron beam lithography was used to

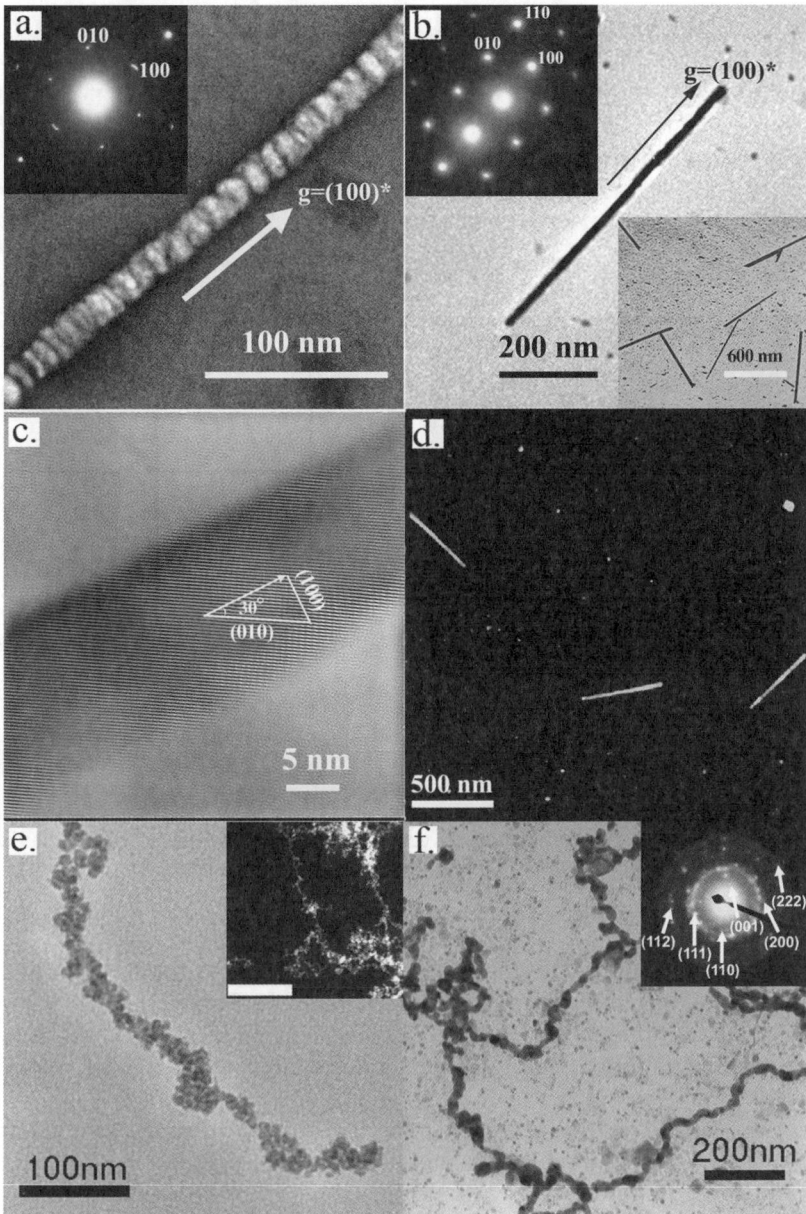

Figure 10.4 Electron microscopy images of ZnS, CdS, CoPt and FePt nanowires synthesized by phage templates. (a) Dark-field diffraction contrast image of pre-annealed ZnS crystals nucleated and oriented along the phage template. (b) An individual ZnS single-crystal nanowire formed after annealing. (c) A lattice image of a single ZnS nanowire; 0.33 nm corresponds to the (010) planes in wurtzite ZnS crystals. (d) Image of CdS nanowires. (e) CoPt nanowires. (f) FePt nanowires. Reproduced with permission from reference 13.

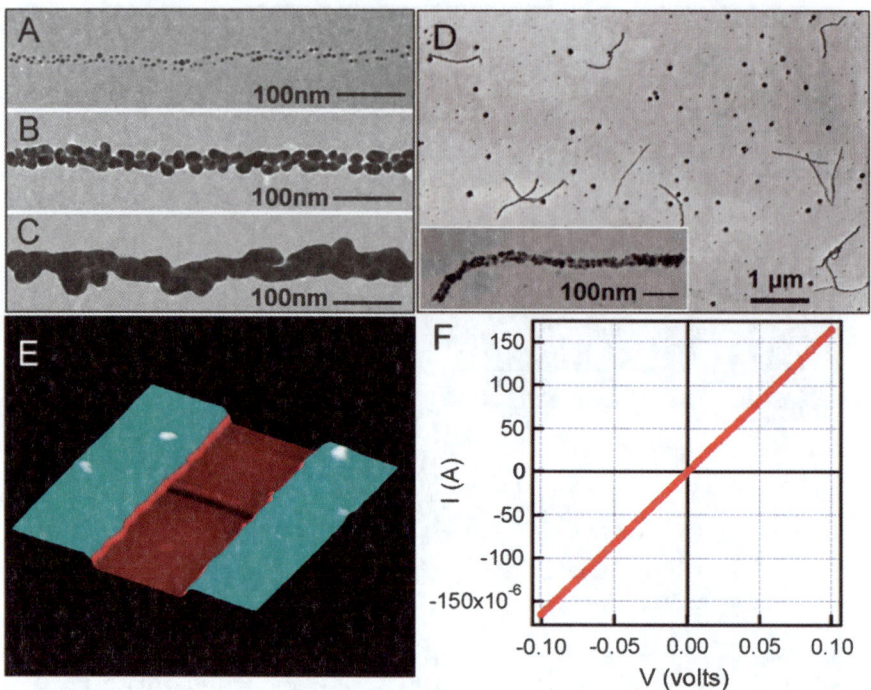

Figure 10.5   (a–c) TEM images of the growth of gold nanowires templated by AuNP arrays on pVIII proteins. (a) Before deposition. (b) 3 min deposition. (c) 5 min deposition. (d) Gold nanowires synthesized directly from the nucleation of gold from solution. (e) AFM image of a gold nanowire from (c) immobilized between a two-terminal device. (f) *I–V* behavior of a gold nanowire. Reproduced with permission from reference 28.

make electrical contacts to the synthesized gold nanowires (Figure 10.5e). The measured *I–V* curve shows a linear relation in Figure 10.5f, in which a resistance of approximately 588 Ω could be estimated. This work shows a promising bottom-up assembly of phage templated materials for electronic applications.

Non-engineered virions can also be used to synthesize and assemble inorganic nanomaterials. For instance, Avery *et al.* synthesized Rh, Ru and Pd nanowires on wild-type (WT) M13 virions by non-specific electrostatic interactions between cationic aqueous metal complexes and the exposed anionic carboxylate groups of the pVIII subunit, followed by reduction.[55] These WT phage can also template the assembly of metal nanoparticles along the virion side walls.[55] The non-engineered virions lack material specificity and therefore can be used to form a variety of materials. Therefore, with or without genetic modification, filamentous virions can serve as bio-templates for synthesizing nanowires or assembling nanoparticles into ordered arrays.

10.5 Synthesis and Assembly of Inorganic Materials on a Self-assembled Phage Scaffold

Filamentous virions can also self-assemble into ordered structures that can be treated as a scaffold for the synthesis and assembly of inorganic materials. Three typical assembled structures include liquid crystalline structures, solid-phase films and 1D bundled fibers.[40,57–61] The virions are able to form liquid crystalline structures because of their monodispersity and rod-like shape. These liquid crystals are an attractive experimental system for fundamental studies because the length, linear charge density and surface chemistry of the virions can be readily altered *via* genetic modification.[40] Different phases of liquid crystalline structures can be accessed by changing the virion concentration. With increasing concentration, the virions progress through an ordered series of phases: isotropic ($< 5\,mg\,ml^{-1}$), nematic (10–$20\,mg\,ml^{-1}$), cholesteric (20–$80\,mg\,ml^{-1}$), smectic A and smectic C ($> 100\,mg\,ml^{-1}$).[10,40] Phage-based thin films can be fabricated on multilayered polyelectrolyte films with layer-by-layer (LBL) deposition.[57–59] 1D fibers of engineered virions can be fabricated through the wet-spinning or electrospinning.[60,61]

Lee *et al.* fabricated a highly ordered composite material composed of pIII engineered M13 virions and ZnS nanocrystals through a liquid crystal template assembly.[19] First, virions displaying a ZnS-binding peptide (CNNPMHQNC) on pIII were affinity-selected from a type 3 random peptide library and amplified (Figure 10.6a). $ZnCl_2$ and Na_2S solutions were then added to a phage pellet, resulting in a virion–ZnS crystal liquid crystalline suspension (Figure 10.6a). Polarized optical microscopy (POM) showed the smectic phase (Figure 10.6b) and cholesteric phase (not shown) at concentrations of 127 and $76\,mg\,ml^{-1}$, respectively.

A highly ordered phage–ZnS film was made by slow evaporation of a suspension of the phage and ZnS nanoparticles. As evaporation progressed, the virion concentration slowly increased, leading to ordered transitions through isotropic, nematic, cholesteric and smectic phases and finally to a highly ordered film. Parallel aligned phage that have almost right-angles between the adjacent directors on the free surface of the phage–ZnS film were imaged by atomic force microscopy (AFM) (Figure 10.6e). In a photoluminescent image of the film (Figure 10.6c), a pattern of 1 µm fluorescent lines corresponding to the ZnS nanocrystals could be discerned, demonstrating the existence of ordered ZnS crystals in the film. This work suggests a new general method for arranging inorganic crystals via a phage-based liquid crystal template.

Yoo and co-workers found that the interdiffusion of two oppositely charged weak polyelectrolytes could lead to the floating of phage on the surface, forming an ordered monolayer of virions atop a cohesive polyelectrolyte multilayer.[57–59] Briefly, thin layers of linear polyethylenimine (LPEI) and poly(acrylic acid) (PAA), with the positively charged LPEI layer on top, were first prepared on a silicon substrate. Negatively charged virions were then randomly deposited on the LPEI surface through electrostatic interaction (Figure 10.7a, step 1, and b). Next, LPEI and PAA were repeatedly and

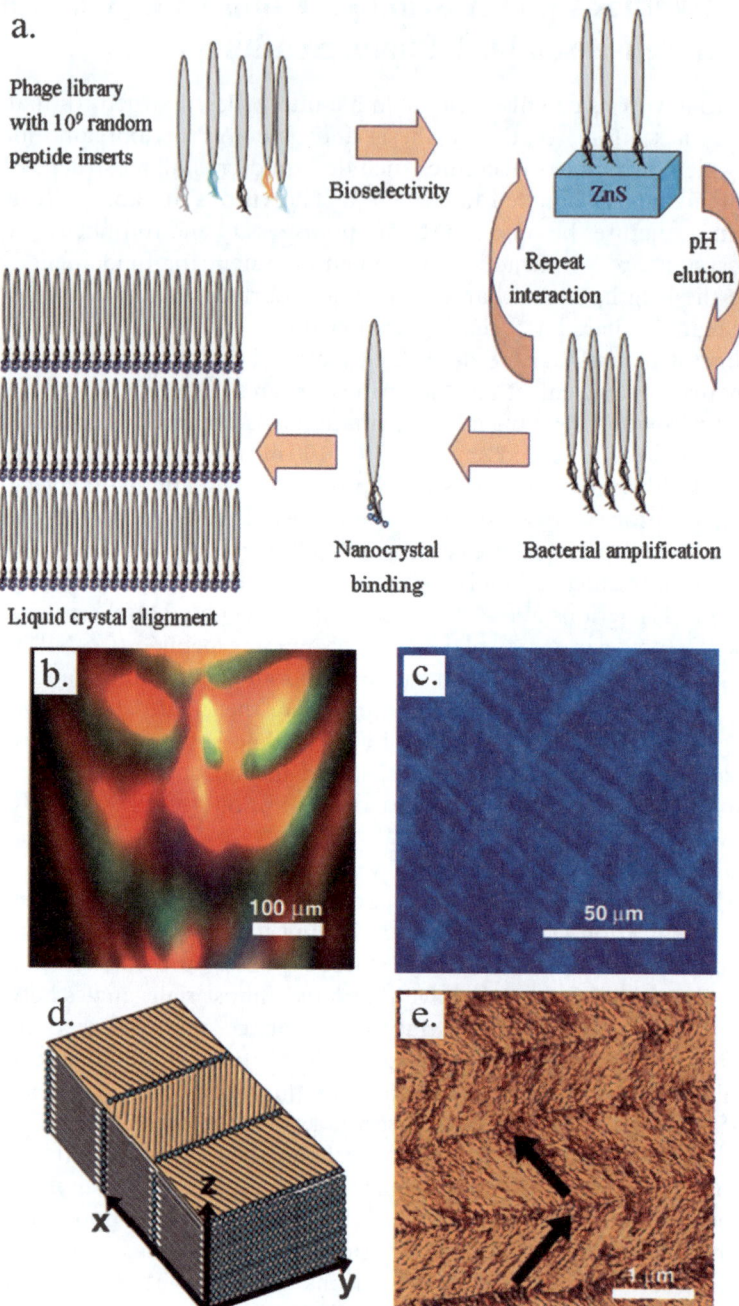

Figure 10.6   (a) Schematic diagram of the process for fabricating the phage–ZnS liquid crystal structure. (b) POM image of a smectic phage–ZnS suspension. (c) Photoluminescent image of the phage–ZnS composite film (excitation at 350 nm). (d) Phage–ZnS composite film structure. (e) AFM image of the free surface of phage–ZnS film. Reproduced with permission from reference 19.

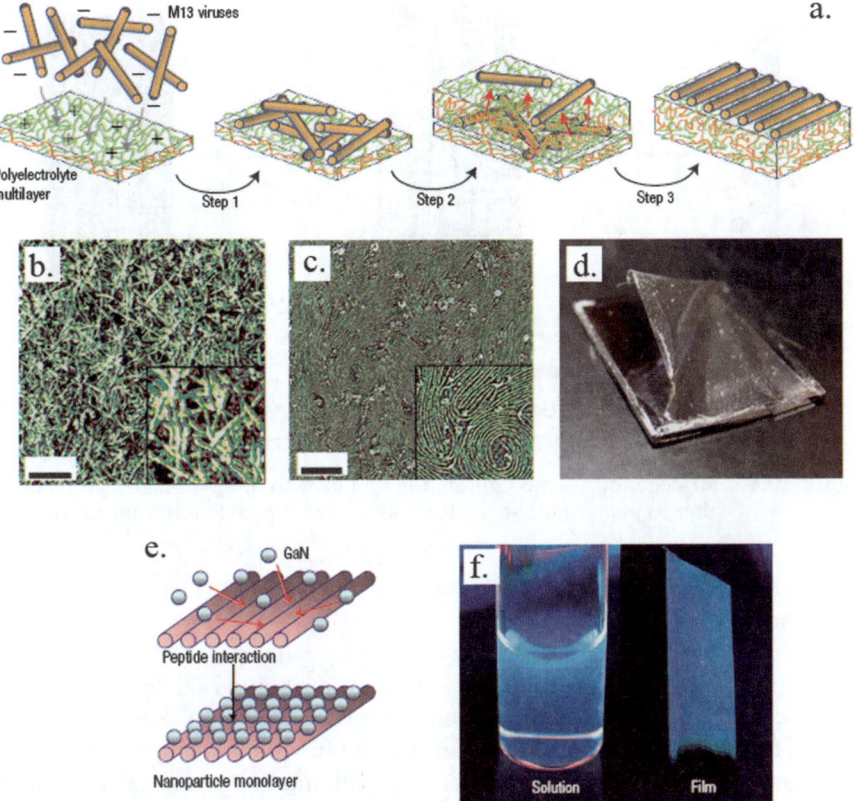

Figure 10.7   (a) Schematic strategy of the phage monolayer assembly. (b) AFM image of the randomly deposited M13 phage on LPEI–PAA multilayers. (c) AFM image of the final product – close packed monolayer of M13 phage. (d) A free-standing phage film after more than 100 repeats of alternate adsorptions of LPEI–PAA. (e, f) When a solution of GaN nanoparticles was incubated with a monolayer film assembled from pVIII engineered GaN-targeting M13 phage, a densely packed GaN film could be prepared. Reproduced with permission from reference 59.

alternately adsorbed on the top phage layer of the thin LPEI–PAA–phage composite, driving the virions to the surface via competitive electrostatic interactions (Figure 10.7a, step 2). Finally, mutually repulsive interactions between neighboring virions induced spontaneous repositioning of the virions into an ordered monolayer atop the cohesive polyelectrolyte multilayer structure (Figure 10.7a, step 3, and c). If more than 100 repeated cycles of alternate adsorptions of LPEI and PAA were able to be achieved, a free-standing film assembled by phage could be made (Figure 10.7d). When a solution of pre-synthesized GaN nanoparticles was incubated with a monolayer film assembled from virions displaying a GaN-targeting peptide on pVIII, a densely packed GaN film could be prepared (Figure 10.7e) with corresponding fluorescence

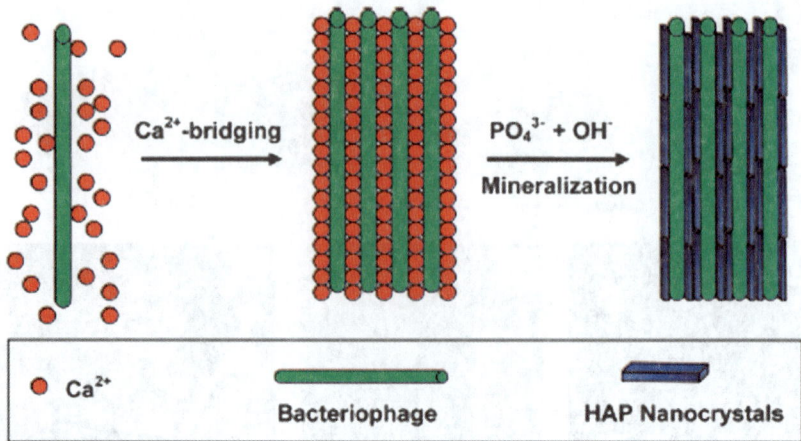

Figure 10.8 Strategy for oriented nucleation and growth of HAP nanocrystals within aligned phage nanofibers. Reproduced with permission from reference 62.

(Figure 10.7f). This method can also be applied to form 2D films made by other inorganic nanoparticles.[59]

1D phage-based fibers have been prepared by Chiang *et al.* through a slow wet-spin of a concentrated virion solution into glutaraldehyde.[61] Glutaraldehyde molecules cross-link neighboring virions, leading to stable micro-scale fibers. They found that assembly of virions into functional nanomaterials could be accomplished either before or after wet-spinning the virions into 1D fibers. To illustrate the first approach, they synthesized fibers with optical properties by mixing CdSe quantum dots with phage before wet-spinning. To illustrate the second approach, they were able to assemble AuNPs densely along 1D fibers that had already been created by wet-spinning.

We recently found that the calcium ions can trigger the self-assembly of WT fd virions into bundles. The bundle can serve as a Ca source and react with phosphate ions to form a bundle decorated with nanocrystalline bone minerals with preferred crystal orientation (Figure 10.8).[62] The resultant mineral-decorated bundles can serve as a building block to form higher-order structures for building bone biomaterials. Therefore, with or without the help of other molecules, 1D and 2D phage scaffolds can be assembled by individual virions to induce on them the synthesis and assembly of inorganic materials.

10.6 Applications of Phage-templated Nanomaterials

Although only a few applications of phage-based nanomaterials have been reported so far, their potential advantages are attracting increasing attention and hold the prospect of important future developments. In this section, we outline applications of phage-based nanomaterials, in the hope of inspiring novel thinking in phage-based nanoscience.

A phage-based lithium-ion battery with high performance has been fabricated.[56] In this work, $FePO_4$-based materials were chosen for the positive electrode because of their low cost, improved safety and low toxicity. Lithium metal was chosen for the negative electrode. The cathode was improved by using engineered virions that displayed EEEE on the N-termini of pVIII subunits and DMPRTTMSPPPR on the N-termini of pIII subunits (Figure 10.1, pIII + pVIII double display). The EEEE peptides enabled the virions to template amorphous anhydrous $FePO_4$ (cathode material) on them, thus reducing both the ionic and electronic paths along the particles. The DMPRTTMSPPPR peptide binds carbon nanotubes, allowing the virions to be cross-linked with single-walled carbon nanotubes (SWNTs) to enhance local electronic conductivity (Figure 10.9a). The phage–$FePO_4$–SWNT cathode showed a capacity of $134 \, mA \, h \, g^{-1}$ at a high discharge rate of $3 \, C$, much better than the record ($80 \, mA \, h \, g^{-1}$ at $3 \, C$) (Figure 10.9b). Moreover, virtually no capacity fade was observed, even after 50 cycles (Figure 10.9d). This application highlights two key advantages of phage nanomaterials: (1) engineered virions can efficiently template the synthesis of nanomaterials on them leading to functional nanowires; and (2) the double-display technique endows virions with the ability to nucleate one functional material while simultaneously binding another.

We have fabricated nanocomposite films by alternating deposition of negatively charged AuNPs and positively charged virions that display RRRR peptides on their side walls.[33] First, a quartz slide was precoated with two layers of anionic poly(vinyl sulfate) (PVS) and cationic poly(diallyldimethylammonium chloride) (PDDA). The layer-by-layer assembly of negatively charged AuNPs and positively charged M13 phage was achieved by alternating immersion of the quartz slide into a tetraArg-M13 solution and AuNP solution for 30 min. The surface plasmon resonance (SPR) spectrum of the film was tunable in response to environmental humidity (Figure 10.10c). When the relative humidity (RH) was decreased, the λ_{max} of the SPR spectra was correspondingly red shifted and this shift was reversed when the RH was increased. A linear dependence on RH was found within the range from 19 to 86%. Therefore, this phage–AuNP film exhibiting unique humidity-dependent SPR spectra could be used as a phage-based humidity sensor by means of a spectrophotometer. A plausible explanation for the humidity dependence is that loss of water flattens the phage–AuNP film, thus reducing the AuNPs spacing (Figure 10.10a). The reduced spacing in turn causes a red shift of the λ_{max} of the SPR spectra. Conversely, when the phage–AuNP film moistens, the AuNPs spacing increases (Figure 10.10 b) and λ_{max} shifts to the blue.

Phage have also been incorporated in biomedical materials for tissue engineering and regenerative medicine applications. Recently, a 3D tissue culture based on magnetic levitation of cells has been achieved using a phage-based hydrogel.[63] The hydrogel was fabricated over cells seeded on a plate by using gold nanoparticles, magnetic iron oxide nanoparticles and phage displaying the peptide CDCRGDCFC, a ligand for αv integrins (Figure 10.11b). After several rounds of washing, only cell-binding or cell-penetrating hydrogels were left. When a magnetic field was applied, these magnetized hydrogels caused the cells

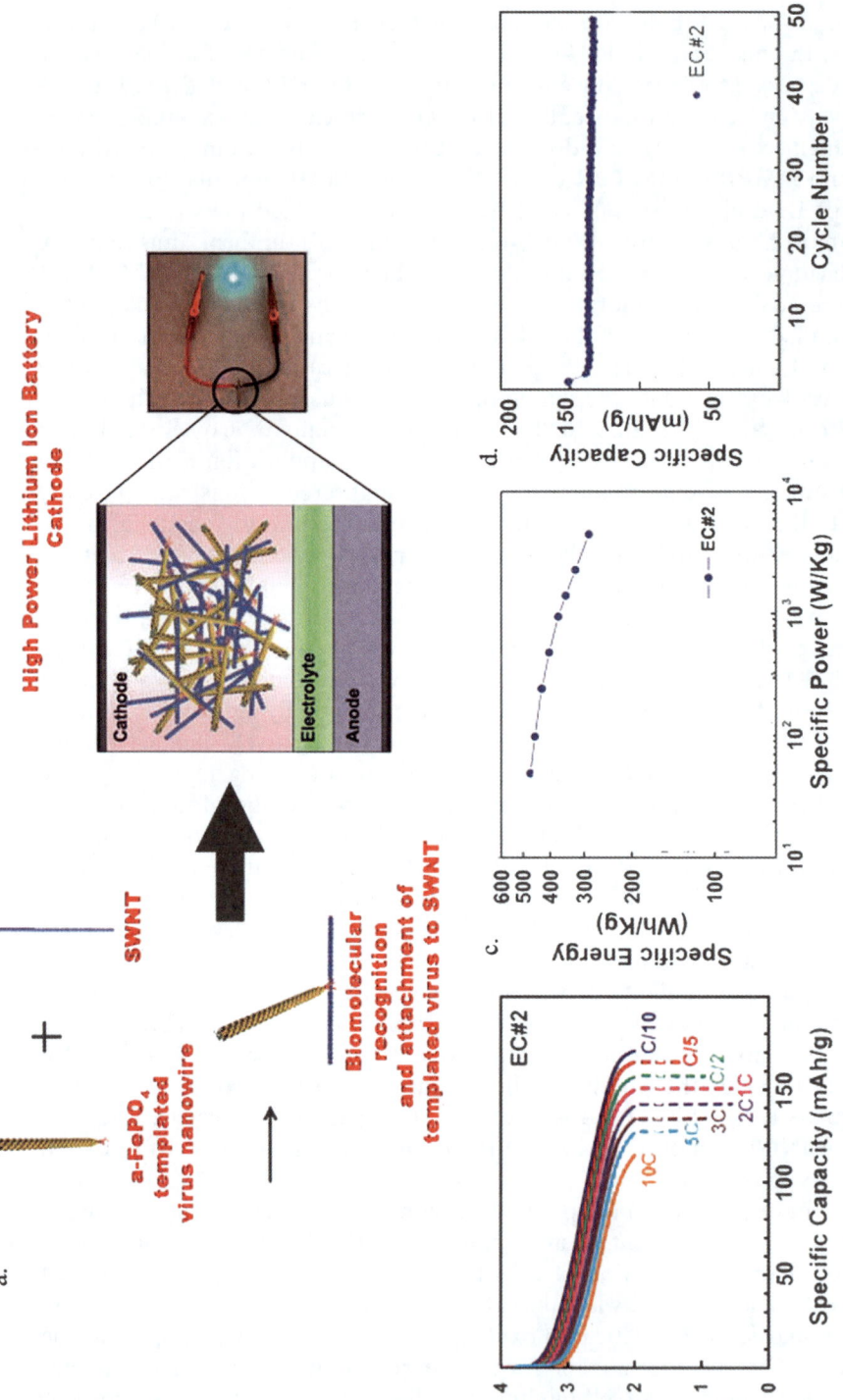

Figure 10.9 (a) Schematic diagram for fabricating lithium-ion battery cathodes using double-displayed M13 phage. This battery can power a green LED. (b, c) The electrochemical performance of a phage–FePO$_4$–SWNT nanocomposite. (b) First discharge curves at different rates. (c) Ragone plot representing rate performance in terms of specific power *versus* energy. (d) Capacity retention for 50 cycles at a rate of 1 C. Reproduced with permission from reference 56.

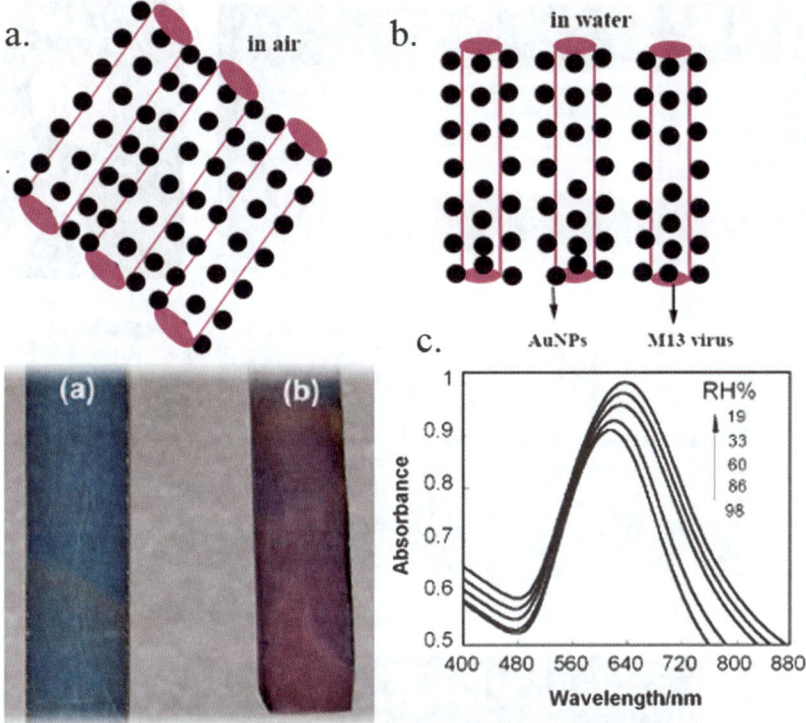

Figure 10.10 (a) When the phage-AuNP film was dried in air, the loss of water resulted in flattening of the AuNP-wrapped M13 phage, leading to reduced AuNPs spacing. (b) Similarly, when the phage–AuNP film was wetted in water, the gain of water resulted in increased AuNPs spacing. (c) SPR spectra of the film were tunable responding to the RH. Reproduced with permission from reference 33.

to rise to the air–medium interface and self-assemble into multicellular structures (Figure 10.11a and c). Side-by-side comparison of human glioblastoma cells cultured in the magnetic levitation system with orthotopic human tumor xenografts from immunodeficient mice showed both morphological and molecular similarities (Figure 10.11d). This *in vitro* 3D culturing may thus become a complementary and cheaper substitute for the generation and maintenance of human brain tumor xenografts in immunodeficient mice. Moreover, confrontation culture assays between human glioblastoma cells and normal astrocytes in their system showed evidence of invasiveness of the former cells, suggesting the potential of this methodology for analysis of brain tumor invasiveness. In short, this powerful hydrogel, based on the elegant combination of the cell-binding ability of engineered M13 phage and the magnetic control from magnetic iron oxide nanoparticles, opens up new approaches to problems in cell biology.

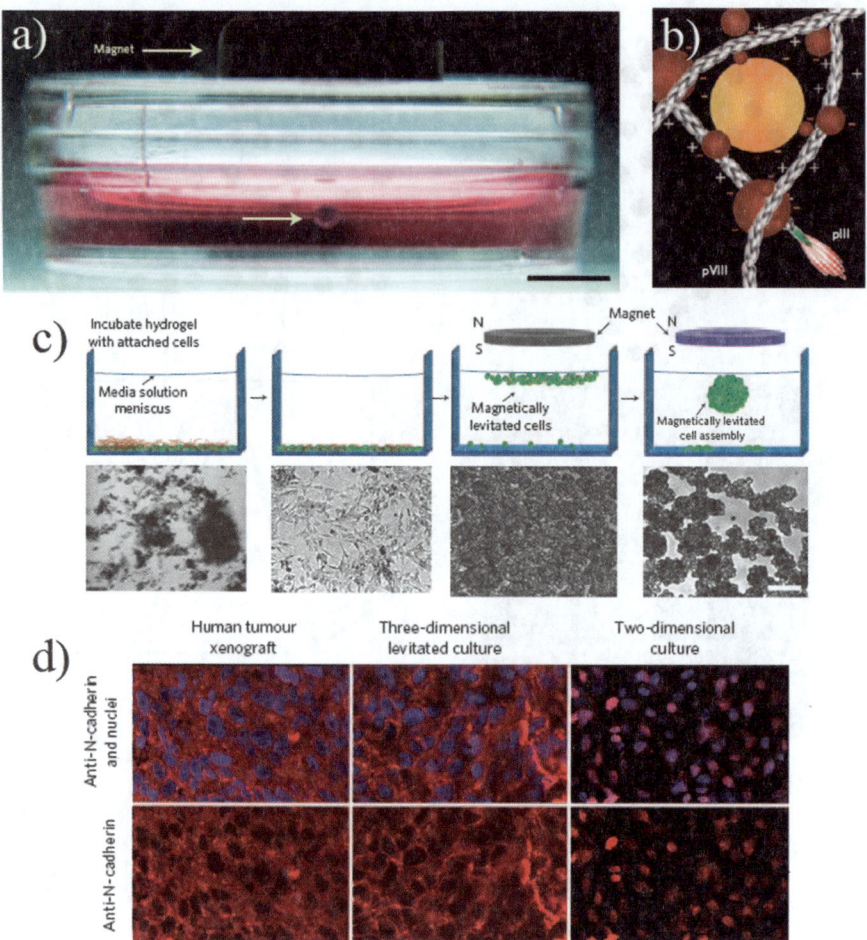

Figure 10.11   (a) Cells rise to the air–medium interface and self-assemble into multicellular structures in a magnetic field. (b) Scheme of electrostatic interactions of gold and magnetic iron oxide nanoparticles with M13 phage. (c) 3D cell culture with magnetic-based levitation. (d) Side-by-side comparison of the human glioblastoma cells cultured in the magnetic levitation system with orthotopic human tumor xenografts from immunodeficient mice. Reproduced with permission from reference 63.

Our laboratory is trying to develop hydroxyapatite–phage scaffolds for bone regeneration studies. Hydroxyapatite (HAP) is the main inorganic component of bone. In order to make an HAP-based bone regeneration scaffold, engineered phage can be employed to assemble HAP nanorods. First, HAP-binding phage were affinity selected from a landscape phage library (Figure 10.3). The HAP-binding phage co-assembled with pre-synthesized HAP nanorods to form a 3D porous scaffold with an alternating binding pattern (Figure 10.12) (our unpublished data). The scaffold with 15–20 μm holes is not water soluble and

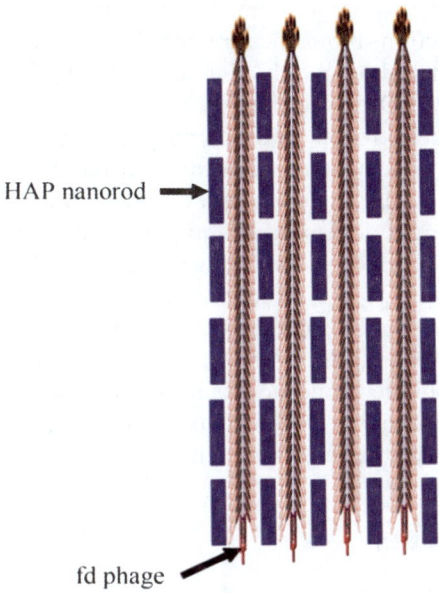

HAP nanorod ➤

fd phage ➤

Figure 10.12 Schematic diagram for fabricating HAP–phage-based scaffold.

can maintain its morphology for several years. It is now being tested for cytotoxicity and for the ability to seed bone cells. This kind of scaffold is a promising candidate for practical bone regeneration surgery in the future.

10.7   Summary and Outlook

With the help of well-established phage display and affinity selection techniques, filamentous phage are able to serve directly as biotemplates for synthesizing inorganic nanowires or assembling nanoparticles into ordered arrays. The templating is achieved *via* two methods. First, engineered phage displaying peptides that bind the target material specifically are able to template the synthesis and assembly of that material. Second, nonspecific electrostatic interactions between inorganic precursors and the exposed anionic carboxylate groups of the thousands of pVIII subunits can generate nanowires without the need for genetic engineering. Besides direct templating, individual phage can self-assemble into higher-order structures such as 1D fibers or 2D thin films to induce the synthesis and assembly of inorganic functional materials. The resultant highly ordered inorganic functional materials have many unique properties that have or will have practical applications in various fields, such as high-performance batteries, sensitive biosensors and biomedical hard tissues.

In the future, the field of phage-based inorganic assemblies can be further advanced by exploring the integration of phage with new inorganic functional materials, leading to more applications in nanoelectronic circuitry, electrochemical energy storage, photovoltaic devices, sensor development, environmental protection, biomedical material fabrication and so on. Because phage and nanomaterials can self-assemble into higher-order structures of increased complexity, 3D devices based on such structures are envisioned for applications in nanoelectronics and energy production and storage. Because phage is a biological nanostructure that can bear target-specific peptides on the surface through phage display techniques, real biological and medical applications of phage–inorganic assemblies are possible. Towards this direction, more efforts need to be made in the following respects. First, integration of the target specificity of phage and functions of inorganic nanomaterials can be designed to develop novel biological or chemical sensors with enhanced target specificity. Second, the complex of phage with target specificity and inorganic nanoparticles with unique properties such as photothermal gold nanoshells or magnetic iron oxide nanoparticles can be used to deliver drugs or genes to target cells or tissues for disease treatment.

Acknowledgements

We thank the National Science Foundation, National Institutes of Health, Department of Defense Breast Cancer Research Program and Oklahoma Center for the Advancement of Science and Technology for financial support.

References

1. M. C. Daniel and D. Astruc, *Chem. Rev.*, 2004, **104**, 293–346.
2. E. H. Sargent, *Adv. Mater.*, 2005, **17**, 515–522.
3. D. Tasis, N. Tagmatarchis, A. Bianco and M. Prato, *Chem. Rev.*, 2006, **106**, 1105–1136.
4. A. H. Lu, E. L. Salabas and F. Schuth, *Angew. Chem. Int. Ed.*, 2007, **46**, 1222–1244.
5. W. J. Liang, M. P. Shores, M. Bockrath, J. R. Long and H. Park, *Nature*, 2002, **417**, 725–729.
6. D. L. Klein, R. Roth, A. K. L. Lim, A. P. Alivisatos and P. L. McEuen, *Nature*, 1997, **389**, 699–701.
7. Y. Huang, X. F. Duan, Y. Cui, L. J. Lauhon, K. H. Kim and C. M. Lieber, *Science*, 2001, **294**, 1313–1317.
8. M. Lazzari and M. A. Lopez-Quintela, *Adv. Mater.*, 2003, **15**, 1583–1594.
9. R. Varghese and H. A. Wagenknecht, *Chem. Commun.*, 2009, 2615–2624.
10. A. Merzlyak and S. W. Lee, *Curr. Opin. Chem. Biol.*, 2006, **10**, 246–252.
11. S. Tang, C. B. Mao, Y. R. Liu, D. Q. Kelly and S. K. Banerjee, *IEEE Trans. Electron Devices*, 2007, **54**, 433–438.
12. A. M. Belcher, C. B. Mao, D. J. Solis, B. D. Reiss, S. T. Kottmann, R. Y. Sweeney, G. Georgiou and B. Iverson, *Abstr. Pap. Am. Chem. Soc.*, 2004, **228**, U542–U542.

13. C. B. Mao, D. J. Solis, B. D. Reiss, S. T. Kottmann, R. Y. Sweeney, A. Hayhurst, G. Georgiou, B. Iverson and A. M. Belcher, *Science*, 2004, **303**, 213–217.

14. C. B. Mao, C. E. Flynn, A. Hayhurst, R. Sweeney, J. F. Qi, G. Georgiou, B. Iverson and A. M. Belcher, *Proc. Natl. Acad. Sci. USA*, 2003, **100**, 6946–6951.

15. J. Sharma, R. Chhabra, A. Cheng, J. Brownell, Y. Liu and H. Yan, *Science*, 2009, **323**, 112–116.

16. C. X. Lin, Y. G. Ke, Z. Li, J. H. Wang, Y. Liu and H. Yan, *Nano Lett.*, 2009, **9**, 433–436.

17. B. Cao and C. Mao, *Langmuir*, 2007, **23**, 10701–10705.

18. S. R. Whaley, D. S. English, E. L. Hu, P. F. Barbara and A. M. Belcher, *Nature*, 2000, **405**, 665–668.

19. S. W. Lee, C. B. Mao, C. E. Flynn and A. M. Belcher, *Science*, 2002, **296**, 892–895.

20. S. W. Lee, S. K. Lee and A. M. Belcher, *Adv. Mater.*, 2003, **15**, 689–692.

21. C. B. Mao, J. F. Qi and A. M. Belcher, *Adv. Funct. Mater.*, 2003, **13**, 648–656.

22. C. E. Flynn, C. B. Mao, A. Hayhurst, J. L. Williams, G. Georgiou, B. Iverson and A. M. Belcher, *J. Mater. Chem.*, 2003, **13**, 2414–2421.

23. C. E. Flynn, S. W. Lee, B. R. Peelle and A. M. Belcher, *Acta Mater.*, 2003, **51**, 5867–5880.

24. K. T. Nam, B. R. Peelle, S. W. Lee and A. M. Belcher, *Nano Lett.*, 2004, **4**, 23–27.

25. J. P. Ni, S. W. Lee, J. M. White and A. M. Belcher, *J. Polym. Sci. Part B Polym. Phys.*, 2004, **42**, 629–635.

26. S. Jaffar, K. T. Nam, A. Khademhosseini, J. Xing, R. S. Langer and A. M. Belcher, *Nano Lett.*, 2004, **4**, 1421–1425.

27. R. Y. Sweeney, C. B. Mao, X. X. Gao, J. L. Burt, A. M. Belcher, G. Georgiou and B. L. Iverson, *Chem. Biol.*, 2004, **11**, 1553–1559.

28. Y. Huang, C. Y. Chiang, S. K. Lee, Y. Gao, E. L. Hu, J. De Yoreo and A. M. Belcher, *Nano Lett.*, 2005, **5**, 1429–1434.

29. S. K. Lee, D. S. Yun and A. M. Belcher, *Biomacromolecules*, 2006, **7**, 14–17.

30. S. Bhaviripudi, J. Qi, E. L. Hu and A. M. Belcher, *Nano Lett.*, 2007, **7**, 3512–3517.

31. K. T. Nam, R. Wartena, P. J. Yoo, F. W. Liau, Y. J. Lee, Y. M. Chiang, P. T. Hammond and A. M. Belcher, *Proc. Natl. Acad. Sci. USA*, 2008, **105**, 17227–17231.

32. K. T. Nam, *Science*, 2008, **322**, 44–44.

33. A. H. Liu, G. Abbineni and C. B. Moo, *Adv. Mater.*, 2009, **21**, 1001–1005.

34. K. T. Nam, D. W. Kim, P. J. Yoo, C. Y. Chiang, N. Meethong, P. T. Hammond, Y. M. Chiang and A. M. Belcher, *Science*, 2006, **312**, 885–888.

35. G. R. Souza, D. R. Christianson, F. I. Staquicini, M. G. Ozawa, E. Y. Snyder, R. L. Sidman, J. H. Miller, W. Arap and R. Pasqualini, *Proc. Natl. Acad. Sci. USA*, 2006, **103**, 1215–1220.

36. G. P. Smith and V. A. Petrenko, *Chem. Rev.*, 1997, **97**, 391–410.
37. V. A. Petrenko and V. J. Vodyanoy, *J. Microbiol. Methods*, 2003, **53**, 253–262.
38. J. T. Stubbs, K. P. Mintz, E. D. Eanes, D. A. Torchia and L. W. Fisher, *J. Bone Miner. Res.*, 1997, **12**, 1210–1222.
39. A. A. Sawyer, D. M. Weeks, S. S. Kelpke, M. S. McCracken and S. L. Bellis, *Biomaterials*, 2005, **26**, 7046–7056.
40. Z. Dogic and S. Fraden, *Curr. Opin. Colloid Interface Sci.*, 2006, **11**, 47–55.
41. S. W. Lee, B. M. Wood and A. M. Belcher, *Langmuir*, 2003, **19**, 1592–1598.
42. J. X. Tang and S. Fraden, *Liq. Cryst.*, 1995, **19**, 459–467.
43. Z. Dogic and S. Fraden, *Langmuir*, 2000, **16**, 7820–7824.
44. J. W. Kehoe and B. K. Kay, *Chem. Rev.*, 2005, **105**, 4056–4072.
45. V. A. Petrenko, G. P. Smith, X. Gong and T. Quinn, *Protein Eng.*, 1996, **9**, 797–801.
46. J. I. Pascual, J. Mendez, J. Gomezherrero, A. M. Baro, N. Garcia, U. Landman, W. D. Luedtke, E. N. Bogachek and H. P. Cheng, *Science*, 1995, **267**, 1793–1795.
47. M. S. Dresselhaus, G. Chen, M. Y. Tang, R. G. Yang, H. Lee, D. Z. Wang, Z. F. Ren, J. P. Fleurial and P. Gogna, *Adv. Mater.*, 2007, **19**, 1043–1053.
48. A. I. Boukai, Y. Bunimovich, J. Tahir-Kheli, J. K. Yu, W. A. Goddard and J. R. Heath, *Nature*, 2008, **451**, 168–171.
49. A. I. Hochbaum, R. K. Chen, R. D. Delgado, W. J. Liang, E. C. Garnett, M. Najarian, A. Majumdar and P. D. Yang, *Nature*, 2008, **451**, 163–165.
50. K. Y. Arutyunov, *Physica C*, 2008, **468**, 272–275.
51. X. Y. Zhao, C. M. Wei, L. Yang and M. Y. Chou, *Phys. Rev. Lett.*, 2004, **92**, 236805.
52. X. W. Teng, W. Q. Han, W. Ku and M. Hucker, *Angew. Chem. Int. Ed.*, 2008, **47**, 2055–2058.
53. A. Hochbaum and P. Yang, *Chem. Rev.*, 2009.
54. L. Cademartiri and G. A. Ozin, *Adv. Mater.*, 2009, **21**, 1013–1020.
55. K. N. Avery, J. E. Schaak and R. E. Schaak, *Chem. Mater.*, 2009, **21**, 2176–2178.
56. Y. J. Lee, H. Yi, W. J. Kim, K. Kang, D. S. Yun, M. S. Strano, G. Ceder and A. M. Belcher, *Science*, 2009, **324**, 1051–1055.
57. P. J. Yoo, N. S. Zacharia, J. Doh, K. T. Nam, A. M. Belcher and P. T. Hammond, *ACS Nano*, 2008, **2**, 561–571.
58. P. J. Yoo, K. T. Nam, J. Park, A. M. Belcher and P. T. Hammond, *Abstr. Pap. Am. Chem. Soc.*, 2005, **230**, U3588–U3588.
59. P. J. Yoo, K. T. Nam, J. F. Qi, S. K. Lee, J. Park, A. M. Belcher and P. T. Hammond, *Nat. Mater.*, 2006, **5**, 234–240.
60. S. W. Lee and A. M. Belcher, *Nano Lett.*, 2004, **4**, 387–390.
61. C. Y. Chiang, C. M. Mello, J. J. Gu, E. C. C. M. Silva, K. J. Van Vliet and A. M. Belcher, *Adv. Mater.*, 2007, **19**, 826–832.
62. F. Wang, B. Cao and C. Mao, *Chem. Mater.*, 2010, **22**(12), 3630–3636.

63. G. R. Souza, J. R. Molina, R. M. Raphael, M. G. Ozawa, D. J. Stark, C. S. Levin, L. F. Bronk, J. S. Ananta, J. Mandelin, M. M. Georgescu, J. A. Bankson, J. G. Gelovani, T. C. Killian, W. Arap and R. Pasqualini, *Nat. Nanotechnol.*, 2010, **5**, 291–296.

64. E. Estephan, C. Larroque, F. J. G. Cuisinier, Z. Balint and C. Gergely, *J. Phys. Chem. B*, 2008, **112**, 8799–8805.

65. R. R. Naik, S. J. Stringer, G. Agarwal, S. E. Jones and M. O. Stone, *Nat. Mater.*, 2002, **1**, 169–172.

66. R. R. Naik, S. E. Jones, C. J. Murray, J. C. McAuliffe, R. A. Vaia and M. O. Stone, *Adv. Funct. Mater.*, 2004, **14**, 25–30.

67. U. O. S. Seker, B. Wilson, S. Dincer, I. W. Kim, E. E. Oren, J. S. Evans, C. Tamerler and M. Sarikaya, *Langmuir*, 2007, **23**, 7895–7900.

68. M. Sarikaya, C. Tamerler, A. K. Y. Jen, K. Schulten and F. Baneyx, *Nat. Mater.*, 2003, **2**, 577–585.

69. K. I. Sano and K. Shiba, *J. Am. Chem. Soc.*, 2003, **125**, 14234–14235.

70. Y. J. Li, G. P. Whyburn and Y. Huang, *J. Am. Chem. Soc.*, 2009, **131**, 15998 + .

71. R. R. Naik, L. L. Brott, S. J. Clarson and M. O. Stone, *J. Nanosci. Nanotechnol.*, 2002, **2**, 95–100.

72. E. E. Oren, C. Tamerler, D. Sahin, M. Hnilova, U. O. S. Seker, M. Sarikaya and R. Samudrala, *Bioinformatics*, 2007, **23**, 2816–2822.

73. D. J. H. Gaskin, K. Starck and E. N. Vulfson, *Biotechnol. Lett.*, 2000, **22**, 1211–1216.

74. C. M. Li, G. D. Botsaris and D. L. Kaplan, *Cryst. Growth Des.*, 2002, **2**, 387–393.

75. M. B. Dickerson, S. E. Jones, Y. Cai, G. Ahmad, R. R. Naik, N. Kroger and K. H. Sandhage, *Chem. Mater.*, 2008, **20**, 1578–1584.

76. M. Umetsu, M. Mizuta, K. Tsumoto, S. Ohara, S. Takami, H. Watanabe, I. Kumagai and T. Adschiri, *Adv. Mater.*, 2005, **17**, 2571–2575.

77. B. D. Reiss, C. B. Mao, D. J. Solis, K. S. Ryan, T. Thomson and A. M. Belcher, *Nano Lett.*, 2004, **4**, 1127–1132.

78. G. Ahmad, M. B. Dickerson, Y. Cai, S. E. Jones, E. M. Ernst, J. P. Vernon, M. S. Haluska, Y. Fang, J. Wang, G. Subrarnanyarn, R. R. Naik and K. H. Sandhage, *J. Am. Chem. Soc.*, 2008, **130**, 4–5.

79. G. Ahmad, M. B. Dickerson, B. C. Church, Y. Cai, S. E. Jones, R. R. Naik, J. S. King, C. J. Summers, N. Kroger and K. H. Sandhage, *Adv. Mater.*, 2006, **18**, 1759–1763.

80. M. T. Klem, D. Willits, D. J. Solis, A. M. Belcher, M. Young and T. Douglas, *Adv. Funct. Mater.*, 2005, **15**, 1489–1494.

81. M. D. Roy, S. K. Stanley, E. J. Amis and M. L. Becker, *Adv. Mater.*, 2008, **20**, 1830–1836.

82. M. Gungormus, H. Fong, I. W. Kim, J. S. Evans, C. Tamerler and M. Sarikaya, *Biomacromolecules*, 2008, **9**, 966–973.

83. S. J. Segvich, H. C. Smith and D. H. Kohn, *Biomaterials*, 2009, **30**, 1287–1298.

84. M. B. Dickerson, R. R. Naik, M. O. Stone, Y. Cai and K. H. Sandhage, *Chem. Commun.*, 2004, 1776–1777.
85. S. Q. Wang, E. S. Humphreys, S. Y. Chung, D. F. Delduco, S. R. Lustig, H. Wang, K. N. Parker, N. W. Rizzo, S. Subramoney, Y. M. Chiang and A. Jagota, *Nat. Mater.*, 2003, **2**, 196–200.
86. D. Kase, J. L. Kulp, M. Yudasaka, J. S. Evans, S. Iijima and K. Shiba, *Langmuir*, 2004, **20**, 8939–8941.
87. Y. Morita, T. Ohsugi, Y. Iwasa and E. Tamiya, *J. Mol. Catal. B Enzymatic.*, 2004, **28**, 185–190.

CHAPTER 11

Phage Vaccines and Phage Therapy

KAREN MANOUTCHARIAN

Departamento de Biología Molecular y Biotecnología, Instituto de Investigaciones Biomédicas, UNAM, México DF, México

11.1 Introduction to Phage

Although vaccination is considered the most effective preventive or therapeutic intervention and can in favorable cases achieve complete eradication of a pathogen, the overall success rate in the field of vaccines is low and out of proportion to the huge financial and human resources invested in vaccine development. There are no effective vaccines against tuberculosis, leprosy, HIV, HCV, most parasitic diseases and cancer. Obviously, there is a need for new concepts and non-standard approaches to resolve the basic problems in the field of vaccines.[1,2]

Modern vaccinology deals with the most complicated cases of human and veterinary diseases and veterinary medicine, since effective vaccines against pathogens and diseases amenable to control by the immune system of the host have already been developed. Historically, vaccines have prevented the spread of many infectious diseases, such as smallpox, polio and measles, and are among the most effective public health interventions. Antigenic variation appears to be a fundamental mechanism by which pathogens avoid immune clearance and is one of the major obstacles to the development of vaccines against the pathogens and diseases with genetic variability.

RSC Nanoscience & Nanotechnology No. 17
Phage Nanobiotechnology
Edited by Valery A. Petrenko and George P. Smith
© Royal Society of Chemistry 2011
Published by the Royal Society of Chemistry, www.rsc.org

The bacteriophage (phage) – bacterial viruses that can be found in water, soil, plants, animals and humans – represent an inexhaustible reservoir of biomaterials for nanomanipulations. In particular, unique structural and functional features of both wild-type and recombinant bacteriophage (described in detail in other chapters) attest to their value as tools for the development of new molecular vaccines and therapeutic agents. Although exploration of phage as vaccine carriers is just getting under way, there is already a large body of studies indicating that they are promising and sometimes unique alternatives to existing standard vaccine platforms. Furthermore, in the light of the recent dramatic global increase in antibiotic-resistant bacterial infections, the application of naturally occurring or genetically modified phage for antibacterial therapy in humans and animals is an attractive alternative to chemical antibiotics. This chapter summarizes results of experiments using phage as vaccine components and discusses the immunogenic properties of this new class of vaccine carrier.

11.2   Phage Immunogens

The bacteriophage or phage, which are simple bacterial viruses, have already made important contributions in the field of vaccines and have been successfully used for the discovery of novel immunogens, therapeutic agents/targets and as effective vaccine carriers. Vaccines based on phage can be classified within all types of vaccines except whole pathogen-based ones. Phage display technology permits the isolation of target-specific peptides/proteins both *in vitro* and *in vivo*, the generation of DNA vaccine- and drug-delivery nanoparticles and nanodevices of biomedical or technical importance (as discussed in several other chapters). This simple methodology relies on the expression of fusion peptides or proteins on the bacteriophage surface, while the DNA encoding them is contained in the phage genome. The most frequently used display systems, based on filamentous phage, λ, T7 and T4 phage, permit the generation of very large random phage-displayed libraries of peptides, antibody fragments (scFv and Fab), functional protein domains such as enzymes, hormones or proteins encoded by cDNA and genomic DNA with complexities up to 10^{11}. The existence of a huge amount of published data describing the application of phage display for the discovery of novel genes, proteins or peptides of therapeutic importance along with the strong immunogenic properties of phage particles make this technology an invaluable tool and technological platform in the field of modern vaccines and molecular therapeutics.

While safety, efficacy and cost effectiveness are basic requirements for any vaccine, the identification of an effective immunogen is the first and decisive step for successful vaccine development. The immunogen must be a sufficiently strong inducer of protective immune responses. The potency of vaccines depends on several structural, chemical and biological properties of immunogen(s) included in vaccine composition, and an easy access/interaction of immunogen(s) with APCs as an initial step of immune response is critical for any vaccine to be successful.

Although there is a general belief that all licensed vaccines induce protective antibody responses, this suggestion can hardly be considered correct since at the time of development of these vaccines T-cell immune responses were not known. Today, cellular immune responses, particularly cytotoxic T lymphocytes (CTLs), are considered principal components of potential vaccines against many infectious diseases caused by intracellular pathogens, such as HIV, HCV, influenza, malaria and tuberculosis, and also against cancer.[2]

Interestingly, the immunogenic properties of phage were reported almost at the same time as when the first phage-displayed peptide library (PDPL) was reported.[3] Thus, the expression of *Plasmodium falciparum*-derived small peptides fused to fd phage major coat protein (cpVIII, 2700 copy numbers) and the display of larger peptides of this pathogen on the surface of hybrid phage (10–30% of cpVIII copies) were performed.[4] The hybrid phage was able to induce malaria-specific Abs in rabbits. The same group engineered the phage carrying multiple copies of peptide sequences from the V3 loop of the HIV-1 surface glycoprotein gp120 fused to cpVIII of the phage.[5] This phage was recognized by human HIV + sera and elicited high titers of HIV-neutralizing sera in mice. An interesting strategy was developed by imposing conformational constraints on a peptide epitope from *Chlamydia trachomatis* to improve its ability to elicit Abs in mice that cross-react with native Ag.[6] To achieve this, an epitope peptide-related phage display library was generated where the epitope's contact residues were randomized and most native-like peptides were isolated by affinity selection. These early studies demonstrated the superiority of phage immunogens over synthetic peptides. The latter are poor immunogens and need to be coupled to a carrier molecule or supplemented with an adjuvant, whereas phage particles *per se* are highly immunogenic and capable of activating helper T cells without any adjuvant.[7]

Immunogenic properties have also been shown for other phage. For example, a liver stage malaria epitope was displayed on RNA-free MS2 phage capsids that were able to induce both Ab and T-cell responses with significant upregulation of interferon-γ in mice.[8] Also, several immunodominant regions of the major capsid protein of hamster polyomavirus (HaPyV) were expressed in the context of RNA bacteriophage coat protein-derived virus-like particles (VLPs) capable of inducing virus-specific Abs in mice and rabbits.[9]

Another recently reported display platform based on bacteriophage T4 allows a multicomponent Ag display. Several anthrax toxin proteins were fused to T4 Hoc (highly antigenic outer capsid protein, 155 copies) and were efficiently displayed on T4 capsid using a defined *in vitro* assembly system.[10] Interestingly, all of the 155 Hoc binding sites can be occupied by one Ag or they can be split among two or more Ags, resulting in recombinant phage carrying multicomponent Ags. Immunization of mice with such phage elicited strong Ag-specific Abs and also lethal toxin neutralization titers. Up to 229 anthrax toxin complexes, equivalent to a total of 2400 protein molecules and a mass of about 133 MDa, were anchored on a single 120×86 nm T4 capsid particle, making it the highest density display reported on any virus.[11]

Among the physico-chemical characteristics of phage particles contributing to their immunogenic potential, the size and defined structure of the Ag surface display perhaps are of the greatest importance. The size of phage nanoparticles not only makes them attractive compounds for APCs targeting both MNC class I and class II peptide loading compartments,[12] but also allow direct targeting of immunologically important tissues/organs, as was recently shown by efficient targeting of Qβ phage-derived VLPs (with a diameter of about 30 nm) to the lymph nodes (LNs) of mice.[13] Furthermore, it was shown in that study that nanoparticles target distinct DC populations which may have important implications for vaccine design. In an earlier study, *in vivo* targeting of fd phage displaying an organ-specific peptide sequence, isolated by *in vivo* biopanning, to LNs resulted in enhancement of the Ab response in mice, which may have broad applications in the development of vaccines, production of Abs and immunotherapy.[14] Also, *in vivo* biopanning with a filamentous phage-displayed peptide library, carried out first in mice then in cancer patients (a B-cell malignancy), resulted in the identification of peptide motifs that localized to different organs, which may have broad implications for the development of targeted therapies.[15] In that study, disconnection of the patient from a life-support system followed short-term intravenous infusion of the phage library into the patient and multiple representative tissue biopsies were carried out. In addition, geometric features of phage such as size and shape, which are different from standard man-made particulates exhibiting a spherical shape, may open the path to new design solutions for systemically administered targeted particulates.[16] The presentation of Ag in a repetitive dense array, as shown by conjugation of Aβ peptide to VLPs or Qβ, allowed the generation of immunogens eliciting balanced Ab responses with negligible T-cell responses against Aβ and considered as new and potentially more effective vaccine candidates for Alzheimer's disease.[17] This type of Ag presentation has an influence not only on the immunogenicity but also on the antigenicity of displayed Ags, as was shown by comparing the dependence of Ab affinity and specificity on presentation of gp23 hexamers of T4 phage capsid in repetitive *versus* monomeric form.[18]

11.3   Epitope Discovery with Phage Libraries and Phage Vaccines

An important challenge in the development of epitope vaccines and therapeutic agents is to isolate antigenic determinants or to find small molecules capable of mediating a desired immunological or biological effect, respectively. In this respect, it is worth mentioning that the unique capability of the biopanning procedure using PDPLs or phage display cDNA, genomic DNA and gene fragment libraries permits not only the identification of useful natural genes, protein domains and epitopes, but also allows the isolation of peptide mimotopes showing functional similarity to natural Ag-derived sequences but bearing no sequence homology with natural Ag. Importantly, the PDPLs permit the identification of not only mimotopes of natural linear epitopes but also peptide sequences that mimic conformational epitopes formed by amino acids distant from each other in primary structure of natural Ag and brought together on the folded molecule. Notably, the mapping of antigenic determinants recognized by monoclonal or polyclonal disease- or pathogen-specific antibodies using phage display is the only experimental tool permitting the identification of linear mimotopes of conformational epitopes. Phage display-derived mimotopes can mimic even natural epitopes of non-protein origin, such as carbohydrates and lipids. Thus, mice were immunized with M13 phage carrying antigenic mimics of hepatitis B virus envelope protein (HBsAg) leading to induction of an HBsAg-specific Ab response, indicating the feasibility of a mimotope–vaccine platform.[19] The same authors also compared humoral immune responses induced in mice by these mimotopes in various molecular contexts: phage cpIII and cpVIII, recombinant human H ferritin, HBV core peptide and multiple antigenic peptides (MAPs). It was shown that phage-displayed mimotopes were the best immunogens and induced the most reproducible and potent responses.[20] Furthermore, it was shown also that HBsAg and HCV phage-displayed mimotopes are strong immunogens when applied in mice by intranasal or intragastric administration, indicating the usefulness of phage for the development of orally effective vaccines.[21]

Phage display was also used to generate immunogens as HIV vaccine candidates based on mimotopes capable of both recognizing the sera from HIV-infected subjects and inducing HIV-neutralizing Abs in mice.[22] In another study, HIV-1 epitope-specific CTL responses were induced in HLA-A2 transgenic mice by immunization with recombinant M13 phage carrying viral reverse transcriptase-derived CTL epitope.[23] A consensus sequence profile was derived from hypervariable region 1(HVR1) of HCV, another pathogen with high genetic polymorphism, displayed on phage and used in a biopanning against sera from infected individuals. Phage bearing mimotopes with the highest cross-reactivity with patients' sera were injected into mice, resulting in the generation of Abs recognizing a panel of natural HVR1 variants.[24] In addition, a sequence pattern responsible for the observed cross-reactivity was identified showing that these findings could be important for the development of a vaccine against HCV. Furthermore, a whole epitope profile of complex pathogens can be obtained through the screening of PDPLs with hyperimmune serum, as demonstrated by the isolation of immunogenic mimotopes of *Mycoplasma hyopneumoniae*, the etiologic agent causing swine enzootic pneumonia.[25]

Currently, several web tools are available that facilitate the analysis of sequences generated by phage display useful for the exploration of protein–protein interaction sites and networks, and also the development of new drugs, vaccines and therapeutics. One such instrument is MIMOX, a free web tool that provides a simple interface to align a set of mimotopes and provides a statistical method to derive the consensus sequences and map these sequences to the corresponding antigen structure and search for all of the clusters of residues that could represent the native epitope.[26]

Despite the success in the isolation of vaccine candidate mimotopes from random PDPLs, only a small fraction of antigenic peptides selected from such libraries have immunogenic fitness, while half of the peptides isolated from natural phage display libraries were immunogenic showing that epitope discovery from these types of libraries is a promising route to subunit vaccines and to small molecule-based therapeutics.[27] Although peptides representing CTL and Th epitopes are important components in epitope vaccine approaches, their efficacy is limited due to the poor immunogenicity of small synthetic peptides. However, expression of such peptides on phage offers a solution to this issue, as shown in a recent study where HCMV MHC II-restricted epitopes displayed on fd phage were able to enhance T-cell responses *in vitro* after efficient processing of hybrid phage by human APCs.[28]

The application of mimotopes seems promising also for cancer vaccine studies, as demonstrated by the isolation of peptide mimics of the epitope recognized by anti-oncogenic protein Her-2/neu monoclonal antibody (MoAb).[29] It may be suitable for the formulation of a breast cancer vaccine. Another selected mimotope of human high molecular weight melanoma-associated Ag (HMW-MAA) after coupling to tetanus toxoid was capable of inducing native Ag-specific Abs in rabbits with anti-tumor *in vitro* activity.[30] In another study, human anti-idiotypic (anti-Id) scFvs were selected against F(ab')$_2$ fragments of trastuzumab, a humanized anti-HER-2/neu MoAb, and used for immunization of mice. The induced antibodies were able to bind HER-2/neu, showing that such mini-Abs could be used as an anti-Id-based vaccine formulations in patients bearing HER-2/neu-positive tumors.[31]

Another intriguing possibility is the application of phage display-derived mimotopes in allergy, which can be employed both for the investigation of allergen–IgE interactions, since in many cases the etiologic agents are unknown in such autoimmune disorders, and for finding safe and efficient novel vaccines.[29] The proof of this was shown in an early study in which mice immunized with phage mimotopes isolated from a phage library after biopanning against purified IgE specific for the major birch pollen allergen Bet V1 produced IgG molecules able to block IgE binding to allergen.[32]

Despite the existence of data showing the high immunogenic potency of phage, surprisingly, only in the late 1990s were attempts made to use recombinant bacteriophage as vaccines. Detailed studies were reported showing that phage, acting as particulate antigens, can access both MHC I and II pathways and were therefore capable of inducing strong humoral and cellular immune responses. Thus, a recombinant phage displaying a B-cell epitope of glycoprotein G of human respiratory syncytial virus was used as a vaccine, conferring protection against virus infection in mice.[33] Then, a new type of immunogen was generated by expressing the Ig VH domain with all three complementarity-determining regions (CDRs) replaced by *Taenia crassiceps* T-cell epitope on M13 phage, and it was shown for the first time that phage immunization elicits protective cellular immune responses in mice susceptible to pathogen challenge.[34] The induction of protective Ag-specific cellular immune responses in pigs vaccinated in controlled conditions with recombinant phage carrying three *Taenia solium*-derived peptides and an Ag was shown by the same group.[35] Pigs are the only intermediate hosts of *Taenia solium*, a parasite causing neurocysticercosis, a common parasitic disease of the central nervous system (CNS) that seriously affects human health on a worldwide scale. Recently, strong protection of pigs vaccinated with the above-mentioned phage vaccines against cysticercosis was demonstrated in a large randomized field trial that included more than 1000 rural pigs. These results illustrate for the first time the application of recombinant phage as practical vaccines.[36]

There are two other earlier studies related to phage vaccines, a report showing protection of neonatal mice against streptococcal infection using maternal immunization with phage-displayed anti-Id Ab scFv fragment,[37] and a study demonstrating for the first time that peptides discovered by phage display selection can be used in the context of phage particles to confer protective Ab-mediated immunity against an infectious agent, in this case a herpes simplex virus (HSV-2) challenge in a murine model.[38]

11.4 Autoimmune Disorders

As mentioned above, phage display technologies hold promise as tools for the development of vaccines and therapeutics to treat complex autoimmune disorders. Thus, a novel active immunization approach against TNF-α was described recently, which resulted in the induction of high titers of therapeutically active autoantibodies. Immunization of mice with virus-like-particles (VLPs) of Qβ covalently linked to either soluble TNF-α protein or a 20 amino acid peptide derived from its N-terminus induced specific antibodies that protected from clinical signs of inflammation in a murine model of rheumatoid arthritis.[39] Similarly, mice immunized with either IL-1α or IL-1β, chemically cross-linked to VLPs of Qβ phage, generated antibodies that efficiently neutralized the binding of the respective IL-1 molecules to their receptors *in vitro* and their pro-inflammatory activities *in vivo*.[40] In the collagen-induced arthritis model, both vaccines strongly protected mice from inflammation and degradation of bone and cartilage, and vaccination with IL-1β strongly protected the mice from arthritis in the T- and B-cell-independent collagen antibody transfer model. These data indicate that these type Qβ phage-based constructs might become efficacious and cost-effective new treatment options for rheumatoid arthritis and other systemic IL-1-dependent inflammatory disorders.

11.5 Cancer

In an important recent study, several tumor Ags, identified using a proteomics-based approach, were expressed on the surface of T7 phage and shown to trigger specific immune responses in mice following oral immunization.[41] These immune responses inhibited tumor growth and metastasis of the mammary 4T1 adenocarcinoma cell line, indicating that surface display of tumor Ags can provide an effective strategy for mucosal cancer vaccines. In addition, arrayed phage-displayed tumor Ags could be useful as a serum-based screening test for the detection of tumor Ags. Earlier, a CTL epitope of melanoma Ag MAGE-A1 displayed on fd filamentous phage was used to immunize mice, resulting in protection against tumor growth, growth control of established tumors and prolonged survival of tumor-bearing mice.[42] The inhibition of tumor growth mediated by CTL activity was also shown in another study in which mice were immunized with fd phage expressing MAGE-A3-derived peptides.[43]

11.6   Neurological Disorders

The nanotubular structure of filamentous phage particles indicates a capacity of such particles to penetrate the CNS, thus providing novel opportunities for the treatment of diseases that affect the CNS. Intranasal administration of filamentous phage carrying an anti-amyloid-beta peptide (Aβ) antibody fragment into Alzheimer's APP transgenic mice permitted *in vivo* targeting of Aβ plaques.[44] Thus, genetically engineered phage proved to be an efficient and non-toxic delivery vector to the brain, which was later confirmed by studies showing that intranasal administration of filamentous phage displaying cocaine-binding scFvs can sequester cocaine in the brains of rats, leading to significant blocking of psychomotor effects of cocaine challenge.[45] Also, immunization of mice with EFRH peptide from the N-terminal region of Aβ displayed on phage led to an improvement in cognitive functions and alleviated amyloid pathology in a transgenic model of Alzheimer's disease.[46]

11.7   Other Diseases

Phage-based vaccines have been successfully applied against diverse diseases such as systemic candidiasis and foot-and-mouth disease. In the first study, a phage-displayed epitope derived from the *Candida albicans* heat shock protein 90 induced both Ab and CTL immune responses and a reduction of colony-forming units in the kidneys of vaccinated mice.[47] In the other case, FMDV capsid protein and proteinase 3C were expressed on the T4 phage surface SOC site. Mice immunized with these recombinant phage either orally or by subcutaneous injection were 100% protected against lethal challenge.[48] Interestingly, data regarding the first human filamentous phage vaccine trial with a small group of multiple myeloma patients can be found on the APALEXO Biotechnologie GmbH web site, showing that phage vaccination can induce tumor-specific immune responses with the potential to exert beneficial effects on patients.[49]

Finally, M13 phage carrying HIV-related mimotopes were tested as HIV vaccine candidates in a recent study.[50] Rhesus macaques immunized with a mixture of such phage experienced lower levels of peak viremia and were protected from progression to AIDS-like illness.[50]

11.8 Antibacterial Therapy

Although there is only a single published report describing the immunization of HIV-infected patients with phage phiX174 for the evaluation of lymphocyte function *in vivo*,[51] the use of lytic bacteriophage discovered over 85 years ago as antibacterial therapeutic agents in humans was common in several European countries and in the USA for decades before the antibiotic era.[52,53] However, the Eastern European practice of phage therapy has received little attention in the West.[53] Due to the increasing prevalence of antibiotic-resistant microbes worldwide, there is interest in and an urgent need for the development of antibacterial phage. These highly virulent and resistant nosocomial bacterial infections are rampant in hospitals everywhere. In US hospitals alone, more than 2 000 000 patients succumb to infectious diseases every year and over 90 000 die, as compared with a yearly mortality rate of 15 000 in the early 1990s.[54] Today, several companies are working with both naturally occurring and genetically modified phage to combat drug-resistant bacteria.[52]

Typically, antibacterial phage preparations are mixtures of different phage of a wide host range capable of infecting and killing many bacterial species and strains, including brucellae, enterococci, pathogenic strains of *Escherichia coli*, *Mycobacterium tuberculosis*, klebsiellae, salmonellae, staphylococci, streptococci, *Shigella* spp., *Vibrio cholerae* and *Yersinia* spp. Notably, 90% of staphylococci, the pathogen responsible for 15% of all bacterial infections, are penicillin resistant and 40% are resistant to methicillin.[54] Bacteriophage therapy can be considered as an alternative treatment and prevention option against enterococci that contribute significantly to patient mortality and morbidity, and also healthcare costs.[55]

Phage therapy is especially interesting for medical care in the developing world. According to the World Health Organization (WHO), in developing countries infections and parasitic diseases are responsible for the death of 20 million people per year. Every year about 8 million children under five years old die of acute respiratory tract infections linked to bacteria such as *Streptococcus pneumonia* or *Haemophilus influenza* type B or of diarrhea-related diseases caused by bacteria such as *Shigella* spp., *Vibrio cholerae* and several types of *E. coli*. Several epidemics, especially in Africa, were caused by strains resistant to known antibiotics, and the availability of advanced antibiotics is often limited by their high cost. On the other hand, phage therapy has proven to be effective against many infections and in cases of serious burns, where a significant cause of death within the first 2 days is *via* infection by *Pseudomonas aeruginosa*.[56]

Along with therapy using living phage, filamentous M13 phage in display systems and non-replicating phage are also effective, for example, for the treatment of *Helicobacter pylori* and *P. aeruginosa*, respectively. In addition to phage particles *per se*, purified phage-encoded peptidoglycan hydrolase (lysine) is also reported to be effective for the treatment of bacterial infectious diseases caused by Gram-positive bacteria such as *Streptococcus pyogenes*, *S. pneumoniae*, *Bacillus anthracis* and group B streptococci.[57] Phage are also promising agents for the prevention or control of infectious biofilms that form spontaneously on both inert and living systems. Clinically relevant biofilms and especially those associated with the use of medical devices can be controlled by using phage-encoded polysaccharide lyase, as has been used to treat *P. aeruginosa* biofilms in cystic fibrosis patients by aerosol administration.[58]

Due to concerns that mass lysis of phage can be problematic for phage medicine (this issue remains under debate), phage carrying molecules that kill bacteria without lysis have been developed. Some drawbacks for the application of phage *in vivo*, such as rapid removal of the phage from the body and the presence of toxins in phage preparations, can be overcome by isolating long-circulating variants of phage using a serial passage technique and by further purifying phage preparations to diminish toxin levels, respectively.

There are several problems in phage therapy that need to be resolved: (i) inactivation of administered phage or lysine by a neutralizing antibody and allergic reactions to them, (ii) the appearance of mutants resistant to phage and (iii) the capture and transfer of bacterial toxin genes by phage. However, multiple possibilities exist to overcome these problems; for example, phage or lysines with different antigenicities or with low immunogenicities could be prepared, or, due to the co-evolution of phage with bacteria, new phage for modified targets could be isolated, or the problem of capture of bacterial toxin or antibiotic-resistant genes by phage may be overcome by the selection of suitable phage that do not have natural generalized or specialized transduction abilities or by construction of genetically modified mutant phage.[57] It is worth noting that in a study describing a minimal toxicity from administration of a phage random peptide library in mice, there is a statement that based on these preclinical data the US Food and Drug Administration (FDA) has approved the implementation of human clinical trials with this technique,[59] although currently there are no related published data. However, the FDA has recently approved a mixture of phage as a food additive that can be used in processing plants for spraying on to ready-to-eat meat and poultry products to protect consumers from the potentially life-threatening bacterium *Listeria monocytogenes*.[60] While this approval does not imply that the application of phage even as oral vaccines will be authorized automatically, the generation of more preclinical and clinical safety/efficacy phage vaccine/therapy data will result, it is to be hoped, in the acceptance of phage-based preparations by regulatory organizations for human and veterinary use in the near future. Interestingly, to enable residents of the USA to have more ready access to phage therapy, the company Phage International recently merged with the Georgian Phage Therapy Center and opened a phage therapy center in Tijuana, Mexico.[61] Another reason contributing to the general reluctance to develop and use phage therapy in the West is the difficulty in obtaining clear intellectual property rights. It is obvious that without patent protection on products there is no real incentive for large pharmaceutical companies to go through the stringent and expensive clinical trials needed for drug approval.[53,62] Many phage companies rather plan to commercialize phage products for agricultural applications where regulations are less stringent.

Although the therapeutic targets of phage are usually bacteria, there have been suggestions that they can also interact with immune cells causing immunomodulatory effects, can inhibit platelet adhesion to fibrinogen and antibody production and even inhibit tumor growth.[62] Furthermore, there are data suggesting that phage can interfere with viral infections. For the latter phenomena, several modes of action have been proposed. Bacteriophage-derived nucleic acids may inhibit virus infection, phage and viruses may be competing for cellular receptors for viral cell entry or phage may induce or display antibodies with antiviral activity. As our current supply of antiviral drugs is fairly limited, the application of phage as therapeutic agents against viruses would be an attractive and cost-effective alternative.[62]

11.9 Conclusion

There are several yet unexplored fields of vaccines where phage can make important contributions at multiple levels, starting from vaccine discovery, development, delivery and storage. For example, the identification and application of epitopes that are immunogenic in the context of multiple MHC molecules would be of significant advantage, overcoming one of the obstacles to the construction of CTL-based vaccines for chronic viral infections. Hence multiple epitopes with different MHC specificities can be expressed in a single phage particle, permitting coverage of larger portions of populations using a minimal number of vaccine components in large-scale vaccination programs. On the other hand, genome-scale epitope screening approaches, particularly the identification of epitopes that induce cellular immune responses, and small molecule-based drug discovery applying combinatorial random peptide or genome-based phage libraries might additionally accelerate the vaccine discovery process. The application of combinatorial approaches in conjunction with phage display techniques will be critical for the success of vaccines against antigenically variable pathogens/diseases. In particular, we can expect the generation of novel immunogens representing structural/molecular mimics of pathogen-derived immunodominant epitopes or protein domains displayed on phage or VLPs capable of inducing protective antibodies, or the construction of novel vaccines based on the incorporation of antigenic/genetic variability of pathogens or cancer cells in the context of phage particles. The diversity of applications and success of phage display are due to its simplicity and flexibility along with the possibilities of very cheap, large-scale production of phage particles by recovering them from infected bacterial culture supernatants as nearly 100% homogeneous preparations free of cellular components. Phage are easy to manage, they resist heat and many organic solvents, chemicals and other stresses, and, importantly, phage particles are highly immunogenic and do not require adjuvant. Furthermore, phage do not require a cold chain (requirement to store vaccines at refrigerated or frozen temperatures), which equates to lower transport and storage costs. Considering these points, recombinant phage should be viewed as promising vaccine discovery tools and vaccine delivery vectors, and it is even worth considering the possibility of replacing the delivery systems of known vaccines currently in use with phage particles as vaccine carriers.

Although there is no guarantee that phage display and phage will resolve all problems in vaccine and drug development, they are offering, at least qualitatively, new tactics and strategies for the discovery and development of novel vaccines, therapeutics and drugs. In summary, there is a general opinion among many experts that phage vaccines, therapeutics and phage therapy could very soon make important contributions in modern medicine.

References

1. I. Benhar, *Biotechnol. Adv.*, 2001, **19**, 1.
2. P. E. Jensen, *Nat. Immunol.*, 2007, **8**, 1041.
3. J. K. Scott and G. P. Smith, *Science*, 1990, **249**, 386.
4. J. Greenwood, A. E. Willis and R. N. Perham, *J. Mol. Biol.*, 1991, **220**, 821.
5. V. F. de la Cruz, A. A. Lal and T. F. McCutchan, *J. Biol. Chem.*, 1988, **263**, 4318.
6. G. Zhong, G. P. Smith, J. Berry and R. C. Brunham, *J. Biol. Chem.*, 1994, **269**, 24183.
7. K. Manoutcharian, G. Gevorkian, A. Cano and J. C. Almagro, *Curr. Pharm. Biotechnol.*, 2001, **2**, 217.
8. K. G. Heal, H. R. Hill, P. G. Stockley, M. R. Hollingdale and A. W. Taylor-Robinson, *Vaccine*, 1999, **18**, 251.
9. T. Voronkova, A. Grosch, A. Kazaks, V. Ose, D. Skrastina, K. Sasnauskas, B. Jandrig, W. Arnold, S. Scherneck, P. Pumpens and R. Ulrich, *Viral Immunol.*, 2002, **15**, 627.
10. S. B. Shivachandra, Q. Li, K. K. Peachman, G. R. Matyas, S. H. Leppla, C. R. Alving, M. Rao and V. B. Rao, *Vaccine*, 2007, **25**, 1225.
11. Q. Li, S. B. Shivachandra, S. H. Leppla and V. B. Rao, *J. Mol. Biol.*, 2006, **363**, 577.
12. M. Gaubin, C. Fanutti, Z. Mishal, A. Durrbach, P. De Berardinis, R. Sartorius, G. Del Pozzo, J. Guardiola, R. N. Perham and D. Piatier-Tonneau, *DNA Cell Biol*, 2003, **22**, 11.
13. V. Manolova, A. Flace, M. Bauer, K. Schwarz, P. Saudan and M. F. Bachmann, *Eur. J. Immunol.*, 2008, **38**, 1404.
14. M. Trepel, W. Arap and R. Pasqualini, *Cancer Res.*, 2001, **61**, 8110.
15. W. Arap, M. G. Kolonin, M. Trepel, J. Lahdenranta, M. Cardo-Vila, R. J. Giordano, P. J. Mintz, P. U. Ardelt, V. J. Yao, C. I. Vidal, L. Chen, A. Flamm, H. Valtanen, L. M. Weavind, M. E. Hicks, R. E. Pollock, G. H. Botz, C. D. Bucana, E. Koivunen, D. Cahill, P. Troncoso, K. A. Baggerly, R. D. Pentz, K. A. Do, C. J. Logothetis and R. Pasqualini, *Nat. Med.*, 2002, **8**, 121.
16. P. Decuzzi, R. Pasqualini, W. Arap and M. Ferrari, *Pharm. Res.*, 2008.

17. B. Chackerian, M. Rangel, Z. Hunter and D. S. Peabody, *Vaccine*, 2006, **24**, 6321.
18. W. Baschong, L. Hasler, M. Haner, J. Kistler and U. Aebi, *J. Struct. Biol.*, 2003, **143**, 258.
19. A. Folgori, R. Tafi, A. Meola, F. Felici, G. Galfre, R. Cortese, P. Monaci and A. Nicosia, *EMBO J.*, 1994, **13**, 2236.
20. A. Meola, P. Delmastro, P. Monaci, A. Luzzago, A. Nicosia, F. Felici, R. Cortese and G. Galfre, *J. Immunol.*, 1995, **154**, 3162.
21. P. Delmastro, A. Meola, P. Monaci, R. Cortese and G. Galfre, *Vaccine*, 1997, **15**, 1276.
22. G. Scala, X. Chen, W. Liu, J. N. Telles, O. J. Cohen, M. Vaccarezza, T. Igarashi and A. S. Fauci, *J. Immunol.*, 1999, **162**, 6155.
23. P. De Berardinis, R. Sartorius, C. Fanutti, R. N. Perham, G. Del Pozzo and J. Guardiola, *Nat. Biotechnol.*, 2000, **18**, 873.
24. G. Puntoriero, A. Meola, A. Lahm, S. Zucchelli, B. B. Ercole, R. Tafi, M. Pezzanera, M. U. Mondelli, R. Cortese, A. Tramontano, G. Galfre and A. Nicosia, *EMBO J.*, 1998, **17**, 3521.
25. W. J. Yang, J. F. Lai, K. C. Peng, H. J. Chiang, C. N. Weng and D. Shiuan, *J. Immunol. Methods.*, 2005, **304**, 15.
26. J. Huang, A. Gutteridge, W. Honda and M. Kanehisa, *BMC Bioinformatics*, 2006, **7**, 451.
27. L. J. Matthews, R. Davis and G. P. Smith, *J. Immunol.*, 2002, **169**, 837.
28. C. Ulivieri, A. Citro, F. Ivaldi, D. Mascolo, R. Ghittoni, D. Fanigliulo, F. Manca, C. T. Baldari, G. L. Pira and G. Del Pozzo, *Immunol. Lett.*, 2008, **119**, 62.
29. A. Riemer, O. Scheiner and E. Jensen-Jarolim, *Methods*, 2004, **32**, 321.
30. S. Wagner, C. Hafner, D. Allwardt, J. Jasinska, S. Ferrone, C. C. Zielinski, O. Scheiner, U. Wiedermann, H. Pehamberger and H. Breiteneder, *J. Immunol.*, 2005, **174**, 976.
31. M. Coelho, P. Gauthier, M. Pugniere, F. Roquet, A. Pelegrin and I. Navarro-Teulon, *Br. J. Cancer*, 2004, **90**, 2032.
32. E. Ganglberger, K. Grunberger, B. Sponer, C. Radauer, H. Breiteneder, G. Boltz-Nitulescu, O. Scheiner and E. Jensen-Jarolim, *FASEB J.*, 2000, **14**, 2177.
33. N. Bastien, M. Trudel and C. Simard, *Virology*, 1997, **234**, 118.
34. K. Manoutcharian, L. I. Terrazas, G. Gevorkian, G. Acero, P. Petrossian, M. Rodriguez and T. Govezensky, *Infect. Immun.*, 1999, **67**, 4764.
35. K. Manoutcharian, A. Diaz-Orea, G. Gevorkian, G. Fragoso, G. Acero, E. Gonzalez, A. De Aluja, N. Villalobos, E. Gomez-Conde and E. Sciutto, *Vet. Immunol. Immunopathol.*, 2004, **99**, 11.
36. J. Morales, J. J. Martinez, K. Manoutcharian, M. Hernandez, A. Fleury, G. Gevorkian, G. Acero, A. Blancas, A. Toledo, J. Cervantes, V. Maza, F. Quet, H. Bonnabau, A. S. de Aluja, G. Fragoso, C. Larralde and E. Sciutto, *Vaccine*, 2008, **26**, 2899.
37. W. Magliani, L. Polonelli, S. Conti, A. Salati, P. F. Rocca, V. Cusumano, G. Mancuso and G. Teti, *Nat. Med.*, 1998, **4**, 705.

38. A. M. Grabowska, R. Jennings, P. Laing, M. Darsley, C. L. Jameson, L. Swift and W. L. Irving, *Virology*, 2000, **269**, 47.
39. G. Spohn, R. Guler, P. Johansen, I. Keller, M. Jacobs, M. Beck, F. Rohner, M. Bauer, K. Dietmeier, T. M. Kundig, G. T. Jennings, F. Brombacher and M. F. Bachmann, *J. Immunol.*, 2007, **178**, 7450.
40. G. Spohn, I. Keller, M. Beck, P. Grest, G. T. Jennings and M. F. Bachmann, *Eur. J. Immunol.*, 2008, **38**, 877.
41. M. Shadidi, D. Sorensen, A. Dybwad, G. Furset and M. Sioud, *Int. J. Oncol.*, 2008, **32**, 241.
42. J. Fang, G. Wang, Q. Yang, J. Song, Y. Wang and L. Wang, *Vaccine*, 2005, **23**, 4860.
43. R. Sartorius, P. Pisu, L. D'Apice, L. Pizzella, C. Romano, G. Cortese, A. Giorgini, A. Santoni, F. Velotti and P. De Berardinis, *J. Immunol.*, 2008, **180**, 3719.
44. D. Frenkel and B. Solomon, *Proc. Natl. Acad. Sci. USA*, 2002, **99**, 5675.
45. M. R. Carrera, G. F. Kaufmann, J. M. Mee, M. M. Meijler, G. F. Koob and K. D. Janda, *Proc. Natl. Acad. Sci. USA*, 2004, **101**, 10416.
46. B. Solomon, *Vaccine*, 2007, **25**, 3053.
47. G. Wang, M. Sun, J. Fang, Q. Yang, H. Tong and L. Wang, *Vaccine*, 2006, **24**, 6065.
48. Z. J. Ren, C. J. Tian, Q. S. Zhu, M. Y. Zhao, A. G. Xin, W. X. Nie, S. R. Ling, M. W. Zhu, J. Y. Wu, H. Y. Lan, Y. C. Cao and Y. Z. Bi, *Vaccine*, 2008, **26**, 1471.
49. www.apalexo.com.
50. X. Chen, G. Scala, I. Quinto, W. Liu, T. Chun, J. S. Justement, O. J. Cohen, T. C. Vancott, M. Iwanicki, M. G. Lewis, J. Greenhouse, T. Barry, D. Venzon and A. Fauci, *Nat. Med.*, 2001, **7**, 1225.
51. I. Fogelman, V. Davey, H. D. Ochs, M. Elashoff, M. B. Feinberg, J. Mican, J. P. Siegel, M. Sneller and H. C. Lane, *J. Infect. Dis.*, 2000, **182**, 435.
52. K. Thiel, *Nat. Biotechnol.*, 2004, **22**, 31.
53. J. R. Clark and J. B. March, *Trends Biotechnol.*, 2006, **24**, 212.
54. www.phage-biotech.com.
55. S. Koch, M. Hufnagel and J. Huebner, *Expert Opin. Biol. Ther.*, 2004, **4**, 1519.
56. www.bioline.org.br.
57. S. Matsuzaki, M. Rashel, J. Uchiyama, S. Sakurai, T. Ujihara, M. Kuroda, M. Ikeuchi, T. Tani, M. Fujieda, H. Wakiguchi and S. Imai, *J. Infect. Chemother.*, 2005, **11**, 211.
58. J. Azeredo and I. W. Sutherland, *Curr. Pharm. Biotechnol.*, 2008, **9**, 261.
59. D. N. Krag, S. P. Fuller, L. Oligino, S. C. Pero, D. L. Weaver, A. L. Soden, C. Hebert, S. Mills, C. Liu and D. Peterson, *Cancer Chemother. Pharmacol.*, 2002, **50**, 325.
60. L. Bren, *FDA Consumer Mag.*, 2007, January–February.
61. www.phageinternational.com.
62. R. Miedzybrodzki, W. Fortuna, B. Weber-Dabrowska and A. Gorski, *Virus Res.*, 2005, **110**, 1.

Subject Index